Author's Note, May 2010

Welcome to this updated second edition of my #1 bestseller *Air Con*. Rather than completely re-writing the book to change data references, I've left the first part of the book intact, more or less just as it was when we first published it a year ago. It was designed with a logical easy-to-read flow to guide you through the pro's and con's of global warming, and I didn't want to mess with that.

Apart from occasional references to the then upcoming 2009 Copenhagen climate talks, the book is timeless and just as up to date on the science of global warming as it was last year.

So instead of radical surgery, I've built on the existing foundation by adding the updated developments in a new 60 page section spanning Chapters 20 through 22.

I had originally intended to confine the update to Climategate, but then the UN IPCC report fell apart and a raft of new studies lending weight to sceptical arguments were published. The good news is it makes this book utterly comprehensive and easy to use as a reference or to discuss with friends and colleagues.

Need to know about disappearing islands and rising sea levels? Ocean acidification? Melting glaciers? It's all in here.

For those of you who'd like to now be kept abreast of daily developments in the climate change debate, sign up to our Twitter account and you'll be alerted every time we post something new:

http://twitter.com/investigatemag

Thanks for taking an interest in this issue. The battle over climate change is ultimately a battle to control the resources and citizens of the world. We are all affected, and we all need to get informed.

Ian Wishart
1 May 2010

For our daughter Sophia

NEED THE FACTS ON CLIMATE CHANGE?

AIR CON

CLIMATEGATE EDITION

2010

Ian Wishart

HOWLING AT THE MOON PUBLISHING LTD

First edition published April 2009
Second edition published May 2010
Howling At The Moon Publishing Ltd
PO Box 188
Kaukapakapa 0843, NEW ZEALAND

www.ianwishart.com
email: editorial@investigatemagazine.com

ISBN 978-0958240161

Typeset in Adobe Garamond Pro and Frutiger
Cover concept: Ian Wishart, Heidi Wishart, Bozidar Jokanovic
Book Design: Bozidar Jokanovic
Cover images: front, Duncan Walker/iStockphoto
back, Imageshop/Corbis

To get another copy of this book airmailed to you anywhere in the world, or to purchase a fully text-searchable digital edition, visit our website:

WWW.IANWISHART.COM

LEGAL NOTICE: Criticisms of individuals in this book reflect the author's honest opinion, for reasons outlined in the text or generally known at the time of writing

Contents

Prologue

The Sky Is Falling

"Chicken Little was in the woods one day when an acorn fell on her head. It scared her so much she trembled all over. She shook so hard, half her feathers fell out. "Help! Help!" she cried. "The sky is falling! I have to go tell the king!"

– Act I, The Story of Chicken Little

The story of Chicken Little is the story of our times. Frankly, it should be a Hollywood blockbuster. Hang on, Al Gore's already produced it. How silly of me. The tale of the chook that misinterpreted a natural event bears striking similarities to some modern roosters currently globetrotting the planet in a cloud of contrails and jetfuel, warning all of us the sky is warming, and the acorn trees are dying.

Despite two decades of hype and dire warnings though, the public are fast getting heartily sick of hearing about global warming and how all our grandkids are going to need flippers and a snorkel just to go downtown of a Friday evening. A recent Pew research poll found a staggering 41% of Americans now believe global warming has been "exaggerated" by its believers – the highest level of doubt recorded. On a list of the top 20 concerns for the public, global warming came in, well, 20th.

But don't be fooled. Just because the public aren't taking it seriously anymore doesn't mean vast armies of lobby groups, climate change believers and politicians won't sign the new Copenhagen Treaty at the end of the year, selling your families into financial slavery to the UN and its multinational carbon-trading minions for the next century.

There is an awful lot riding on the global warming industry. Vast fortunes stand to be made by some on the inside, and where there's money there's power and greed close behind. Would you be willing to bet next month's wages that the Copenhagen Treaty won't go ahead, just because the public increasingly don't believe any more? I wouldn't take that wager.

It'll take more than a stake through the heart and a necklace of garlic to keep global warming believers from achieving their stated goals over the next few years.

Which is why I've written *Air Con*. I wanted to go on a voyage of discovery about global warming and to follow the evidence where it leads, using my experience as an investigative journalist. I explain, as we go through, what I believe about it and why, so I won't distract you with the detail at this point.

For the record, because I know the believers will work themselves into a lather over this book, I have not sought, been offered, nor accepted, a cent from "Big Oil", "Big Industry", "The Cigarette Smoking Man" or in fact anyone else, to write this book and publish it. Heidi and I own an independent book publishing, magazine and newspaper business which has published dozens of books by various authors over the years, many of them bestsellers. In fact, the publishing empire was born when major publishing houses were threatened behind the scenes with massive lawsuits if they handled my first book, *The Paradise Conspiracy*, on the international tax haven dealings of some major corporates. A security sweep at my office inside Television New Zealand confirmed the phone lines had been bugged, my home suffered break-ins, copies of the manuscript were taken and there was an attempt on my life when the brakes on my vehicle were sabotaged – an incident that prompted the Mitsubishi dealership to call in police when they found the damage. Surrender, however, was not an option: the house was mortgaged, money was borrowed from family, and a fully-fledged publishing company emerged with a predatory wolf as its logo and a take-no-prisoners attitude to match. *Air Con* is just the latest in that fiercely independent pedigree.

Planet Earth is in the grip of Chicken Little Syndrome, and frankly that makes this not just an important story but a compelling story. If humans are really the major cause of global warming via greenhouse gas emissions, and the planet will keel over and die if we don't do something urgently, then I'd be among the first to be storming the barricades of complacency. But if significant warming (catastrophic) is not actually happening, or if it is but it is caused by Nature, not humans, then frankly I don't see the point in throwing the world's economy down the plughole if it's not going to stop the warming. King Canute found he could not order the tide to stop coming in. If global warming is a planetary tide that we have no control over, why waste money and time trying to stop it?

As you go through this book, you're going to be reading some science. Nothing to be afraid of, in fact it's essential if you have any hope of finding out whether you've been conned by the global warming industry.

To try and make that journey a little easier, here are some basic tools to help you understand the main claims people like Al Gore make about climate change. This is what they believe:

Global Warming Theory: (usually called 'climate change' by believers in order to cover all eventualities). In simple terms, the theory says that by burning fossil fuels like oil and coal (hydrocarbons), humans have unleashed waste gases like Carbon Dioxide (CO_2) and Methane into the atmosphere in such huge amounts that the planet is warming up uncontrollably.

Why? Because Carbon Dioxide (CO_2) absorbs the sun's heat when it is reflected off the planet's surface back into the air, and traps that heat in our atmosphere instead of letting it escape into space.

What happens next? As the atmosphere begins to warm because of this trapped heat being stored by gases like CO_2 and Methane, the extra heat melts icecaps and causes extra evaporation from the oceans. This evaporation turns into water vapour (steam, humidity, clouds) which increasingly blanket the planet and cause even more warming, evaporation and humidity, and in turn more ice melt.

How did this theory arise? Changing weather patterns have been with us forever, but our ability to study them is only quite recent. In the 1980s a group of scientists who suspected humans were polluting Earth to the point of sickness started discussing their beliefs. The US Congress heard testimony about "possible" global warming in 1988, and by 1990 the United Nations had already assembled a team of experts to report on the new threat of global warming. If that sounds suspiciously fast for a bureaucracy to move you'd be right. As you'll see later in this book, not much happens behind the scenes on global warming that has anything to do with coincidence. By 1992 the United Nations had convened a massive "Earth Summit" in Rio de Janeiro, Brazil, to talk about the future of the planet.

In 1995 the United Nations issued its second report on Climate Change via the group known as the UN Intergovernmental Panel on Climate Change (IPCC for short) – a select group of climate scientists who'd been invited to contribute.

The seeds planted at the 1992 Rio conference bloomed into the Kyoto Protocol of 1997, which you've heard so much about, and the UN Intergovernmental Panel on Climate Change issued its third report on the problem in 2001, with a fourth bite at the cherry in 2007. Each report was scarier than the last.

The Jargon: Boy, the global warming believers love their jargon. It's part of the process of keeping the rest of us confused and content to let the "experts" handle things. Sadly, these experts are working for people who want to steal your money. You're going to have to get down and get dirty in order to see what they're up to. They use a lot of acronyms like "IPCC" to cover for horrendously long and ponderous titles – most of these you can check in the Index at the back of the book if you forget what they mean. Let's take a look at some of the other key phrases, however:

Anthropogenic: It means 'human-caused'. Essentially, believers argue that global warming is entirely human-caused because of the greenhouse gases we've emitted through pollution.

Climate Change: This is now the official phrase to describe global warming, because it is so vague it can actually describe any weather or climate event. "Global Warming" doesn't sound convincing to the public when it's snowing outside.

Greenhouse Effect: This is a way of comparing runaway global warming on Earth to the heat and humidity inside a greenhouse. The only problem is that – unlike a greenhouse – Earth doesn't have glass walls or a roof, so heat actually can and does escape. Global warming believers know this, but they use "greenhouse effect" to paint an easy mental picture in the public's mind even if it isn't technically true.

Forcings: A word to describe things that "force" temperature increases on Earth. An obvious culprit is the sun (solar forcing), but global warming believers prefer to claim that carbon dioxide (CO_2 forcing) is now responsible.

Feedbacks: Things that kick in after an initial 'forcing' to create merry havoc and cause even more warming. Examples might be melting permafrost, which releases more methane and carbon dioxide to join the party, or water vapour which supposedly cloaks the planet in humidity and creates a speed-up in the warming/melting process (thus releasing more carbon dioxide from melting ice and permafrost and so on).

Summary: If it's true it's one of the biggest crises mankind has faced, because the Earth will indeed begin to warm up uncontrollably and sea levels will rise in catastrophic fashion. If it's true.

There are, however, very good reasons to believe it is utterly wrong and that swallowing the lie that global warming is caused by humans will cost you, personally, thousands of dollars a year and cause the collapse of western civilisation (something the Green movement has been hoping for for decades).

This steamroller machine to introduce carbon taxes and emissions cuts at the end of this year has been sold to the public, the media and politicians as utterly essential, regardless of the cost, because "the science is settled", and we are doomed if we don't act now.

History is littered with the bodies of stupid people who did things in haste because their leaders assured them it was "utterly essential, regardless of the cost". This book is a direct challenge to that kind of mentality, primarily showing that not only is the science not settled, but that the current panic over climate change has been deliberately provoked to cover a deeper agenda. I have bent over backwards in this book to quote peer-reviewed studies by leading scientists that show a healthy and ongoing debate about the reality and extent of global warming. I'm sure some of my critics will attempt to cherry pick their lines of attack, but the totality of the studies leaves little room for doubt that warming is largely natural in origin, and probably overestimated because of the dubious quality of international surface temperature sensors and sea level gauges.

After reading *Air Con*, if you hear anyone on a blogsite, in a newspaper or on TV trying to denigrate sceptics as just "deniers", and assuring you "the science is settled", I suspect like me you'll be rolling all over the floor laughing so hard it'll take you a week to stand up. The science may be many things, but settled it isn't. And when there's so much of your future and your family's future at stake in the outcome of this debate, the last thing you need to hear are meaningless, patronising platitudes.

This book is a weapon against spin and propaganda. Use it well.

Chapter One

The Science Is Settled

"We've got to ride the global warming issue. Even if the theory is wrong, we will be doing the right thing in terms of economic and environmental policy"

– Timothy Worth, Council on Foreign Relations, 1990

If you believe 95% of the news media reports, Planet Earth is dying because of pollution from us evil humans. One report out in January of this year warns we're now facing a "sea level rise of one metre within 100 years", and notes that this is "three times higher than predictions from the UN's Intergovernmental Panel on Climate Change (IPCC)".[1] On the other hand, if we're going to hell in a handcart, another report at least provides some cold comfort – we won't have to go without frozen confectionery as the planet heats up!

"Edible antifreeze saves ice cream", screeched the headline in *New Scientist* magazine.[2]

Woohoo! Now *that's* the kind of practical response to global warming we like to see. Why not drop the whole icecream thing and just go with flavoured antifreeze in a cone?

If anyone epitomises global warming on a popsicle stick, it's former US Vice President Al Gore, whose "documentary" *An Inconvenient Truth* broke box office records and scared political leaders the world over when it was released a couple of years ago. I use scare-quotes around documentary because a British court found Gore's film was factually incorrect in a number of areas, and ruled that schools

1 "Sea Level Rise of One Meter Within 100 Years", *Science Daily*, 11 January 2009

2 *New Scientist,* 11 January 2008, http://www.newscientist.com/article/dn13178-edible-antifreeze-promises-perfect-ice-cream.html

wanting to show the film to kids should teach classes about all the points Gore got wrong.[3]

But if the old adage about never letting the facts get in the way of a good story is true, then Gore is a master storyteller. Take this transcript of a US National Public Radio broadcast:[4]

"Even once-sceptical Republicans are coming over to Gore's side – and it seems the debate has shifted from arguing whether there is a climate crisis to disagreement over how to fix it.

" 'The science is settled', Gore told the lawmakers. Carbon-dioxide emissions – from cars, power plants, buildings and other sources – are heating the Earth's atmosphere.

"Gore said that if left unchecked, global warming could lead to a drastic change in the weather, sea levels and other aspects of the environment. And he pointed out that these conclusions are not his, but those of a vast majority of scientists who study the issue."

As Nazi propagandist Josef Goebbels once helpfully pointed out to the public relations industry, "If you tell a lie big enough and keep repeating it, people will eventually come to believe it".

The biggest lie in Al Gore's comments is this: "The science is settled". It's what practitioners of the dark arts of public manipulation refer to as a "lizard brain" phrase, that parks itself deep into the subconscious of listeners, thanks to a comforting appeal to authority figures (scientists), and an assurance they've got it right (settled). Lizard brains are where instinctive, knee jerk reactions are generated often before the person consciously realises.

Around the world, politicians pushing a global warming barrow picked up on the "science is settled" phrase and its variant, "consensus".

"These days the scientific consensus is almost total – climate change is coming ready or not. The sceptics are receding, just like the Arctic ice-sheet," exclaimed New Zealand's Minister for the Environment Pete Hodgson confidently.[5]

Journalist Andrew Bolt, writing in the *Herald Sun*, tried to challenge Australia's Climate Change Minister Penny Wong with some inconvenient facts:[6]

---

3 "Al Gore's inconvenient judgment", *Times of London*, 11 October 2007, http://business.timesonline.co.uk/tol/business/law/corporate_law/article2633838.ece

4 NPR, 21 March 2007, "All things considered", http://www.npr.org/templates/story/story.php?storyId=9047642

5 *NZ Herald*, 16 February 2005, Op/ed page

6 http://blogs.news.com.au/heraldsun/andrewbolt/index.php/heraldsun/comments/im_sorry_but_i_tried/

"At the Melbourne Press Club today, I asked Climate Change Minister Penny Wong the following question (from memory):

" 'The IPCC, the UN's climate change body, in its February report said it had detected human-induced global warming in just one 25 year period, up to around 1998. Since 1998, the consensus of the four bodies that measure the world's temperature is that the world has not warmed. It has not warmed for a decade, and over the past couple of years has actually cooled. Minister, how many more years of no-warming will it take before you accept that the global warming theory on which you've based your huge carbon cutting scheme is actually wrong? One more year of no-warming? Five years? Or 15 years?'

"I'm afraid that Ms Wong did not answer my question. Then again, she did not question the premise, either. Instead she said she had to go with the 'consensus' science...But I suspect this is the question that will start to haunt her, because the facts – and an honest answer – are so deadly to her cause," writes Bolt.

"Consensus", too, seems to be order of the day in the journal *Foreign Policy*, with this claim from environmental writer Bill McKibben: "The science is settled, and the damage has already begun."[7]

McKibben's comments are worth further examination, because as you read this book you will see how utterly wrong he is.

"Scientists Are Divided?" poses McKibben rhetorically, before offering an immediate rebuke: "No, they're not. In the early years of the global warming debate, there was great controversy over whether the planet was warming, whether humans were the cause, and whether it would be a significant problem. That debate is long since over."

Really? You can only put such a claim, made early this year, down to sheer *chutzpah* and old fashioned arrogance; it must be, because surely McKibben wasn't unaware of this announcement when he wrote his dismissive comments:

"More than 31,000 scientists have signed a petition denying that man is responsible for global warming," reported Britain's *Daily Telegraph* in mid 2008.[8] "The academics, including 9,000 with PhDs, claim that greenhouse gases such as carbon dioxide and methane are actually beneficial for the environment."

7 *Foreign Policy*, January/February 2009 issue, http://www.foreignpolicy.com/story/cms.php?story_id=4585

8 *Daily Telegraph*, 30 May 2008, http://www.telegraph.co.uk/news/worldnews/2053842/Scientists-sign-petition-denying-man-made-global-warming.html

There is no similar list of even 500 scientists who've put their names to a document supporting, like McKibben or Al Gore or Penny Wong, claims that humans are definitely causing global warming. In other words, far from a "consensus" or "settled", the numbers of scientists prepared to publicly and visibly speak out against global warming hysteria seriously outnumber those who publicly list themselves as believers.

Keep repeating the lie often enough…

"Although the details of future forecasts remain unclear," continues McKibben, "there's no serious question about the general shape of what's to come.

"Every national academy of science, long lists of Nobel laureates, and in recent years even the science advisors of President George W. Bush have agreed that we are heating the planet. Indeed, there is a more thorough scientific process here than on almost any other issue: Two decades ago, the United Nations formed the Intergovernmental Panel on Climate Change (IPCC) and charged its scientists with synthesizing the peer-reviewed science and developing broad-based conclusions. The reports have found since 1995 that warming is dangerous and caused by humans. The panel's most recent report, in November 2007, found it is 'very likely' (defined as more than 90 percent certain, or about as certain as science gets) that heat-trapping emissions from human activities have caused 'most of the observed increase in global average temperatures since the mid-20th century.'

"If anything, many scientists now think that the IPCC has been too conservative – both because member countries must sign off on the conclusions and because there's a time lag. Its last report synthesized data from the early part of the decade, not the latest scary results, such as what we're now seeing in the Arctic."

McKibben, a former journalist turned environmentalist, is undoubtedly a nice enough guy, but he's such an earnest global warming believer that even Al Gore looks up to him:

"When I was serving in the Senate, Bill McKibben's descriptions of the planetary impacts... made such an impression on me that it led, among other things, to my receiving the honorific title 'Ozone Man' from the first president Bush," said Gore.[9]

---

9 http://en.wikipedia.org/wiki/Bill_McKibben

Well, now we know who to blame.

Politically, global warming is big news. Huge news. Not only is it the secular Armageddon story of our times, but it's also tailor-made for TV and Hollywood. Fortunes will, and already have, been made terrifying the public in the name of infotainment. This is *Fear Factor*, *Storm Chasers* and *Who Wants To Be A Millionaire?* all rolled into one.

For politicians, up until the world economic crash, global warming was seen as the "must have" issue on campaign tickets, to make it appear that the candidate or their party was listening to public [read "media-generated"] concerns.

This media/political trendsetting exercise is a curious but effective parasitical relationship, somewhat akin to the remora fish that attaches itself to sharks. The media, who can broadcast sexy pictures of cute polar bears, and crashing glaciers, love the hype and excitement and sense of urgency and drama that comes with these "end of the world" kind of stories. The politicians thrive on providing the solutions to media scare stories, and therefore looking like heroes and saviours. If the Bible's Book of Revelation is *Left Behind* for intellectuals, then Al Gore's movie *An Inconvenient Truth* is *Left Behind* for idiots.

And there's something so very messianic about Al 'the Baptist' Gore and Barack 'The One' Obama's tag team mission to change the world.

"Al Gore is lauded by the Academy Awards, by the Hollywood elitists, by the United Nation's activists, by the far left, by the environmental extremes," notes Republican sceptic Senator James Inhofe. "Katie Couric calls him the secular saint. Oprah Winfrey calls him the Noah of our time."[10]

The insatiable media appetite for this kind of news meant, particularly in the early days, that they sought out people prepared to give them "bad news" soundbites. Those scientists prepared to speak up became high profile, and universities and institutions wanting to get in on the act provided research funding, at the same time approaching philanthropist donors and governments seeking money to pay for the grants to global warming researchers. The scarier the hypothesis, the easier it was to sell another round of "further

10 James Inhofe, keynote speech, CPAC conference, 2 March 2007, http://www.conservative.org/pressroom/2007/speech_inhofe.asp

study needed", and pretty soon hordes of academics, bureaucrats and journalists found themselves gainfully employed fulltime on the story of the century.

And what a story it's turned out to be. In March this year, British economist Lord Nicholas Stern tried to ratchet up the fear factor just a touch by suggesting it is no longer a two degree rise in temperature we face, but "six, seven degrees Centigrade" by the end of this century, bringing with it "deserts spreading across much of southern Europe, collapses in crop yields, rivers drying and perhaps billions of people being forced to leave their homes," the *Times of London* reported.[11]

"What would be the implication of that?" Stern asked rhetorically. "Extended social, extended conflict, social disruption, war essentially, over much of the world for many decades. This is the kind of implication that follows from temperature increases of that magnitude."

Stern's threats are valid only if the temperature rise is real, and not merely the result of scaremongering hot air. Given that Planet Earth is currently cooling slightly, Lord Stern's a pretty brave man making that kind of statement. But then, he's preaching to the choir of fellow warming believers.

Science, like other disciplines, is influenced by the market. If the popular tide is flowing one way, it's easier to get funding if you go with the flow and give the audience what they want. It's not so easy trying to challenge herd-think, and it's even harder getting research cash to carry out what many regard as flogging a dead horse.

There are politicians the world over – and I know some of them – who will stand up in their respective parliaments and congresses on any given day and channel Al Gore ad nauseum in debates intended for public consumption. Privately, those same politicians will admit they think anthropogenic [human-caused] global warming is a crock.

"But what can we do?" one told me. "If we are not seen to be doing something we'll suffer a backlash from voters or other governments that could hurt our trade opportunities."

The Emperor has no clothes, but politically few have the guts to shout it out.

---

11 "Nicholas Stern: politicians have no idea of the impact of climate change", *Times of London*, 12 March 2009, http://www.timesonline.co.uk/tol/news/environment/article5895518.ece

At stake, however, are a number of things. And those stakes are now even higher thanks to the global financial collapse. If we are wrong about the extent of global warming, or if we are wrong about what causes it, vast amounts of taxpayer money – trillions of dollars potentially – will be thrown down a black hole for nothing.

That's money that could be spent feeding the poor, providing good healthcare around the world, improving education opportunities, easing the pressure in world troublespots.

Instead, that cash is being earmarked for multinational corporates in emissions trading schemes, and governments through carbon taxes.

But will paying a carbon tax, or offsetting a carbon emission with buying a carbon credit, really make a blind bit of difference to the problem, or just create a new level of international bureaucrats and money-traders? That's the question scientists, politicians, the news media and the public need to be asking.

Republican senator James Inhofe made an interesting point a couple of years back. He told delegates to a conference what the four biggest tax increases were in the US over the past four decades.[12]

"First was the Revenue Expenditures Control Act 1968, $35 billion; the crude oil windfall profit tax, $23 billion; the Budget Act of 1990, $29 billion; and then Bill Clinton's tax increase of $32.3 billion. Now these are the largest tax increases in the recent history of America.

"If you put this next to what the tax increase would be if we were to pass something like Kyoto," said Inhofe, "it would be a $300 billion tax increase. It would be 10 times greater than the Bill Clinton tax increase of 1993."

In dollar terms, every American household would be forking out an average of US$2,750 a year in extra taxes to help prevent global warming. And that's assuming the Kyoto Protocol actually makes a difference.

One of the reasons the US did not ratify Kyoto is because its own climate change scientists didn't think it would work. It was simply a feel-good scheme to fool the public. Tom Wigley, then of the National Centre for Atmospheric Research in the US, reported to the Clinton administration in 1998 that even if all industrialised nations in the world adopted the Kyoto Protocol and followed it to

12 The CPAC 2007 conference, Washington DC, 2 March 2007. *Congressional Quarterly Transcriptions.*

the letter, it would reduce global temperatures in 2050 by only 0.07 degrees Celsius,[13] or 7/100ths of one degree. That's a figure so low it doesn't even begin to emerge from the margin of error, it is actually scientifically undetectable. Every year at the moment we're seeing climate swings of 0.5°C up or down based on natural variations. The Kyoto figure is meaningless.

The US Energy Information Administration ran its own figures, and declared that by 2010 the cost of to the US of ratifying Kyoto would be between $100 billion and $400 billion. President Clinton, naturally, realised Kyoto was a lame duck and it sank without trace in Washington.

Activists and pressure groups seized on US reluctance to sign up to Kyoto and rolled out a PR blitz, as these comments from KyotoUSA show:[14]

"Global warming is the most serious threat facing the planet today. Studies conducted by the world's most respected climate scientists demonstrate that we must act collectively and immediately to make significant reductions in the amount of greenhouse gases that we are releasing into the atmosphere. Our failure to act now will result in catastrophic and irreversible consequences for all life on this planet.

"Countries around the globe are attempting to address this problem. In an agreement known as the Kyoto Protocol, most industrialized countries have agreed to cut their greenhouse gas emissions. Unfortunately, the United States steadfastly refuses to ratify the Kyoto Protocol and actively works against the efforts of the world community to reach its full implementation."

Downunder, one of Australasia's well known left wing bloggers, Russell Brown, also showed little appreciation for Kyoto's ineffectiveness at a practical, temperature-reducing level.[15]

"Speaking of global warming, the Australian government, in a fit of greed and denial, has confirmed its decision to ignore the Kyoto Protocol in case it costs some money. We hear so much about how the US and Australia aren't ratifying that it's easy to forget that 70 other countries already have."

There's an old saying, "truth is not determined by majority vote".

---

13 Marlo Lewis op/ed column, "All cost, no benefit", *Tech Central Station*, 20 July 2005. http://cei.org/gencon/019,05329.cfm

14 http://www.kyotousa.org/

15 Russell Brown's Hard News, 7 June 2002, http://www.nznews.net.nz/hardnews/2002/20020607.html

Just because a whole bunch of people (or in this case governments) believe something, doesn't necessarily make it correct. Whilst the appearance of long lists of eminent names is relevant to clarifying whether debate on an issue is "settled" or not, big lists don't actually prove the empirical truth of a claim.

Yet the Left very quickly turned Kyoto into a debate about 'doing our bit', and singled out countries they felt were not 'doing their bit' – like the US and Australia. The debate should never have been about everyone pitching in, it should have been about achieving a goal. And in real terms, reducing temperatures, Kyoto would achieve nothing no matter how hard we worked.

Brown's native New Zealand, under the leadership of then prime minister and now UN appointee Helen Clark, was one of the first countries to ratify Kyoto in a blaze of publicity. Clark was the first leader outside Europe to initiate an emissions trading scheme. But (there's always a 'but'), Helen Clark's legacy is a salutary lesson to the rest of the world about the flash over substance involved in political solutions to global warming.

"Helen talked of reducing net emissions to zero," wrote blogger David Farrar.[16] "Kyoto is about getting them back to 1990 levels. But surely Helen managed to at least keep them constant? Nope. From 1999 to 2006, this is the net increase (including offsetting with land use and forests) for various countries:

1. Sweden -61.8% [a 61.8% reduction]
2. Norway -31.8%
3. Estonia -23.4%
4. Monaco -21.4%
5. Finland -9.2%
6. France -6.3%
7. Belgium -5.3%
8. Hungary -4.6%
9. Slovakia -4.5%
10. Poland -4.3%
11. Denmark -3.4%
12. Netherlands -3.2%
13. United Kingdom -2.6%

16 http://www.kiwiblog.co.nz/tag/carbon_emissions

14. Germany -2.0%
15. European Community -0.9%
16. Portugal +0.9% [a 0.9% increase in emissions]
17. Japan +0.9%
18. United States +0.9%
19. Italy +2.7%
20. Ireland +3.0%
21. Liechtenstein +3.9%
22. Iceland +5.3%
23. Bulgaria +6.2%
24. Greece +7.0%
25. Australia +8.2%
26. Czech Republic +8.6%
27. Switzerland +8.8%
28. Canada +11.0%
29. New Zealand +12.0%
30. Spain +18.0%
31. Turkey +33.3%

"Only two industrialised countries (excluding those who are below their Kyoto targets) have a worse record than New Zealand under Helen Clark. We are also at 33% over our 1990 Kyoto target. The US is only 14% over, and Australia 7% over. The United Kingdom is 16% under.

"If Labour try to claim any sort of moral high ground on climate change, just remember these facts. Helen Clark's record was one of the worst in the world on carbon emissions," concluded blogger David Farrar.

The icing on the cake is that despite New Zealand being one of the worst performers in the world, even worse than the US which pointedly did *not* ratify Kyoto, the United Nations last year named Helen Clark as a "Champions of the Earth" award winner for her work on preventing climate change.[17]

"New Zealand's Prime Minister Helen Clark – whose country will host World Environment Day this year with the theme 'Kick the Habit: Towards a Low Carbon Economy!' – [is] among the seven

17 United Nations Environment Programme release, "Climate Change Links 2008 Champions of the Earth Award Winners", 28 January 2008 http://www.unep.org/Documents.Multilingual/Default.asp?DocumentID=525&ArticleID=5738&l=en

environmental achievers chosen for this year's awards, the United Nations Environment Programme announced today.

"The Champions of the Earth prize, which will be given out at a ceremony in Singapore on 22 April, recognizes individuals from each region of the world who have shown extraordinary leadership on environmental issues."

It's kind of like awarding Zimbabwe's Robert Mugabe a Nobel Peace Prize because he allowed his soldiers to use their swords to steal other people's ploughshares in furtherance of agricultural policy. Arguably, George W. Bush would have been a worthier recipient than the New Zealander – at least his country substantially outperformed Helen Clark's when it came to actually reducing carbon emissions. But this is the problem with the global warming industry. They're so busy scratching each other's backs and conspiring to get funding and awards for the "in crowd" that they're all mouth and no action.

The irony in this case is that Helen Clark has just been appointed to the number three job in the United Nations, heading the powerful UN Development Programme with a $5 billion budget to promote a range of international projects including climate change policy and social engineering, among other things.

Environmental and other lobby groups (collectively referred to as 'Non Governmental Organisations' or NGOs in UN-speak), which generally have very close relationships with the news media, are another sector who do very well out of environmental scare stories. For them it's not awards or the lure of a cushy United Nations job, but money. The more punters they spook, the more donations they get and the more NGO funding and international conferences they can tap into.

Everyone's a winner on the global warming gravy train except, perhaps, the public and the polar bears. The polar bears lose because, in truth, none of the hand-wringing and tax gouging is actually going to change the course of global warming if it's happening. And the public lose because they're the patsies who'll ultimately bear the entire financial cost through higher food and fuel prices, and higher taxes, all of which flow through every other aspect of the economy.

That's why we have to be absolutely sure that anthropogenic global warming is taking place, before we commit serious dollars and time to possible solutions. Proof of ordinary, natural global warming is not good enough. If it's natural, there's probably no hope of preventing it and we're on for the ride whether we like it or not. And of course if no abnormal global warming is happening at all, human or otherwise, then we can all rest easy, take a paracetamol and get some sleep before turning to the real issues that face our civilisation in the 21st century.

Don't misunderstand me: there are very good reasons to reduce pollution and protect the environment, and I understand that most who believe in global warming are genuine in their reasons. I respect that, but I don't respect the people who have conned you.

I only quoted you the most famous *line* from Nazi propaganda specialist Josef Goebbels at the start of this chapter, but so you can appreciate what this book is about, I'll now print the entire quote. Read it carefully:

"If you tell a lie big enough and keep repeating it, people will eventually come to believe that. The lie can be maintained only for such time as the State can shield the people from the political, economic and or military consequences of the lie. It thus becomes vitally important for the State to use all of its powers to repress dissent, for the truth is the mortal enemy of the lie, and thus by extension, the truth is the greatest enemy of the State."

In simple terms, propaganda fools the public by lying to them, and then using every available tool to marginalise opponents, write them off as "deniers" and "fools", and passing laws to reinforce belief in the lie. It is not just the lie itself, but the efforts put into squashing any opposition to the lie, that are the mark of state propaganda.

But truth is the mortal enemy of the lie, and in this book I'm aiming to put some balance back in the global warming debate by exposing some genuinely inconvenient truths.

You've probably detected from my tone I'm a global warming sceptic. I haven't always been, but I am now. However, I'm also an investigative journalist who gets professional satisfaction from testing the evidence. Therefore, throughout this book, you'll find me looking for the best evidence in favour of global warming generally, and human-induced global warming specifically, because I personally want to test that evidence for myself. I *want* to know whether global warming is real, and whether it's caused by humans.

So should you. After all, it's your money they're after. So let's find out.

Chapter Two

Is The Greenhouse Effect Such A Bad Thing?

"In searching for a new enemy to unite us, we came up with the idea that pollution, the threat of global warming, water shortages, famine and the like would fit the bill...all these dangers are caused by human intervention...the real enemy, then, is humanity itself"

– ***The Council of the Club of Rome, 1991***[18]

It might seem like a silly question, but it isn't. At the centre of this debate is the claim that CO_2 emissions are causing a "greenhouse effect", where Earth will heat up uncontrollably and we will all be forced to live on the tops of mountains because of rising sea levels. The next few chapters will address all of this, but let's start with the greenhouse effect. While most of us have been brainwashed by the media to believe greenhouse gases are "bad", we wouldn't be here without them. Part of getting you to be fearful of climate change involves making you fearful of carbon dioxide and the greenhouse effect. In this chapter, I want to show you why Earth needs a greenhouse effect in order for life to exist, and anyone who tells you differently is lying to you.

You'll also find out in this chapter a little bit more of the science behind carbon dioxide (CO_2), which you'll need to know if you want to break free of the brainwashing and save your wallet from being raided.

Is it really that terrible if Earth heats up a little? Let me explain what a "greenhouse effect" actually does by showing you what life might be like if we didn't have one.

---

18 *The First Global Revolution: A Report by the Council of the Club of Rome* by Alexander King & Bertrand Schneider, Pantheon Books, New York, 1991

Throughout the solar system, there are examples of planets and moons whose *lack* of a greenhouse effect is a killer. Mars is a good example. Without enough gravity, Mars couldn't hold on to its atmosphere or its water vapour and its greenhouse gases escaped. It's effectively dead, even allowing for the possibility that microbes pinged into space by an asteroid strike on Earth have landed on Mars[19] and found a niche producing methane. Those bacteria can produce methane till the cows come home but they'll never create a greenhouse effect on Mars because the methane slips away into space.[20]

The Moon is another excellent example of what happens with no greenhouse effect. During the lunar day, average surface temperatures reach 107°C, while the lunar night sees temperatures drop from boiling point to 153 degrees below zero. No greenhouse gases mean there's no way to smooth out the temperatures on the moon.

On Earth, greenhouse gases filter some of the sunlight hitting the surface and reflect some of the heat back out into space, meaning the days are cooler, but conversely the gases insulate the planet at night, preventing a lot of the heat from escaping. The moon doesn't have this protection. If it's getting sunlight it's scalding hot. Without sunlight, it's cold enough to turn a human into a Popsicle in seconds.

Earth without a greenhouse effect would be lethal.

Greenhouse gases on Earth perform another service – they actually help cool the planet down! When they heat up near the surface (such as with sunlight on the ground), they carry that heat high up into the atmosphere on what meteorologists refer to as "convective towers". The heat then dissipates into space at night. Some estimates suggest the benefit from this circulation of heat-absorbing gases through the air is enough to reduce Earth's surface temperature by about 60°C.[21] In other words, without greenhouse gases, ironically, it would be far hotter in most cities than the Sahara desert.

It is sometimes said that Earth has a twin – the planet Venus – and Venus has often been raised as Earth's global warming "scarecrow" – an example of what happens with a runaway greenhouse effect.

---

19 In just the same way that Martian rocks have been found on Earth, rock debris from asteroid craters on Earth has undoubtedly found its way to Mars after being flung into space carrying dirt and other organic matter. If life is found on Mars, scientifically the best odds are that we'll discover they are descendants of earth bugs.

20 On Earth, our stronger gravity holds onto methane longer, but even so the gas takes only 8.5 years to dissipate out of the atmosphere, mostly by reacting with oxygen and hydrogen, which helps create water vapour, but also through stratospheric loss into space.

21 http://www.junkscience.com/Greenhouse/

So does Venus, which is closer to the sun than we are, really show us what's in store if global warming takes hold?

It is true that Venus is covered in clouds and greenhouse gases, many of them belched from volcanoes. But it is not really accurate that those gases created the greenhouse problem on Venus. Temperatures on the green planet run to 480°C – hot enough to melt lead and destroy satellite probes sent in. But what most people don't realise is that Venus doesn't rotate on its axis (creating day and night) as much as Earth does. A Venusian day is slightly longer than a Venusian year, meaning the sun beats down continuously, slowly roasting the planet while it turns on its axis once a year. It's the ultimate case of sunburn, barring that endured by Mercury.

Undoubtedly the gases in Venus' atmosphere have accentuated the blistering solar barrage, but the problem in the first place appears to be the slow rotation, which overheats the surface and would have turned any existing water into steam (water vapour, perhaps the most significant of the greenhouse gases).

Another big problem for Venus is that it doesn't have plate tectonics – the system of floating continental crusts that creates fault lines and earthquakes on Earth. Down here, the continents move on a kind of conveyor belt, one edge constantly slipping underground where it meets another continental plate, and back into the earth's mantle, while on the other side of the continental plate new crust is being generated from out of the bowels of the earth.

While earthquakes can be terribly tragic for those caught in them, they're much milder than the way Venus deals with pressures from underground. Every 500 million years or so, Venusian continents simply tip up like the lid on a boiling pot in a vast upwelling from beneath, and slide down into the molten core like the Titanic going down. Lock, stock and all smoking barrels. These events release vast amounts of heat into the atmosphere and, of course, if Venus had an ocean it would be like pulling the plug out when an entire continent tipped up and disappeared.

As if all that wasn't enough for poor old Venus, the sun has become steadily hotter by about 30% over the past few billion years, and there's little escaping the obvious conclusion: Venus is what happens when the sun gets too hot for an already crippled planet to handle.

Global warming believers are confident Venus provides valid lessons for Earth, however.

"Understanding all of this will help us pin down when Venus lost its water," says David Grinspoon, one of the team responsible for the Venus Express satellite probe.[22] "That knowledge can feed into the interpretation of climate models on the Earth because although both planets seem very different now, the same laws of physics govern both worlds."

Except, perhaps, for that whole lack of a 24 hour day thing, and the physics of continental renewal being utterly and completely different. It is unlikely that Venus offers much to science at all in regards to the greenhouse effect on Earth. But good luck to Grinspoon and his team, and good on them for trying. Regardless of the nebulous linkage to terrestrial warming, the Venus Express project provides useful information about Venus.

In all of this, then, we can see how greenhouse gases are crucial for life on earth to exist because of their ability to dissipate and transfer heat, as well as their insulative effects at night. But the greenhouse gases have other benefits too.

Carbon dioxide (CO_2) is the black sheep of greenhouse theory. According to global warming believers, it's this colourless gas that is responsible for our plight this century. CO_2, however, makes up only 385 parts per million in the atmosphere or, expressed another way, only 0.038% of the concoction we call "air". Pre-industrialisation last century, CO_2 levels were sitting at 280ppm, so they've increased by around 35% over the past century or so. By far the biggest player in global warming is, however, water vapour. Up to 4.0% of air is water vapour, steam, call it what you like. Not only is water vapour[23] vastly more prevalent than CO_2, it is also directly responsible for 90 to 95% of Earth's greenhouse effect.[24]

"The remaining portion [5 to 10% influence] comes from carbon dioxide, nitrous oxide, methane, ozone and miscellaneous other minor greenhouse gases," reports the website Junkscience.com.[25]

"As an example of the relative importance of water, it should be

---

22 European Space Agency news release, March 2008, carried at http://www.dailygalaxy.com/my_weblog/2008/03/climate-change.html

23 Water vapour is used in a general sense here and includes clouds. For the record, clouds are believed to account for around 20% of the greenhouse effect, and water vapour 70%, to reach the combined value of 90% (or up to 95%, depending on how you calculate the split between heating of the troposphere (lower atmosphere) and stratosphere (upper atmosphere))

24 Freidenreich and Ramaswamy, "Solar Radiation Absorption by Carbon Dioxide, Overlap with Water, etc", *Journal of Geophysical Research* 98 (1993):7255-7264

25 http://www.junkscience.com/Greenhouse/

noted that changes in the relative humidity on the order of 1.3-4% are equivalent to the effect of doubling CO_2."

Why then, if CO_2 emissions (both human and natural) are only responsible (on a best case scenario from global warming believers) for between 5 and 10% of global warming, are we wasting time targeting carbon emissions at all? Why, if water vapour is the biggest culprit, aren't we targeting that instead?

The cynical answer is because targeting carbon is a money-making exercise. It is something our civilisation uses plenty of, and it can be easily taxed. Wheeler dealers can make cash out of carbon credits and emissions schemes, and governments and the United Nations can very easily identify and isolate carbon users, so as to be seen to be doing something.

Taxing water vapour, on the other hand, is impossible. It isn't sold in fuel bowsers, and in fact the vast bulk of atmospheric water is a direct result of natural processes. Trying to tax water vapour is almost as useless as taxing air itself.

In global warming theory, water vapour is the elephant in the room that no global warming believers really want to talk about. And when they do, they talk around it and deflect back to carbon – such as this piece from a Green website:[26]

"However, just because water vapour is the most important gas in creating the natural greenhouse effect does not mean that human-made greenhouse gases are unimportant. Over the past ten thousand years, the amounts of the various greenhouse gases in the Earth's atmosphere remained relatively stable until a few centuries ago, when the concentrations of many of these gases began to increase due to industrialization, increasing demand for energy, rising population, and changing land use and human settlement patterns. Accumulations of most of the human-made greenhouse gases are expected to continue to increase, so that, over the next 50 to 100 years, without control measures, they will produce a heat-trapping effect equivalent to more than a doubling of the pre-industrial carbon dioxide level."

If you spotted the inaccuracies, such as the claim that gases began to increase "a few centuries ago…due to industrialisation", good on you. For the record, industrialisation began in the late nineteenth

26 http://www.greenfacts.org/studies/climate_change/l_3/climate_change_8.htm

century and is not believed by reputable scientists to have begun having any global warming impact until well into the 20th century. Any greenhouse gas increases "a few centuries ago" were entirely natural in origin.

Then there's this, slightly more sophisticated (but ultimately flawed) offering from a global warming believer at *ScienceBlogs*:[27]

"There is some truth in this, insofar as water vapour is a greenhouse gas, but ... you will discover it is important to understand the difference between a forcing and a feedback. Carbon dioxide and methane are forcings – they *cause* the planet to warm. Water vapour is a feedback. The amount of water in the air is a *consequence* of temperature, and it in turns results in more warming. There's nothing we can do about feedbacks, but we can control anthropogenic forcings."

Well, yes, and no. There is a certain amount of question-begging going on there, given that the sun has been the primary "forcer" up until now, and increased solar radiation can and does result in increased evaporation from oceans, lakes and rivers, thus creating more water vapour which then cloaks the planet in warm humidity. Carbon dioxide doesn't have to be a significant ignition factor in that equation at all, and in fact might well be a by-product, a feedback, itself, rather than a "forcer". Studies have shown[28] that as oceans warm, CO_2 is released, so if the sun is the initial forcer, it figures that everything else (CO_2, water vapour and methane) follows.

Increased solar radiation also thaws tundra and ice, which in turn releases methane and CO_2 into the air along with steam.

However, let's tackle one of the other assumptions in this green spin – human-caused carbon dioxide emissions. You would think from reading it that humans have been pouring copious amounts of the stuff into the atmosphere sufficient to knock Earth off its perch. Yet in reality, of all the carbon dioxide emissions into the atmosphere every year, only about 3.4% is caused by humans. All our cars, all our jet planes, factories, cattle, Google searches – all of it – amounts to just 3.4% of world carbon dioxide emissions every year. And don't take my word for it, that's from the UN's Intergovernmental Panel on Climate Change (UN IPCC).

---

27 Blogger James Hrynyshyn, "Island of Doubt", http://scienceblogs.com/islandofdoubt/2009/01/dear_john.php?utm_source=sbhomepage&utm_medium=link&utm_content=channellink

28 "Role of deep sea temperature in the carbon cycle during the last glacial", P Martin, D Archer, D Lea, *Paleoceanography*, VOL. 20, PA2015, doi:10.1029/2003PA000914, 2005

The rest of the carbon dioxide comes from forests, non-vegetative organic life, volcanoes and oceans. Of humanity's 3.4%, only around half (1.7%) actually goes into the atmosphere. The rest is soaked up by plants and other surface level "sinks" that utilise carbon.

Yet despite the fact we contribute only 1.7% of the atmospheric carbon dioxide emissions, global warming believers are utterly convinced it's *our* 1.7% tipping the planet towards environmental Armageddon, not the 98.3% of carbon dioxide emissions that nature throws up. Remember, carbon dioxide in total probably only accounts for 10% of the total greenhouse effect (the majority of the heat coming from water vapour), so on that math human-induced carbon dioxide accounts for 0.17% of total global warming annually. One estimate[29] says that humans account for only 2.5% of the *total* greenhouse effect since industrialisation began in the late 1800s. In other words, 2.5% of total global warming since then (that's 2.5% of a 0.5°C rise in world temperature, or 0.01 degrees) can be laid at our doorsteps.[30]

Yet despite this, global warming believers remain utterly convinced… you get the picture. It's almost as if belief in global warming is primarily a religious experience, a common faith where those rebelling against the perceived evils of modern civilisation can gather and worship. Forget the fact that their good book doesn't really support their beliefs, and that their high priest is Al Gore, because apparently facts don't matter, and the inconvenient fact that Gore's house uses enough power each month to electrify the entire Darfur region in the Sudan is *especially* irrelevant to those of the faith.

The bitter irony is that while many global warming believers would be the first to argue for a separation of church and state, they don't see anything wrong in trying to get the public to pay for this green religious belief. Indeed, in schools across the Western world, your taxes are being used to fund compulsory indoctrination classes into this belief system.

---

29 "The Real Inconvenient Truth", http://www.junkscience.com/greenhouse/

30 I was staggered to read, from NASA GISS scientist Gavin Schmidt on his blog RealClimate, a claim that "The correct statement is that CO_2 is around 30% higher than it was in the pre-industrial period, and <u>all</u> of that rise is due to human emissions." All of it? Really? No CO_2 released for natural reasons at all? All of the warming since 1700AD hasn't released *any* extra CO_2 from ice, tundra, soil or oceans? I thought that was the whole argument of the greenhouse effect. You can't melt your cake and keep it frozen at the same time. Either warming releases CO_2, or it doesn't. Extra CO_2 release is supposed to be a 'feedback' of warming. It's been warming for a couple of hundred years now. To take Schmidt at his simplistic word is to argue that the greenhouse theory doesn't actually work

It would be more helpful, probably, if your children were being taught the benefits of carbon dioxide. Far from being "bad", CO_2 is crucial for all life, and crucial for humanity's survival in particular.

Examples? *Junkscience* has a few:

"Estimates vary, but somewhere around 15% seems to be the common number cited for the increase in global food crop yields due to aerial fertilization with increased carbon dioxide since 1950. This increase has both helped avoid a Malthusian disaster *and* preserved or returned enormous tracts of marginal land as wildlife habitat, land that would otherwise have had to be put under the plough in an attempt to feed the growing global population.

"Commercial growers deliberately generate CO_2 and increase its levels in agricultural greenhouses to between 700ppm (parts per million) and 1,000ppm to increase productivity and improve the water efficiency of food crops far beyond those in the somewhat carbon-starved open atmosphere. CO_2 feeds the forests, grows more usable lumber in timber lots meaning there is less pressure to cut old growth or push into 'natural' wildlife habitat, makes plants more water efficient helping to beat back the encroaching deserts in Africa and Asia and generally increases bio-productivity.

"If it's 'pollution', then it's pollution the natural world exploits extremely well and to great profit. Doesn't sound too bad to us."

In fact, boosting atmospheric CO_2 to 700ppm in enclosed greenhouses is claimed by agricultural companies to boost crop yields by 30%. That's thirty percent more food generated from the same number of plants.

But surely, if we've moved from CO_2 at 280ppm in the 1800s, to 385ppm now, and the boffins are all screaming about runaway global warming, then allowing CO_2 to hit 700ppm would be the end of life as we know it, right?

The truth, as you're about to see, is a lot stranger than science fiction.

Chapter Three

What Cars Did The Dinosaurs Drive?

"If all mankind minus one, were of one opinion, and only one person were of the contrary opinion, mankind would be no more justified in silencing that one person, than he, if he had the power, would be justified in silencing mankind"

– John Stuart Mills, On Liberty

One of the most misleading statistics to be quoted by the global warming industry is the claim that greenhouse gas levels in the atmosphere are higher now than they have been in 650,000 years, based on data obtained from Antarctic ice core samples dating back that far. It's true, but it doesn't give you the truth, if you understand the subtle distinction.

The whole greenhouse effect/global warming argument hinges on convincing you that rising carbon dioxide levels do definitely cause rising temperatures, and that proof of this is that both CO_2 levels and temperatures are currently higher than they've ever been before. This is the claim they beat you around the ears with in news reports every day, this is the reason they want to tax you back to the Stone Age at the end of this year.

As you'll see in this chapter though, the ancient past shows no link between high carbon dioxide levels and high temperatures. In fact, it shows the opposite in many cases. If I can show you that in the past CO_2 levels have been very high while temperatures have been low, then this might prove to you that CO_2 does not cause global warming.

Keep your hand on your wallet, because understanding this is crucial to breaking through the spin. Let's look first at the global warming industry's hype about high greenhouse gas levels:

In its *Summary for Policymakers*, the IPCC's [Intergovernmental

Panel on Climate Change] 2007 *Fourth Assessment Report* gives the example of methane. With 2005 levels of 1,774 parts per billion (1,774 methane molecules per billion molecules of other dry air gases), the IPCC notes that this concentration "exceeds by far the natural range of the last 650,000 years", where methane has been recorded in ice at levels between 320 and 790ppb.[31]

The report helpfully publishes some graphs (one of which is on the facing page) tracking CO2, Nitrous Oxide and Methane levels for the past 10,000 years against rises in surface temperature of the earth (described in Watts per m^2; imagine pointing a 2400 watt electric fan heater at a square metre of your carpet and you'll get the analogy), as if to prove the point.

As Eric Wolff, of the British Antarctic Survey, pointed out a few years ago, "In 440,000 years we have never seen greenhouse gas concentrations, either methane or carbon dioxide, anything like as high as they are today."[32]

Looked at in isolation, the graphs are indeed alarming and it's little wonder politicians around the globe (to whom the IPCC's *Summary for Policymakers* is aimed at) have jumped on the bandwagon declaring "the science is settled".

Who, without scientific training and a finely honed B/S [bad smell] detector, could easily pick holes in graphs that so clearly illustrate the massive rise in greenhouse gas emissions in the modern era?

But the graphs presented by the IPCC to scare politicians and the news media are misleading. They only tell part of the story. Forget the past 10,000 years, or even the past 650,000. The planet is around 4.5 billion years old, and has survived many climate swings.

If we travel back in time to the dinosaur era, the figures tell a new and compelling story. Firstly, the average global temperature was between 22°C and 25°C,[33] a staggering 10° higher than today's average of 12° to 14°C. Remember, the IPCC is predicting temperature increases of between 1.4°C and 5.8°C this century (the wide range mainly the result of disagreement between IPCC scientists as to the real impact of climate change).[34]

31 IPCC, 2007: Summary for Policymakers. In: *Climate Change 2007: The Physical Science Basis. Contribution of Working Group 1 to the Fourth Assessment Report of the Intergovernmental Panel on Climate Change*

32 "Warm spell to last", *The Guardian*, 10 June 2004, http://www.guardian.co.uk/environment/2004/jun/10/research.environment

33 Temperature data from geologist Christopher Scotese's Paleomap Project, http://www.scotese.com/climate.htm

34 http://www.grida.no/publications/other/ipcc_tar/?src=/climate/ipcc_tar/wg1/figspm-5.htm

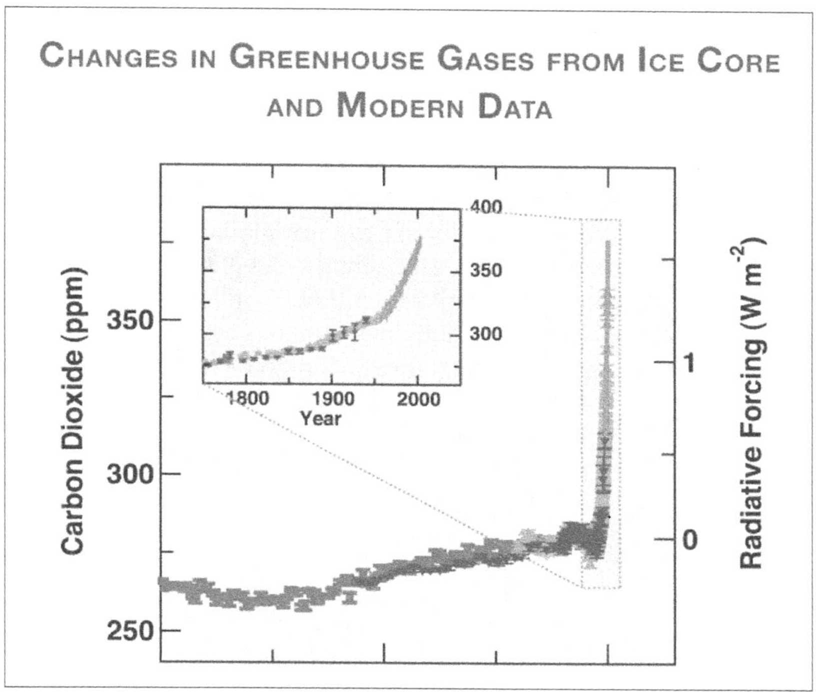

Even in their wildest dreams, the global average temperature in the most scaremongering computer modelling available to the IPCC today, doesn't come close to Earth's average temperature during the dinosaur era.

Yet, the dinosaurs didn't drive cars, operate heavy machinery or use those evil incandescent light bulbs that waste most of their light as heat. So what gives? Were greenhouse gas concentrations higher? Yes, and sometimes no – which is creating a headache for the global warming industry because it creates a problem that's hard to explain: ancient temperatures don't appear to be directly related to CO_2 levels.

Let's take a more detailed look.

Early in the Paleozoic era, which began around 600 million years ago with the Cambrian period, atmospheric CO_2 levels have been pegged at a whopping 7000 parts per million.[35] Now that compares with 2005 levels of 379 ppm. What did the IPCC have to say about modern levels of CO_2?

35 http://www.geocraft.com/WVFossils/Carboniferous_climate.html

"The global atmospheric concentration of carbon dioxide has increased from a pre-industrial value of about 280ppm to 379ppm in 2005. The atmospheric concentration of carbon dioxide in 2005 exceeds by far the natural range over the last 650,000 years (180 to 300ppm) as determined from ice cores."

Again, the IPCC has been misleading, because clearly the data from the last 650,000 years doesn't explain global warming in the dinosaur era. The impression given by the IPCC statement is that CO2 levels have usually been low on Earth, but are now tracking dangerously upward. Well, if they're dangerous after climbing from 280 to 379ppm, imagine how hot Earth must have been with CO2 rampaging through the air at 7000ppm![36]

Here's the twist: despite leaping from 5000ppm to 7000ppm, there was no corresponding jump in temperature. It remained at a tropical 22°C. The sky didn't fall, literally.

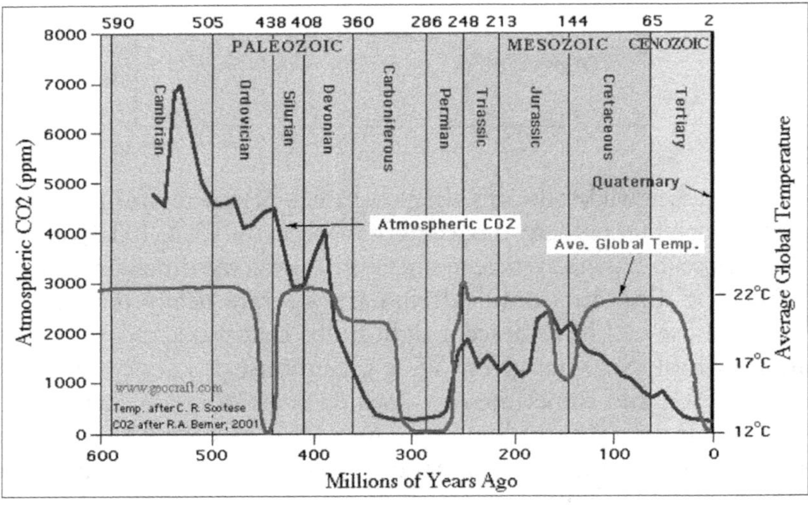

Some 480 million years ago (mya), coming up to the border between the Ordovician and Silurian periods, CO2 had dropped from 7000 down to a more manageable 4000ppm. Not suddenly, but gradually over the previous 100 million years. The temperature stayed steady at 22°C.

36 Carbon data taken from a Yale study, *GEOCARB III: A REVISED MODEL OF ATMOSPHERIC CO2 OVER PHANEROZOIC TIME,* ROBERT A. BERNER and ZAVARETH KOTHAVALA, Department of Geology and Geophysics, Yale University, published *American Journal of Science,* Vol. 301, February, 2001, P. 182–204

Then something strange happened. *CO_2 levels rose* sharply from 4000ppm to 4500ppm, yet suddenly, *world temperatures plummeted* to an average matching today's: 12°C.[37] In plain English, carbon dioxide levels went up but the temperature went down. So much for global warming theory.

Take a mental note here – with CO_2 levels at 4500ppm, compared to 2005's 379ppm (less than one tenth), the average temperature was comparable to Earth today.

You can see why the IPCC doesn't quote Paleozoic temperature data to politicians and policymakers; it simply doesn't fit the global warming scare industry. Even if carbon emissions double in the next hundred years and push atmospheric concentration to nearly 800ppm, they will still only be less than a fifth of the 4500ppm CO_2 level that existed when global temperatures *dropped* by nearly half to 12 degrees.

Embarrassingly for the IPCC, that drop to a planetary average of 12 degrees – the late Ordovician period – was a global Ice Age, *even though CO_2 levels were around 12 times higher than they are now.*

At 438 million years ago, atmospheric CO_2 *dropped* sharply again, from 4500ppm down to 3000ppm, yet the fossil record shows temperatures shot just as sharply back *up* to 22°C.

And let's take another mental note here – the average world temperature tended to hover around 22°C, regardless of whether CO_2 levels were 7000ppm or 3000ppm. So what impact are our current levels of 385ppm likely to have on global climate or the continued viability of Planet Earth?

In fact, the Paleozoic data throws up some more fascinating information. In the past 600 million years there have only been three occasions where Earth's average temperature has been as low as 12°C. You are living in one of those periods. Chronologically, we've just outlined the first of them where, paradoxically, CO_2 levels were extremely high (far higher than anything humans will ever achieve), yet the planet was in an Ice Age. That leaves just one other period in known history where Earth enjoyed the same average climate that it does today: a 45 million year cool spell known as the Late Carboniferous period (so-called because that's when many of the planet's carbon-bearing coalfields date to).

During this time, around 315 million years ago, something spe-

37 We don't actually have a precise average temperature for Earth because there are so many variables, but climates generally average between 12 and 14 degrees Celsius.

cial happened – both the CO_2 levels and the temperature dropped at the same time, in this case to levels similar to those on Earth today.[38] This is the only time in known history where both CO_2 and temperature were simultaneously the same as modern Earth. Previously, as you've seen, the CO_2 levels and temperatures went in different directions.

Although this happy event lasted far longer (45 million years) than our modern era has, nonetheless both CO_2 levels and temperature shot back up at the end of it, just as the main Mesozoic dinosaur era was beginning. The CO_2 levels rose to between 1800ppm and 1200ppm, while the temperature went back to the same old 22°C it had always preferred, even back in the days of 7000ppm of CO_2.

If ever you needed evidence that CO_2 has little to do with temperature increases, this is it. Earth appears capable of regulating its temperatures and greenhouse gas absorption at levels far higher than humans are able to generate.

At around 180 million years ago, CO_2 levels rocketed up from 1200ppm to 2500ppm, coinciding (again) in a big temperature *drop*, from 22°C down to around 16°C. At the border between the Jurassic period (think T-Rex) and the Cretaceous that followed, CO_2 levels began to drop again, while the temperature went back up to 22°C, where it stayed until well after the dinosaurs had become extinct. It is only comparatively recently that global temperatures have dropped again to 12°C, well out of sync with the slide in CO_2 levels that began back in the Jurassic.

So here's a point to ponder:

If Earth, until recently, had an average temperature in the early 20s and CO_2 levels in the thousands of parts per million, rather than the very low 385ppm we have now, is there any possibility that humans reducing carbon emissions will make a blind bit of difference to planetary breathing cycles?

Are human-induced carbon emissions really causing global warming, or is that simply an arrogant over-belief in mankind's ability to significantly alter the planet?

---

38 To be precise, this cooling in the Late Carboniferous actually became an Ice Age, thanks to the placement of the continents at that time which encouraged the growth of massive polar ice-caps and prevented warmer ocean currents from circulating as effectively as they currently do. Technically, Earth should be experiencing an Ice Age at the moment given our global temperature averages, but the current continental layout is allowing warm currents to freely circulate closer to polar regions. The net result? We're in an Ice Age with "benefits"

Here's what former Vice President Al Gore said in *An Inconvenient Truth*:[39]

"When there is more carbon dioxide, the temperature gets warmer, because it traps more heat from the sun inside," he intones on screen, pointing to an ice core graph. "*Carbon dioxide having never gone above 300 PPM* [my emphasis], here is where CO_2 is now [380]. We give off where it has never been as far back as this record will measure."

"Never"? Here's the real inconvenient truth: For most of known history, CO_2 levels have been between four and 18 times higher than they are today, yet Earth, and life, survived. Clearly then, when all the historic data and not just selected portions are analysed, carbon cannot be the cause of runaway global warming that the United Nations and politicians courting the Green vote claim it to be.

And Al Gore's 650,000 year old ice core sample is a frozen drop in the bucket compared to temperature indications from the past 600 *million* years.

So what do we take away from this chapter and how does this information help protect your wallet?

All physicists agree that the laws of physics and chemistry are the same today as they were a billion years ago. There's nothing magic about CO_2 in the 20th century that will make it behave differently than it always has. I've shown you that CO_2 levels rose to record levels in the past, at the same time as temperatures fell. This is directly opposite to what modern computer projections predict. It proves that regardless of the properties of CO_2 in an enclosed glasshouse and under controlled conditions, it doesn't behave *exactly* the same way in the open atmosphere. If it did, temperatures would rise and fall in direct relation to CO_2 levels. But they didn't, and they're not now.

If temperatures are going up, there has to be bigger reason than carbon dioxide, and if carbon dioxide isn't the main culprit there is no moral or practical reason to tax you over it.

The current forecasts and fundamentals of global warming theory are wrong. But then, as the story you're about to read shows, humans have a history of getting climate forecasts wrong.

---

39 http://forumpolitics.com/blogs/2007/03/17/an-inconvient-truth-transcript/

Chapter Four

Fry-up or Freeze-up?

"Laypeople frequently assume that in a political dispute the truth must lie somewhere in the middle, and they are often right. In a scientific dispute, though, such an assumption is usually wrong."

– Paul Erhlich

The debate about whether we've overdue for another Ice Age has raged furiously over the past century, perhaps most famously epitomised by Newsweek's 1970s prediction of an imminent "Little Ice Age".

For the benefit of those who've never read it, here's what *Newsweek's* Peter Gwynne wrote on April 18 1975:

> There are ominous signs that the Earth's weather patterns have begun to change dramatically and that these changes may portend a drastic decline in food production – with serious political implications for just about every nation on Earth. The drop in food output could begin quite soon, perhaps only 10 years from now. The regions destined to feel its impact are the great wheat-producing lands of Canada and the U.S.S.R. in the North, along with a number of marginally self-sufficient tropical areas – parts of India, Pakistan, Bangladesh, Indochina and Indonesia – where the growing season is dependent upon the rains brought by the monsoon.
>
> The evidence in support of these predictions has now begun to accumulate so massively that meteorologists are hard-pressed to keep up with it. In England, farmers have seen their growing season decline by about two weeks since 1950, with a resultant overall loss in grain production estimated at up to 100,000 tons annually. During the same time, the average temperature around the equator has risen by a fraction of a degree – a fraction that in some areas

can mean drought and desolation. Last April, in the most devastating outbreak of tornadoes ever recorded, 148 twisters killed more than 300 people and caused half a billion dollars' worth of damage in 13 U.S. states.

To scientists, these seemingly disparate incidents represent the advance signs of fundamental changes in the world's weather. The central fact is that after three quarters of a century of extraordinarily mild conditions, the earth's climate seems to be cooling down. Meteorologists disagree about the cause and extent of the cooling trend, as well as over its specific impact on local weather conditions. But they are almost unanimous in the view that the trend will reduce agricultural productivity for the rest of the century. If the climatic change is as profound as some of the pessimists fear, the resulting famines could be catastrophic. "A major climatic change would force economic and social adjustments on a worldwide scale," warns a recent report by the National Academy of Sciences, "because the global patterns of food production and population that have evolved are implicitly dependent on the climate of the present century."

A survey completed last year by Dr. Murray Mitchell of the National Oceanic and Atmospheric Administration reveals a drop of half a degree in average ground temperatures in the Northern Hemisphere between 1945 and 1968. According to George Kukla of Columbia University, satellite photos indicated a sudden, large increase in Northern Hemisphere snow cover in the winter of 1971-72. And a study released last month by two NOAA scientists notes that the amount of sunshine reaching the ground in the continental U.S. diminished by 1.3% between 1964 and 1972.

To the layman, the relatively small changes in temperature and sunshine can be highly misleading. Reid Bryson of the University of Wisconsin points out that the Earth's average temperature during the great Ice Ages was only about seven degrees lower than during its warmest eras – and that the present decline has taken the planet about a sixth of the way toward the Ice Age average. Others regard the cooling as a reversion to the "little ice age" conditions that brought bitter winters to much of Europe and northern America between 1600 and 1900 – years when the Thames used to freeze so solidly that Londoners roasted oxen on the ice and when iceboats sailed the Hudson River almost as far south as New York City.

Just what causes the onset of major and minor ice ages remains a mystery. "Our knowledge of the mechanisms of climatic change is at least as fragmentary as our data," concedes the National Academy of Sciences report. "Not only are the basic scientific questions largely unanswered, but in many cases we do not yet know enough to pose the key questions."

> Meteorologists think that they can forecast the short-term results of the return to the norm of the last century. They begin by noting the slight drop in overall temperature that produces large numbers of pressure centres in the upper atmosphere. These break up the smooth flow of westerly winds over temperate areas. The stagnant air produced in this way causes an increase in extremes of local weather such as droughts, floods, extended dry spells, long freezes, delayed monsoons and even local temperature increases – all of which have a direct impact on food supplies.
>
> "The world's food-producing system," warns Dr. James D. McQuigg of NOAA's Center for Climatic and Environmental Assessment, "is much more sensitive to the weather variable than it was even five years ago." Furthermore, the growth of world population and creation of new national boundaries make it impossible for starving peoples to migrate from their devastated fields, as they did during past famines.
>
> Climatologists are pessimistic that political leaders will take any positive action to compensate for the climatic change, or even to allay its effects. They concede that some of the more spectacular solutions proposed, such as melting the Arctic ice cap by covering it with black soot or diverting arctic rivers, might create problems far greater than those they solve. But the scientists see few signs that government leaders anywhere are even prepared to take the simple measures of stockpiling food or of introducing the variables of climatic uncertainty into economic projections of future food supplies. The longer the planners delay, the more difficult will they find it to cope with climatic change once the results become grim reality.

Newsweek wasn't the first. Since the late 1800s, journalists and scientists looking for research funding have been belting out warnings about climate change. Only, back then, they feared a looming ice age.

Some of the headlines look familiar in style:[40]

New York Times, February 24, 1895:

"Geologists Think the World May Be Frozen Up Again"

Los Angeles Times, October 7, 1912:

"Fifth ice age is on the way. Human race will have to fight for its existence against cold."

New York Times, October 7, 1912:

"Prof. Schmidt Warns Us of an Encroaching Ice Age"

Los Angeles Times, June 28, 1923:

40 Headlines List courtesy British author James P. Hogan's "Bulletin Board", http://www.jamesphogan.com/bb/bulletin.php?id=1086

"The possibility of another Ice Age already having started ... is admitted by men of first rank in the scientific world, men specially qualified to speak."

Chicago Tribune, August 9, 1923:

"Scientist says Arctic ice will wipe out Canada."

Time Magazine, September 10, 1923:

"The discoveries of changes in the sun's heat and the southward advance of glaciers in recent years have given rise to conjectures of the possible advent of a new ice age."

New York Times, September 18, 1924:

"MacMillan Reports Signs of New Ice Age"

In the 1930s, a warming period got people's hopes up, and the headlines changed:

GLOBAL WARMING: 1930s-1950s

New York Times, March 27, 1933:

"America in Longest Warm Spell Since 1776; Temperature Line Records a 25-Year Rise"

Time Magazine, January 2, 1939:

"Gaffers who claim that winters were harder when they were boys are quite right.... weather men have no doubt that the world at least for the time being is growing warmer."

Chicago Daily Tribune, November 6, 1939:

"Chicago is in the front rank of thousands of cities [throughout] the world which have been affected by a mysterious trend toward warmer climate in the last two decades."

Time Magazine, 1951:

Noted that permafrost in Russia was receding northward at 100 yards per year.

New York Times, 1952:

Reported global warming studies citing the "trump card" as melting glaciers. All the great ice sheets stated to be in retreat.

U.S. News and World Report, January 18, 1954:

"[W]inters are getting milder, summers drier. Glaciers are receding, deserts growing."

New York Times, February 15, 1959:

"Arctic Findings in Particular Support Theory of Rising Global Temperatures."

But then, suddenly, we were lurching back into an ice age, according to the pundits:

GLOBAL COOLING: 1970s

Washington Post, January 11, 1970:

"Colder Winters Herald Dawn of New Ice Age."

Time Magazine, June 24, 1974:

"Climatological Cassandras are becoming increasingly apprehensive, for the weather aberrations they are studying may be the harbinger of another ice age."

Christian Science Monitor, August 27, 1974:

"Warning: Earth's Climate is Changing Faster than Even Experts Expect", Reported that "glaciers have begun to advance"; "growing seasons in England and Scandinavia are getting shorter"; and "the North Atlantic is cooling down about as fast as an ocean can cool".

New York Times, December 29, 1974:

"[P]resent climate change [will result in] mass deaths by starvation and probably in anarchy and violence."

Science News, March 1, 1975:

"The cooling since 1940 has been large enough and consistent enough that it will not soon be reversed, and we are unlikely to quickly regain the 'very extraordinary period of warmth' that preceded it."

Newsweek, April 28, 1975:

"The Cooling World": "There are ominous signs that the Earth's weather patterns have begun to change dramatically and that these changes may portend a drastic decline in food production – with serious political implications for just about every nation on Earth. The drop in food output could begin quite soon, perhaps only 10 years from now."

International Wildlife, July-August, 1975:

"But the sense of the discoveries is that there is no reason why the ice age should not start in earnest in our lifetime." In this article, *New Scientist* magazine's former editor Nigel Calder intoned: "The threat of a new ice age must now stand alongside nuclear war as a likely source of wholesale death and misery for mankind."

New York Times, May 21, 1975:

"Scientists Ponder Why World's Climate is Changing; A Major Cooling Widely Considered to Be Inevitable"

So what the heck was going on in science labs and newsrooms all over the planet? Was Earth's temperature really rising and falling and rising and falling across the globe over the course of a century, like some kind of planetary deep breathing cycle?

To get answers, we need to ask the right questions. During the 20th century, while weather data was OK (and the data collected since 1850 remains at the core of climatologists' calculations today), for the most part they didn't have hi-tech weather sensors or weather satellites in space.

Although studying the weather has been a key pastime of humankind since the first caveman drew on a wall, we've lacked access to the big picture up to now. And even now, as we try and dig into the past, the picture is confusing.

More recently, Maureen Raymo, a Massachusetts Institute of Technology researcher and associate professor of Earth, Atmospheric and Planetary Sciences, published a paper in the journal *Nature*[41] suggesting that Earth has endured massive climate swings on a number of occasions in the past 1.5 million years. Their data was hauled from Greenland ice core samples and ocean floor sediment cores from around the world.

"These climate swings are so dramatic that if we lived through one today, it would be like New England taking on Miami-like weather within a 25 year period," the website *ScienceDaily* reported[42]. In a southern hemisphere context, that's like Dunedin or Melbourne taking on Brisbane's climate.

The temperature swings they documented showed Earth's average temperature could increase or decrease by as much as 10°C within just a few decades. Those are monster climate changes, on a scale modern humans can only stare at in wonder, yet they fit the same massive changes that took place in the dinosaur era, regardless of CO2 levels.

And again, there were no cars back in the more recent times, in fact there were very few humans fullstop.

With that scale of temperature change, you'd see glacial ice sheets form across Canada and into the northern USA. These massive swings happened so frequently that Raymo's paper notes, "Our results suggest that such millennial-scale climate instability may be a pervasive and long-term characteristic of Earth's climate."

---

41 *Nature*, 16 April 1998

42 Massachusetts Institute Of Technology. "MIT Researcher Finds Evidence Of Ancient Climate Swings." *ScienceDaily* 20 April 1998. http://www.sciencedaily.com/releases/1998/04/980420080212.htm

Could it be that climate swings are mostly, if not entirely, natural phenomena? And if they are, what difference are energy efficient light bulbs and hybrid cars actually going to make? None. Not a whisker. If climate change is natural, there is nothing, repeat, nothing, that President Obama or anyone else can do that is going to change the end destination.

Not that global warming believers agree.

Raymo made the point that after more than 8,000 years of global warming since the last ice age, we're now due to head into a cooling period, "but all bets are off because of global warming from the buildup of carbon dioxide or other greenhouse gases in the atmosphere."

There's still an assumption in that statement, and it is simply this: that modern global warming exists *and* is caused by the build-up of CO2. That's an assumption that still appears to be without hard evidence.[43] And it's the same when it comes to historical causes; as for the events that triggered the ancient massive climate swings, Raymo confessed that science simply doesn't know: "What causes climate variations on this time scale is a black box for scientists right now."

In 2004, as scientists began analyzing the main data from ancient Antarctic ice core samples, one of the lead researchers tried to pour cold water on claims that global warming was good, because it's delaying the onset of a new Ice Age.

"If people say to you: the greenhouse effect is a good thing because we would go into an ice age otherwise, our data says no, we weren't about to go into an ice age," scoffs the British Antarctic Survey's Dr Eric Wolff. "We have another 15,000 years before that was coming."[44]

That was back in 2004. Wolff, and others, reported in the journal *Nature* that the ice core evidence categorically showed we weren't due for an ice age until the year 17,004.

"We have eight examples of how the climate goes in and out of ice ages, and you can learn what the rules are," explained Wolff. "Those are the rules that go into the climate models that predict the future."

Wolff, as you can see, was pretty unequivocal. Global warming bad; ice age not due for another 15,000 years; Antarctic data proves it and gives us clear rules for climate modelling.

---

43 As you'll see later in the book, the question of whether CO2 emissions *cause* warming, or are merely symptoms of warning caused by something else, appears to have been answered by ice core data revealing warming first, *then* higher CO2 levels as a result.

44 "Warm spell to last", *The Guardian*, 10 June 2004, http://www.guardian.co.uk/environment/2004/jun/10/research.environment

Great, but was he right?

At the end of 2008, a conference of the American Geophysical Union was given a startling new take on the Antarctic ice core data: that greenhouse gas emissions are the *only* thing standing between humanity and a new ice age.

Climatologist Stephen Vavrus – based at the University of Wisconsin-Madison's Centre for Climatic Research and the Nelson Institute for Environmental Studies – told delegates that a re-examination of the ice cores had led to a disturbing conclusion: we are indeed due for an ice age right now, but it hasn't happened yet.

"We're at a very favourable state right now for increased glaciation," said John Kutzbach, one of Vavrus' team. "Nature is favouring it at this time in orbital cycles, and if humans weren't in the picture it would probably be happening today."[45]

Berger and Loutre's 2002 paper,[46] however, argues that barring minor fluctuations, earth is in a long term warming period that could last a further 50,000 years, because of the current track of earth's orbit around the sun. Clearly this current "warming" cycle allows for "cooling" within it, because the planet has experienced numerous cool-downs over the past ten thousand years since the last ice age melt, but never cooling enough to return to a full ice age. The overall trend has been warming. But how do they know this natural warming cycle is expected to last?

At the centre of their claim are changes in the earth's orbital path around the sun, earth's axial tilt, and the sun's orbital path around the centre of the solar system. All of these factors change from time to time, and have a big impact on climate change. Geologists have matched past ice ages with regular cyclical changes in the Earth's orbit, known as "Milankovitch cycles".

If you've ever watched a child's spinning top on the floor, you'll recall how the top experiences variations in the direction of its axis (wobble). These wobbles in the top are known as "precessions", and Milankovitch discovered that Earth experiences one of these roughly every 26,000 years.

Continuing the spinning top analogy on the floor, you'll recall how the top's orbits around the floor can also vary from wild loops

45 University of Wisconsin media release, 17 December 2008, http://www.eurekalert.org/pub_releases/2008-12/uow-sde121708.php

46 *Science* 23 August 2002: Vol. 297. no. 5585, pp. 1287 – 1288

to tight circles. Earth, spinning around the sun, moves through a complete orbital variation cycle roughly every 21,000 years.

These variations primarily occur because the planet shares its space with a bunch of other planets, a moon and the sun, all exerting gravitational pull depending on how close they get. These forces can pull the earth closer to or further away from the sun, with the logical result of global warming or global cooling on each occasion.

Then there's variation in the Earth's axial tilt. If you look at a globe, you'll notice earth spins on an axis that is not straight up and down, but off around 22 degrees. It's this axial tilt that gives northern and southern hemispheres their differing seasons. If you look at your spinning top, sometimes it leans over more heavily and sometimes it stands almost upright. In the same way, earth's axial tilt varies between 22.1 degrees of angle and 24.5 degrees, but the variations take an incredible 41,000 years to cycle.

At the moment, the axial tilt angle of 23.44 degrees, and the earth's current position in its orbital variation cycle, mean that southern hemisphere summers and winters are more extreme than northern hemisphere ones because of a 6.8% differential in solar exposure to the hemispheres. The earth is closest to the sun during the southern summer,[47] and furthest away during the southern winter. That's a big impact on weather patterns and the flood of weather stories on the TV news every year, and it has nothing to do with human-induced global warming in the slightest.

It means that, generally, northern hemisphere winters will also be slightly warmer than average because earth is closest to the sun during the northern winter/southern summer.

If you want evidence of the impact orbital cycles can have on planetary climate, "Exhibit A" is the Sahara Desert.[48] Nine thousand years ago, the degree of tilt on Earth's axis was 24.14 degrees (today it has reduced to 23.45), meaning the angle that sunlight struck the Earth was slightly different to the way it is today. Like Australia in the southern hemisphere, the Sahara occupies a similar latitude in the northern hemisphere. It just so happened that back then, the point at

47 This happens around the end of the first week in January. As if to confirm the point, New Zealand's meteorological service announced January 8, 2009 as a "record breaker", with the southern city of Christchurch recording its hottest January temperature ever, of 36°C. Whilst the spin implied global warming, the reality is it would have been helped by the earth's orbit coming closest to the sun that week.

48 "Sahara's abrupt desertification started by changes in Earth's orbit", *ScienceDaily*, 12 July 1999, http://www.sciencedaily.com/releases/1999/07/990712080500.htm

which the Earth came closest to the sun was in July (northern summer), whereas it is currently January (Australian summer).

This combination of orbital positions is believed to have set off a climate change over the Sahara which, between 7,000 years ago and 3,600 years ago, gradually transformed it from grassland with patches of jungle, into the sandtrap we know and love today. Where once there were great lakes and river systems, now only red dust remains (although massive amounts of water remain trapped underground[49]). As the hot northern summers burnt away the grasses and caused bushfires like Australia's today, the Sahara entered a period of runaway climate change.

Now, of course, a similar dry-heat attack is hitting Australia. So it's swings and roundabouts over cyclical periods of thousands or tens of thousands of years. The Sahara, one day, will probably return to lush foliage. But none of us will be alive to see it.

So that's one big factor – orbital changes. Then there's the impact of El Niño/La Niña weather patterns on the globe. Now remember the IPCC is forecasting global temperatures to rise about 1.5°C by the end of the century. Keep that in mind as you digest this next piece of information.

The temperatures on the West Antarctic ice shelf rose between 3 and 6°C from 1939 to 1942, during a major El Niño, then plummeted between 5 and 7°C between 1942 and 1944. That's a six degrees up, 7 degrees down swing over the space of five years, not a century!, and it happened in the Antarctic.[50]

What caused it? To understand that, you first have to learn a little bit more about El Niño.

El Niño is a pressure change in the South Pacific ocean which causes the ocean to warm up near the tropics. This particular weather system has a huge impact on the planet, and it is significant that the 1997 El Niño was one of the most severe in the whole of last century, because it set the scene for a decade of global warming.

According to a graph from Britain's Hadley Centre, 1998 remains the hottest year on Earth since modern measurements began a century ago. Likewise, the other massive El Niño of last century,

49 "New technique dates Saharan groundwater as Million Years old", *ScienceDaily*, 2 March 2004, http://www.sciencedaily.com/releases/2004/03/040302075003.htm

50 "Antarctic Climate: short term spikes, long term warming linked to tropical Pacific", National Center for Atmospheric Research, *ScienceDaily*, 15 August 2008, http://www.sciencedaily.com/releases/2008/08/080812160619.htm

in 1982, sent temperatures in 1983 through the roof, making that year the hottest of all the years between 1970 and 1989.

The 1997/98 El Niño reportedly takes the blame for a 16% die-off of the world's coral reefs, and ongoing problems with bleaching on coral reefs. Repeats of El Niño through the early 2000s helped keep those years among the warmest on record – although TV news reports were mostly attributing these record years to "global warming".

In other words, El Niños have a more dramatic effect on global warming than carbon emissions do, and they're an entirely natural phenomenon.

In 2007, La Niña began to form, which conversely cools the planet down. World temperatures have been falling again since.

But getting back to these weather patterns and their impact on Antarctica. A team of scientists from the US National Centre for Atmospheric Research reported late last year that West Antarctic ice is quite vulnerable to El Niño's grasp and certainly has been for the past hundred years.

"Ice cores reveal that West Antarctica's climate is influenced by atmospheric and oceanic changes thousands of miles to the north," NCAR's David Schneider said in a news release.[51]

That means that melting ice in that region may have more to do with El Niño than it does with global warming. There's certainly enough ice there to cause some problems – at more than a mile thick the West Antarctic sheet, if it melted, could raise global sea levels between 2.5 and five metres, or 8 to 16 feet.

Knowing this, what then are we to make of this 2009 scare story published in the journal *Nature* and covered by news media around the world in January 09?[52]

"Antarctica, the only place that had oddly seemed immune from climate change, is warming after all, according to a new study," reported *Associated Press*.

"For years, Antarctica had been an enigma to scientists who tracked the effects of global warming. Temperatures on much of the continent at the bottom of the world were staying the same or slightly cooling, research indicated."

The new study, it turns out, concludes that "since 1957, the annual temperature for the entire continent of Antarctica has warmed by

51 http://www.ucar.edu/communications/quarterly/summer08/antarctica.jsp

52 *Nature*, Jan 22 2009, Volume 457 Number 7228, cover story

about 1 degree Fahrenheit, but is still 50 degrees below zero. West Antarctica, which is about 20 degrees warmer than the east, has warmed nearly twice as fast," said the study's lead author, University of Washington's Eric Steig.

OK. So the new study shows West Antarctica warming at double the average rate, increasing by nearly 2 degrees Fahrenheit, or just over 1.1 Celsius. Remember how the same area rose in temperature by up to 5 times more in the early 1940s (a 6°C rise), then fell by a similar amount after El Niño wore off?

Call me cynical, but a study showing West Antarctica warming faster than the rest of the ice continent seems entirely consistent with nearby El Niño cycles, perhaps magnifying the effects of any solar activity as well. The West Antarctic Peninsula – that claw reaching up towards the lowest tip of South America – is undeniably warmer than it was 50 years ago, but it's also smack bang in the capture zone of ocean currents affected by El Niño to the north.

It is this same peninsula that has provided all of us with graphic TV images of stark, pristine collapsing ice shelves over the past decade. Despite the hype, however, no scientific study has come anywhere close to proving that the warming is anything other than natural, and regional.

Climate change believers, though, have seized on this January 09 study as conclusive proof of global warming.

"Contrarians have sometimes grabbed on to this idea that the entire continent of Antarctica is cooling, so how could we be talking about global warming?", the *Nature* study's co-author Michael Mann, from Penn State University's Earth System Science Centre, was quoted by AP. "Now we can, so no, it's not true…it is not bucking the trend".

Well maybe, maybe not. For a start, we should be worried about Michael Mann's work. He was the man responsible for the now widely discredited "hockey stick" graph used in Al Gore's *Inconvenient Truth*. He needs something to restore his credibility and continue to bolster belief in global warming.

But Mann's study actually shows East Antarctica cooled from the late 1970s through the 1990s, and satellite measurements over Antarctica don't reveal any warming since 1980.[53] According to a

53 AP report carried by *NZ Herald* (24 Jan 09) and *Detroit Free Press* (22 Jan) http://www.freep.com/article/20090122/NEWS07/901220425

recent University of Bristol study reported in *ScienceDaily*, "the ice mass in East Antarctica has been roughly stable, with neither loss nor accumulation over the past decade."[54]

The vast continent only has a handful of ground-based weather stations, some of which are situated right next to heat-emitting buildings on Antarctic research stations that may be skewing readings.

In other words, there's room for reasonable doubt on the ground temperature readings, and if the satellites can't find any warming, maybe there isn't any.

Because Antarctica is the storehouse for so much of the world's fresh water, the question of whether the ice continent is really melting or not is crucial to the global warming debate. By some estimates, if all the ice on the Antarctic mainland thawed out, sea levels would rise up to 85 metres worldwide, submerging all seaside towns and most major cities deep underwater.

The scientists are agreed, however, that the West Antarctic ice sheet is more vulnerable to climate change than the interior of the continent. And the West Antarctic sheet contains enough water to raise global sea levels by five metres.

What chance is there, however, of a total melt? After all ice melts at 0°C, and the West Antarctic temperature average is still about 34°C below zero. The average across the whole of Antarctica is 46 below.

With atmospheric CO_2 levels of 7000 parts per million in prehistoric times (in contrast to 385ppm now), the entire planet's average temperature was still only 22-24°C, an increase of eight to ten degrees on the current world average. What would it really take in terms of CO_2 emissions and temperature change to unlock Antarctica's ice?

NASA's Drew Shindell, a global warming believer, argues that West Antarctica "will eventually melt if warming like this continues."[55] One of the authors of the *Nature* study, he claims a temperature rise of just 3°C could be the catalyst for that region to enter meltdown.

But we've seen temperature fluctuations of double that on the Western Antarctic ice sheets in the past century, so presumably

54 "Increasing Amounts of Ice Mass Have Been Lost from West Antarctica" *ScienceDaily*, 14 January 2008, http://www.sciencedaily.com/releases/2008/01/080113143438.htm

55 www.newsdaily.com/stories/tre50k5bm-us-antarctica-warming/

there's far more to it than simply plucking a number out of the air and constructing a theory around it to scare small children and Green Party voters with of a night.

One who knows a little about scaring people is Al Gore, who spins a great little yarn about the collapse of West Antarctica's Larsen B ice shelf.[56]

"If you were flying over it in a helicopter, you'd see it 700 feet (213m) tall," says Gore in his movie. "They are so majestic, so massive. In the distance are the mountains, and just before the mountains is the shelf of the continent. This is floating ice, and there is land based ice on the down-slope of those mountains. From here to the mountains is about 20 to 25 miles. They thought this would be stable for about a hundred years, even with global warming. The scientists who study these ice shelves were absolutely astonished when they were looking at these images. Starting in January 31, 2002, in a period of 35 days, this ice shelf completely disappeared. They could not figure out how in the world this happened so rapidly. They went back to figure out where they had gone wrong. That's when they focused on those pools of melting water.

"Even before they could figure out what had happened there, something else started going wrong. When the floating sea-based ice cracked up, it no longer held back the ice on the land. The land-based ice then started falling into the ocean. It was like letting the cork out of a bottle. There's a difference between floating ice and land-based ice. It's like the difference between an ice cube floating in a glass of water, which when it melts doesn't raise the level of water in the glass, and a cube sitting atop a stack of ice cubes, which melts and flows over the edge. That's why the citizens of these Pacific nations had all had to evacuate to New Zealand."

For the record, no Pacific islanders were evacuated to New Zealand as a result of the Larsen B collapse, probably because it caused a sea level rise of about 0.1 millimetres (1/100th of a centimetre) – not enough to cause anyone to scamper up a beach let alone shinny up a palm tree to escape the wave. The world's oldest man, using a Zimmer frame, could outrun a sea level rise of that strength. Al Gore's claim is bogus.

And for what it's worth, although glaciers are melting, they're

56 http://forumpolitics.com/blogs/2007/03/17/an-inconvient-truth-transcript/

not melting much. Australia's national science organisation, the CSIRO, has been keeping close tabs on Antarctic melt in the wake of Gore's movie, just in case they have to evacuate Australians to New Zealand as well. Thankfully, it doesn't look likely.[57]

"Measurements by satellite techniques based on gravity indicate mass loss at a rate of 138 ± 73 billion tonnes per year during 2002-2005, mostly from the West Antarctica Ice Sheet. That is equivalent to a rise in global sea level of 0.4 ± 0.2 mm per year."[58]

Remember, that's not 0.2 cm, that's 0.2 millimetres (2/100ths of a centimetre or 7/1000ths of an inch). If that rate of sea level rise continues, we'll all be enjoying the benefits of a 1.8 centimetre rise in sea levels by the end of this century.

For what it's worth, and in direct contradiction of the early 2009 claim that Antarctica is warming overall, satellite measurements[59] suggest the East Antarctic north of latitude 81[60] gained (not lost) up to 52 billion tons of ice *per year*, between 1992 and 2003. That's around 500 billion tons of ice added during the decade said to have been the warmest in the past hundred years. "A gain of this magnitude," reported the study authors, "is enough to slow sea level rise by 0.12 mm per year".

As for Gore's claim that Larsen B is definitely a victim of global warming, scientists at the front line describe it as regional warming.

"Temperature rises on the Antarctic Peninsula have been five times faster than the global mean, that's 2.5°C since records began in the 1940s," notes British Antarctic Survey glaciologist Dr David Vaughan.[61]

"Geographically this is unusual and it is also unusual in historical terms.[62] There has been no similar warming in the last five hundred years but I am not alarmed about any immediate impact on the rest of the world from the Larsen B collapse. It is still too small in size to cause a rise in sea level or affect the amount of radiation

---

57 http://www.cmar.csiro.au/sealevel/sl_drives_recent.html

58 The same CSIRO report also notes that the estimate conflicts with other data suggesting either a very small loss or an actual net gain in Antarctic ice. But for the sake of running with this point, I'll take the most alarmist estimate at face value.

59 "Snowfall driven Growth in East Antarctic Ice Sheet Mitigates Recent Sea Level Rise", Davis et al, *Science*, 24 June 2005, Vol 308, no 5730, pp. 1898-1901

60 Up till now, satellites have been unable to measure ice mass close to the south pole itself, because of limitations in the orbit paths and instrumentation they carry. A European satellite launching this year is expected to fill in the blanks. It is reasonable to assume that if East Antarctica's northern plains were adding ice mass, so too the colder south polar cap.

61 http://climatex.org/articles/climate-change-info/larsen-ice-shelf-b-antarctic-2002/

62 Regional hotspots are entirely consistent, however, with El Niño and other oscillations.

reflected back into the atmosphere (albedo). However, the event gives scientists like us an indicator of which processes to look at in order to predict what's going on over the next hundred years. This will be relevant to sea level rise."

A much more measured commentary than the former Vice President's.

A study of the Larsen C shelf however, adds even more perspective. Scientists monitoring that sheet don't think the highly unusual warming of the West Antarctic Peninsula is the direct culprit in the Larsen collapses.

"The thought is that warmer ambient temperature – about 2.5 degrees centigrade in the last 50 years – alone isn't breaking up the ice shelf."[63]

Instead, the culprit is thought to be "a slightly warmer ocean assaulting the ice from underneath, carving out huge cavities and channels that thin and weaken the shelf."

Which is the El Niño scenario yet again – warmer ocean currents from the north circulating down.

Indeed, there's supporting evidence in a recent study[64] that backs away from modern global warming as a major player in the Larsen B collapse.

"Ice shelf collapse is not as simple as we first thought," conceded Professor Neil Glasser, after he and a colleague discovered Larsen B had been a disaster waiting to happen, or as they put it, "an ice shelf in distress for decades previously".

"It's likely that melting from higher ocean temperatures, or even a gradual decline in the ice mass of the Peninsula over the centuries, was pushing the Larsen to the brink", said the National Snow and Ice Data Centre's Ted Scambos, who co-authored the paper.[65]

Centuries. Long before humans built factories. And if that's the cause of the Larsen collapse, it's probably the same for the Pine Island Glacier and the rest of West Antarctica, such as the Wilkins ice sheet.

To make things more difficult, there is a raft of assumptions being made about Antarctica which should really be admitted as "guesswork". For example, the Larsen B ice shelf collapse on the peninsula in 2002 was said to be the biggest since the last ice age 12,000 years ago, and the shelf itself was said to date from that time.

---

63 US Antarctic Programme, http://antarcticsun.usap.gov/science/contentHandler.cfm?id=1546

64 "Antarctic ice shelf collapse blamed on more than climate change", *Journal of Glaciology*, referred by *ScienceDaily*, 11 February 2008, http://www.sciencedaily.com/releases/2008/02/080210100441.htm

65 "A structural glaciological analysis of the 2002 Larsen B ice shelf collapse", N Glasser and T Scambos, *Journal of Glaciology*, Vol. 54, No. 184, 2008

The American Institute of Physics wasn't quite so definite, saying only that the Larsen shelf "probably" dated back to the ice age.[66] In truth, we don't actually know. Antarctica was only discovered in the loose sense of the word by Captain Cook in 1772, and no one set foot on Antarctica until the 1820s. It wasn't until 1892 that Captain James Larsen, of ice shelf fame, discovered his ice shelf on the Antarctic Peninsula.[67]

Another study, of the "currently healthy" George VI ice shelf on the Antarctic Peninsula, reveals the ebb and flow of ice shelves is quite common. *ScienceDaily* reported that this particular shelf "experienced an extensive retreat about 9,500 years ago, more than anything seen in recent years. The retreat coincided with a shift in ocean currents that occurred after a long period of warmth."[68]

And yet, in 2009, the George VI shelf is in perfectly good nick.

The shelf is sea ice, similar to that in the Arctic Circle, and it's the Arctic that provides strong evidence against claimed lifespans of ten thousand years for Antarctic sea ice.

In September 2007, European satellite images revealed the fabled "Northwest Passage" – a sea route between the UK and Asia across the top of the Canadian Arctic Circle – opened for the first time in recorded history.[69] In case that sounds impressive, recorded history in regard to the Passage only began in 1978 with satellite monitoring.

That aside, the retreat of sea ice in the Arctic in 2007 was, nevertheless, spectacular. It prompted speculation that the Arctic might not freeze up again, that perhaps the tipping point had been reached on global warming. Somewhere in the region of a million square kilometres of sea ice just melted away.

Yet the following winter, it all came back and the Northwest Passage froze up, just as it has most years. Had we not known, thanks to satellites, that the ice had melted enough to open the passage in 2007 and close it again a few months later, the world might have continued ignorantly believing that the Arctic sea ice in 2009 had been there undisturbed for thousands of years too.

"Global warming seems of little help to explorers," noted the

66 http://www.aip.org/history/climate/xIceShelf.htm

67 A better estimate of Larsen B's age is 2000 years, based on ocean sediments analysed from beneath it, but they simply tell us how long ice has been overhead, not necessarily how thick or thin the ice has been, or whether it has been there continuously.

68 "Antarctic Ice Shelf Retreats Happened Before", British Antarctic Survey, reported in *ScienceDaily*, 28 February 2005, http://www.sciencedaily.com/releases/2005/02/050224115901.htm

69 http://news.bbc.co.uk/2/hi/americas/6995999.stm

K2Climb.Net explorers' website wistfully in January 2009.[70] "Arctic sailor Henk De Velde reported: "As seen on the 2008 late July ice map; a NW passage will be impossible for a small boat this year."

"Eclipse hunters Nike and David Speltz travelled on an Arctic ice breaker for a view of a total solar eclipse. At the exact North Pole, Nike celebrated by taking a swim on the spot. The couple's live images over Contact 4.0 showed an extensive ice shield covering the Arctic sea late summer.

"This – and sat pics showing the same thing – still didn't deter Britain's Lewis Gordon Pugh. In spite of a trailing ice breaker (and media crew), Pugh's attempt to kayak 1,200 km to the North Pole in order to raise awareness for global warming was stuck in thick ice only days after departure."

You've got to love serendipity!

Although the Northwest Passage opened up again late in the northern summer of 2008, by December of that year there were signs the Arctic sea ice was making up lost ground by gaining in extent after hitting its 2007 low.

"Average Arctic sea ice extent for the month of December was 12.53 million square kilometres (4.84 million square miles). This was 140,000 square kilometres (54,000 square miles) *greater* than for December 2007 and 830,000 square kilometres (320,000 square miles) less than the 1979 to 2000 December average," reported the US National Snow and Ice Data Centre in January 2009.

In February this year scientists were embarrassed at the discovery they'd *underestimated* the re-growth of Arctic sea ice by as much as 500,000 square kilometres, because of faulty sensors on the ice.[71] That's an area of extra ice bigger than California and twice as big as New Zealand.

But if the Arctic was still slightly smaller overall in 2008 compared to 1979, the Antarctic was larger. The University of Illinois' Arctic Climate Research Centre posted the results of its own analysis in January 2009, reporting that globally, sea ice coverage in 2008 had returned to nearly the same levels satellites first recorded back in 1979.[72]

Not that this prevents global warming believers from pushing their

70 http://www.k2Climb.net/news.php?id=17941
71 "Arctic ice extent underestimated because of 'sensor drift'", http://news.slashdot.org/article.pl?sid=09/02/19/0420255&from=rss
72 http://www.dailytech.com/Article.aspx?newsid=13834

agenda. The UN's International Polar Year project, set up with the understanding it would support the IPCC's claims about global warming, reported late February 2009 that studies of the polar caps have proved the worst – melting is happening even faster than expected.

Critics of the UN IPCC, like Steve Goddard, spent the next few days rolling all over the floor laughing:[73]

"It was reported last week that the IPY (International Polar Year) released a study claiming that both polar ice caps are melting 'faster than expected'," explained Goddard.

"Given that NSIDC [the US National Snow and Ice Data Centre] *shows Antarctica gaining ice at a rapid pace*, I find myself surprised that IPY would release a study saying exactly the opposite. But then again, an IPY official reportedly forecast that last summer (2008) might have an 'ice free Arctic' [it didn't].

"Columnist George Will reported that overall global sea ice area is normal, and was correct. Dr. Meier [from NSIDC] confirmed that on January 1 global sea ice levels were normal.

"The UIUC [University of Illinois Arctic Centre] graph shows global ice levels well within one standard deviation of the 1979-2000 mean. [NASA's] Dr. [James] Hansen was correct that according to global warming theory, both poles should be losing ice – though we know now it theoretically should be happening more slowly in the Antarctic. Yet 20 years later we actually see the Antarctic gaining ice, which is contrary to Dr. Hansen's theory, contrary to IPY claims, and probably contrary to Steig's questionable temperature analysis," concludes Goddard.

This is one of the most frustrating things about the global warming issue: the scientists pushing it keep singing their tune no matter what the evidence actually shows.

Given that sea ice is much more responsive to temperature changes than glacial ice, i.e., much faster to melt or regrow, the University of Illinois' figures have caused many to wonder whether there's any significant global warming at all, if the ice extent hasn't materially changed in thirty years.

"Each year," wrote *DailyTech's* Michael Asher, "millions of square kilometres of sea ice melt and refreeze. However, the mean ice anomaly – defined as the seasonally-adjusted difference between

73 "Poll and Polar Ice Trends", 2 March 2009 http://wattsupwiththat.com/2009/03/02/poll-and-polar-ice-trends/#more-5955

the current value and the average from 1979-2000, varies much more slowly. That anomaly now stands at just under zero, a value identical to one recorded at the end of 1979, the year satellite record-keeping began.

"Sea ice is floating and, unlike the massive ice sheets anchored to bedrock in Greenland and Antarctica, doesn't affect ocean levels. However, due to its transient nature, sea ice responds much faster to changes in temperature or precipitation and is therefore a useful barometer of changing conditions.

"Earlier this year, predictions were rife that the North Pole could melt entirely in 2008. Instead, the Arctic ice saw a substantial recovery. Bill Chapman, a researcher with the UIUC's Arctic Center, tells *DailyTech* this was due in part to colder temperatures in the region. Chapman says wind patterns have also been weaker this year. Strong winds can slow ice formation as well as forcing ice into warmer waters where it will melt.

"Why were predictions so wrong? Researchers had expected the newer sea ice, which is thinner, to be less resilient and melt easier. Instead, the thinner ice had less snow cover to insulate it from the bitterly cold air, and therefore grew much faster than expected, according to the National Snow and Ice Data Center."

Not only did the ice grow back, it grew back faster than it ever has since satellite monitoring began. In fact, it grew back even faster than it melted in the first place.

Interestingly, it isn't the first time melting sea ice in the Arctic has prompted speculation about global warming. The *New York Times*, back in 1952,[74] carried a story reporting, "We have learned that the world has been getting warmer in the last half century". As evidence, it cited the appearance of cod in the diet of Arctic Inuit people, a fish not previously seen by them before 1920.

Later that same decade, the *Times* again joined the fray on global warming with a headline proclaiming, "Arctic Findings in Particular Support Theory of Rising Global Temperatures."[75] It's in this article that the *Times* noted vanishing sea ice in the Arctic.

"Ice in the Arctic ocean is about half as thick as it was in the late nineteenth century."

By 1969, the *Times* ignored growing ice age fever amongst the rest

74 *New York Times*, 10 August 1952
75 Ibid, 15 February 1959

of the media and quoted polar explorer Colonel Bernt Balchen[76] as warning "the Arctic pack ice is thinning and that the ocean at the North Pole may become an open sea within a decade or two."[77]

As it transpired, Balchen's calculations were out by 20 or so years, but the Northwest Passage did indeed open up. It had been thinning since the mid-1800s, however, prior to the industrial revolution really kicking in, and the area had warmed enough that previously unknown fish like cod were being caught by Eskimos back in the 1920s.

And it would be remiss of me not to pass on the most telling piece of evidence in regard to the Northwest Passage: Roald Amundsen took a ship through it in 1903, as his own diary records.[78]

"The North West Passage was done. My boyhood dream – at that moment it was accomplished. A strange feeling welled up in my throat; I was somewhat over-strained and worn – it was weakness in me – but I felt tears in my eyes. 'Vessel in sight' ... Vessel in sight."

And if that's not enough evidence for you, a Royal Canadian Mounted Police ketch assigned to Arctic patrol duties made regular trips through the Northwest Passage in the 1940s:[79]

"Between 1940 and 1942 *St. Roch* navigated the Northwest Passage, arriving in Halifax harbour on October 11, 1942. *St. Roch* was the second ship to make the passage, and the first to travel the passage from west to east. In 1944, *St. Roch* returned to Vancouver via the more northerly route of the Northwest Passage, making her run in 86 days."

And just to rub dung on the noses of global warming-promoting TV journalists everywhere, the opening of the "fabled" Northwest Passage in September 2007 had been gazumped in real life by another voyage through there in 2000, with no media fanfare, just seven years earlier![80]

"The *St. Roch II* has crossed through the Northwest Passage in just three weeks! ... There was so little ice that most of the trip was smooth sailing except for the occasional iceberg floating by."

76 Norwegian-born Balchen had worked on Amundsen's North Pole exploits, and had extensive Arctic and Antarctic experience. On November 28, 1929, he became the first person to fly a plane over the South Pole. Ironically, this was fifty years to the day before an Air New Zealand DC10 on a sightseeing flight to Antarctica slammed into the side of volcanic Mt Erebus, killing all 257 people on board on November 28, 1979.

77 *New York Times*, 20 February 1969

78 http://www.athropolis.com/links/nwpass.htm

79 http://hnsa.org/ships/stroch.htm

80 http://www.athropolis.com/news/st-roch.htm

That was 2000. What was it global warming believers in the news media were saying in September 2007 again?

"The Arctic's sea ice cover has shrunk so much that the Northwest Passage, the fabled sea route that connects Europe and Asia, has opened up for the first time since records began," wrote the *Observer's* Robin McKie.[81]

"The discovery, revealed through satellite images provided by the European Space Agency, shows how bad the consequences of global warming are becoming in northerly latitudes.

"This northern summer there was a reduction of a million square kilometres in the Arctic's ice covering compared with last year, scientists have found.

"As a result, the Northwest Passage that runs between Canada and Greenland has been freed of the ice that has previously blocked it and that, over the centuries, has frustrated dozens of expeditions that tried to sail northwest and open up a commercial sea route between the Atlantic and the Pacific..."

Well as long as you ignore people sailing through the passage in 1903, the 1940s and 2000, I guess you could run the line that it's been frozen for "centuries".

Environmentalist, and Al Gore's muse, Bill McKibben, is another beating up the significance of the Northwest Passage.[82]

"By the end of the summer season in 2008, so much ice had melted that both the Northwest and Northeast passages were open. In other words, you could circumnavigate the Arctic on open water. The computer models, which are just a few years old, said this shouldn't have happened until sometime late in the 21st century. Even sceptics can't dispute such alarming events."

Memo to McKibben and Gore: get new computer models, and read some history books.

Receding sea ice in the Arctic turns out not to be such a new or rare phenomenon at all, and probably is not caused by humans if it was happening a hundred years ago.

Additionally, scientists have discovered that when sea ice is retreating in the north, it's growing in the south. This doesn't just apply to the obvious fact that when ice is melting in a northern summer, it's

81 Reprinted in the *NZ Herald*, 17 September 2007, http://www.nzherald.co.nz/shipping/news/article.cfm?c_id=317&objectid=10464083&pnum=0

82 "Think Again", *Foreign Policy*, January/February 2009 issue, http://www.foreignpolicy.com/story/cms.php?story_id=4585

winter at Antarctica. A recent review of satellite data[83] has shown that whilst Arctic sea ice has reduced overall, Antarctica's sea ice has increased overall to compensate.

These, then, are some of the cautions you absolutely have to keep in mind when assessing doomsday claims from global warming believers, and *especially* ones you see on the TV news. There are others, even more intriguing, such as the ancient extremes of "Snowball Earth", and "Spa-pool Earth", and that is where we now head in our search for clues to what might happen in the future.

83 Cavalieri, D.J., C.L. Parkinson, and K.Y. Vinnikov. 2003. "30-year satellite record reveals contrasting Arctic and Antarctic decadal sea ice variability", *Geophysical Research Letters* doi:10.1029/2003GL018031.

Chapter Five

The Ice Planet

"Some circumstantial evidence is very strong, as when you find a trout in the milk"

– Henry David Thoreau, 1850

Earth has not always been warm, with green grass, forests and animal life. Six hundred million years ago, it was a very different planet.

At that stage, life as we know it was tiny and mostly bacterial. The "Cambrian Explosion", where all the nearly three dozen types [phyla] of animals, plants and sea life known on Earth today sprang into existence in the fossil record virtually overnight in geological terms, was still nearly a hundred million years in the future. Pre-Cambrian earth was in the grip of the biggest ice age ever recorded.

Geologists refer to this time as "Snowball Earth".

"There is strong geologic evidence of tropical glaciation at sea level during those times," Dr David Pollard of Penn State University's Earth and Mineral Sciences Environmental Institute told delegates at the American Geophysical Union's conference in 2001. "We wanted to determine how low-level tropical glaciers could have formed."

The key point was the discovery of glaciers, in the tropics, at sea level. According to Pollard and his fellow researchers, this could only happen if all of Earth's oceans had iced up, even at the equator.

"During the lead-up to a snowball Earth episode," reported *ScienceDaily*,[84] "the Earth gradually cools because the amount of carbon dioxide in the Earth's atmosphere decreases.[85] Relatively fast weathering of silicate rocks on large tropical landmasses causes this

84 *ScienceDaily*, 31 May 2001, http://www.sciencedaily.com/ releases/2001/05/010529235718.htm

85 Except, of course, for those annoying times in the past when CO2 levels dropped and temperatures went up.

decrease that locks up carbon. As the earth cools, the oceans begin freezing. The high reflectivity of the snow and ice that covers the northern and southern oceans, reflects, rather than absorbs, the sun's heat and further cools the planet. This cooling takes place slowly until the oceans are frozen to about 30 degrees latitude, or from the North Pole down to New Orleans, La. and from the South Pole up to the tip of South Africa."

In other words, there's a tipping point and the oceans essentially become so cold that icy tentacles reach out and freeze all the water that's left.

"This [the freeze-down to 30 degrees latitude] is the coldest that the Earth can get before all the entire ocean surface freezes," Pollard told *ScienceDaily*. "Beyond this, there is no stable point at say 20 or 10-degrees latitude: instead, the ice reflectivity feedback becomes unstable and the system collapses rapidly to a snowball Earth with all oceans ice-covered."

This, then, was Earth in 600 million BC. Frigid, windswept, bleak. The ideal setting perhaps for the second *Star Wars* movie, *The Empire Strikes Back*, only colder. So cold that even the equator had iced up.

What caused it? We really don't know. *ScienceDaily* speculated it was a drop in atmospheric CO_2, but that's based on the fad of the moment. You've already seen how prehistoric Earth's climate bounced around with clear disregard for CO_2 levels.

The sun was around 30% weaker back then, so earth was more vulnerable to iciness. Even so, ice ages account for only about 4% of Earth's history. With the gradual heating up of the sun, a trend towards warmth over time, with fewer cold spells, seems only logical.

There's argument, still, about whether the Earth did, in fact, completely freeze over. Pollard insisted it had, because there was no other rational explanation for the existence of sea level glacier trails in the tropics. But others argue life would have been unlikely to survive a complete freeze-out.

To get around this, Snowball Earth believers speculate that perhaps warm spots near volcanoes remained, allowing life to congregate in those places. They call these hotspots, "refugia".

Countering the Snowball believers are a group of scientists advocating a rival "Slushball" theory: that earth became icy, yes, but not completely frozen over at the equator. In 2005, *ScienceDaily* reported the discovery of ancient lifeforms dating to the "Snowball" period, which proves that whatever ice there was cannot have been thick enough to block out sunlight.[86]

"If there was ice, it had to have been thin enough that organisms could photosynthesize below it or within it," reported lead researcher Alison Olcott, a Ph.D earth sciences student at USC, who examined sediment drill samples for evidence of ancient life.

She and her team found "a complex and productive microbial ecosystem", many of whose members relied on photosynthesis – turning sunlight into energy – to survive. They could not have lived in black seas under thick ice sheets.

Snowball believers hit back, suggesting she might have accidentally drilled in a "refugia" area, but she questioned the odds of that, if refugia areas were supposed to be few and far between. Besides, the drill samples were taken from a wide geographical area, not a small one.

"At what point does an enormous refugium become open ocean?" Olcott asked wryly.

Snowball or Slushball, everyone agrees it was cold.

At the end of 2008, new research published in *Nature Geosciences* shed light on how Earth extricated itself from its self-imposed blizzard.

"The planet's present day greenhouse scourge, carbon dioxide, may have played a vital role in helping ancient Earth to escape from complete glaciation, say scientists in a paper published online today," reported *ScienceDaily*.[87]

According to this latest research, Earth never completely froze over, because of a self-levelling mechanism. They believe that cooling temperatures allowed the oceans to absorb more oxygen out of the atmosphere, which in turn reacted with organic, carbon-based matter in the oceans and turned into CO_2 gas. Those ancient CO_2 emissions allegedly helped turn the Ice Age tide and melt the glaciers.

"In the climate change game," noted Professor Phillip Allen from London's Imperial College, who was the lead author on the study, "carbon dioxide can be both saint and sinner. These days we are so concerned about global warming and the harm that carbon dioxide is doing to our planet. However, approximately 600 million years

86 *ScienceDaily*, 3 October 2005, http://www.sciencedaily.com/releases/2005/10/051003232816.htm

87 *ScienceDaily*, 8 December 2008, http://www.sciencedaily.com/releases/2008/11/081130164511.htm

ago this greenhouse gas probably saved ancient Earth and its basic lifeforms from an icy extinction."

Allen sounds a worthy note of caution to cut through current hysteria.

"There is so much about Earth's ancient past that we don't know enough about. So it is really important that climate modellers get their targets right. They need to build into their calculations a warmer planet, with open oceans, despite lower levels of solar radiation at this time. Otherwise, climate models about the Earth's distant past are aiming for a target that never existed."

OK, so that's the scenario for the ice planet. If it was cold enough to freeze almost everything even up to the tropics and possibly the equator, imagine how cold it must have been at the poles.

Contrast that mental image with our next scenario, "Spa-pool Earth", a planet sweltering with ocean temperatures at the equator of 35-37°C (95-98.6°F). Hotter than a heated swimming pool. As steaming, in fact, as a hot tub. Imagine jumping off a boat in the Caribbean or Fiji, into waters that searing.

The time of this occurrence was 91 million years ago during what's known as the Cretaceous Thermal Maximum.

Now you would think, at a time when the ocean at the equator was so hot you couldn't stay submerged in it for more than a few minutes, that there would be palm trees growing and ukuleles playing way down south in Antarctica.

Apparently not. At least, not entirely. Much of current global warming theory hinges on a belief that if Earth gets too hot Antarctica and Greenland's glaciers will melt, pouring the world's reserves of fresh water into the sea and drowning New York.

But in January 2008, after the most recent IPCC Assessment had already released its gloomy forecast, new research came to light showing that Antarctica had kept a polar ice cap, even when the equatorial oceans were spa pools.

"New research challenges the generally accepted belief that substantial ice sheets could not have existed on Earth during past super warm climate events," reported *ScienceDaily*.[88]

The study, by researchers from Scripps Institution of Oceanography at UC San Diego, has turned up evidence of polar ice caps

88 *ScienceDaily*, 11 Jan 2008, http://www.sciencedaily.com/releases/2008/01/080110144824.htm

at least half the size of those currently in existence, at a time when Earth was at its hottest, climatologically speaking.

As you might recall, the world's average air temperature was a tropical 22⁰C to 25⁰C at this point, compared with the modern average between 12⁰C and 14⁰C.

In proving that ice caps existed, the Scripps Institution team studied tiny marine fossils left on the ocean floor 91 million years ago, and accessible now through specialist drilling samples. What they were looking for was chemical evidence of sea conditions at the time, recorded in the fossils – in this case an isotope of oxygen molecules known as d18O. In all cases, the fossils showed changes consistent with the effects of a large ice cap nearby.

"Until now, it was generally accepted that there were no large glaciers on the poles prior to the development of the Antarctic ice sheet about 33 million years ago," study co-author Richard Norris, a professor of paleobiology, told reporters.

"This study demonstrates that even the super-warm climates of the Cretaceous Thermal Maximum were not warm enough to prevent ice growth."

As verification that their claim of a polar ice cap was accurate, the team pointed to evidence that sea levels fell between 25 and 40 metres (82-131 ft) at the same time, which is what you'd expect if water became landlocked in massive ice floes.[89]

While global warming believers are telling journalists and political leaders that, "trust us, the science is settled", the latest research is suggesting anything but.

Another new study, published after the latest IPCC report has come out (and therefore more up to date), suggests that an ancient time period being used as a poster child for modern climate studies has in fact been badly misunderstood by modern scientists.

"Geoscientists have long presumed that, like today, the tropics remained warm throughout Earth's last major glaciation 300 million years ago," reported *ScienceDaily* late last year.[90] "New evidence, however, indicates that cold temperatures in fact episodically gripped these equatorial latitudes at that time."

---

89 Geological Society of America *Bulletin*, March/April 2004, National Science Foundation study, "Ice sheets caused massive sea level change during late Cretaceous"

90 "Cold And Ice, Not Heat, Episodically Gripped Tropical Regions 300 Million Years Ago", National Science Foundation release reported by *ScienceDaily*, 1 August 2008, http://www.sciencedaily.com/releases/2008/07/080731140227.htm

The study, by Oklahoma University's Gerilyn Soreghan and a research team, found evidence of glaciers in Colorado's Rocky Mountains. A 'no-brainer', you might quip, except that these were not modern glacier trails but ones laid down 300 million years ago when the Rockies were part of the tropics on the supercontinent Pangaea.

According to a National Science Foundation news release, climate model simulations are unable to replicate such cold tropical conditions for that time period.

"We are left with the prospect that what has been termed our 'best known' analogue to Earth's modern glaciation is, in fact, poorly known," *ScienceDaily* quoted Soreghan.

"The evidence we found indicates that glaciers were common at this time, even in tropical latitudes. This calls into question traditional assumptions of long-lasting equatorial warmth in the Late Paleozoic, and raises the possibility of large-scale and unexpected climate change in the tropics during that time."

All of which calls into question current alarmist statements by global warming believers. Either the climate models used by the IPCC are correct, or they're not. And if they're not, then all the political huffing and puffing about emissions caps and carbon taxes and making households pay is nothing more than additional, needless and highly expensive hot air.

Chapter Six

So Why Are We Seeing Global Warming At The Moment?

"Global warming, ozone depletion, deforestation and over-population are the four horsemen of a looming 21st century apocalypse. As the cold war recedes, the environment is becoming the No. 1 international security concern"

– Michael Oppenheimer, Council on Foreign Relations, 1990[91]

I'm so glad you asked this question. You've recently seen the massive scale of temperature changes on Planet Earth in prehistoric times, from the globe completely freezing over, to the oceans being as steamy as spa pools. You've also seen how those fluctuations had little or no correlation with CO_2 levels in the ancient atmosphere.

So we know Earth has a dynamic climate, and we know that climate can change – for reasons we still don't fully understand – on a dime, with massive temperature decreases ushering in ice ages over just a couple of decades.

However, if you listen to global warming believers, there's been a clear and worrying warming trend over the past 150 years that is now threatening the existence of civilisation as we know it.

Take the earnest Bill McKibben again:[92]

"Solving this crisis is no longer an option. Human beings have already raised the temperature of the planet about a degree Fahrenheit. When people first began to focus on global warming (which is, remember, only 20 years ago), the general consensus was that at this point we'd just be standing on the threshold of realizing its

91 "From Red Menace to Green Threat", Michael Oppenheimer, *New York Times*, 27 March 1990

92 Bill McKibben, "Think Again", *Foreign Policy* January/February 2009, http://www.foreignpolicy.com/story/cms.php?story_id=4585&page=2

consequences – that the big changes would be a degree or two and hence several decades down the road. But scientists seem to have systematically underestimated just how delicate the balance of the planet's physical systems really is.

"The warming is happening faster than we expected, and the results are more widespread and more disturbing. Even that rise of 1 degree has seriously perturbed hydrological cycles: Because warm air holds more water vapour than cold air does, both droughts and floods are increasing dramatically. Just look at the record levels of insurance payouts, for instance. Mosquitoes, able to survive in new places, are spreading more malaria and dengue. Coral reefs are dying, and so are vast stretches of forest.

"None of that is going to stop, even if we do everything right from here on out. Given the time lag between when we emit carbon and when the air heats up, we're already guaranteed at least another degree of warming.

"The only question now is whether we're going to hold off catastrophe. It won't be easy, because the scientific consensus calls for roughly 5 degrees more warming this century unless we do just about everything right. And if our behaviour up until now is any indication, we won't," said McKibben.

The rise of just over one degree Fahrenheit, or half a degree Celsius, has happened since 1850. What isn't really addressed by global warming believers is that by making 1850 their "ground zero", they're painting a misleading picture about the significance of recent warming.

Why? Because the late 1800s were the end of a centuries-long cool spell, the peak of which was known as "The Little Ice Age". The planet became very cold and it has only just ended that cycle an eyeblink ago in geological time. Naturally, if you come out of a cold spell, the temperature goes up.

The 20th century might well be warm in comparison with the 19th century, but it is not the warmest on record in human history, not by a longshot.

"A review of more than 200 climate studies led by researchers at the Harvard-Smithsonian Center for Astrophysics has determined that the 20th century is neither the warmest century nor the century with the most extreme weather of the past 1000 years," begins a news release from Harvard.[93]

---

93 "20th Century Climate Not So Hot", 31 March 2003, Harvard-Smithsonian Center for Astrophysics http://www.cfa.harvard.edu/press/archive/pr0310.html

Bet you haven't heard global warming believers telling you *that* on the TV.

"The review also confirmed that the Medieval Warm Period of 800 to 1300 A.D. and the Little Ice Age of 1300 to 1900 A.D. were worldwide phenomena not limited to the European and North American continents. While 20th century temperatures are much higher than in the Little Ice Age period, many parts of the world show the medieval warmth to be greater than that of the 20th century.

"Smithsonian astronomers Willie Soon and Sallie Baliunas, with co-authors Craig Idso and Sherwood Idso (Center for the Study of Carbon Dioxide and Global Change) and David Legates (Center for Climatic Research, University of Delaware), compiled and examined results from more than 240 research papers published by thousands of researchers over the past four decades. Their report, covering a multitude of geophysical and biological climate indicators, provides a detailed look at climate changes that occurred in different regions around the world over the last 1000 years.

" 'Many true research advances in reconstructing ancient climates have occurred over the past two decades', Soon says, 'so we felt it was time to pull together a large sample of recent studies from the last 5-10 years and look for patterns of variability and change. In fact, clear patterns did emerge showing that regions worldwide experienced the highs of the Medieval Warm Period and lows of the Little Ice Age, and that 20th century temperatures are generally cooler than during the medieval warmth'."

It's official, then. Despite what Al Gore said in his movie. In fact, it directly contradicts Gore – who should have known better because this study was published three years before his movie came out.

In the ironically-named *Truth*, Gore deliberately glides over the Medieval Warm Period, and even suggests it matches the CO_2 records.

"They can go back in a lot of these mountain glaciers a thousand years. They constructed a thermometer of the temperature. The blue is cold and the red is warm. I show this for a couple of reasons. Number one the so called sceptics will sometimes say 'Oh, this whole thing is cyclical phenomenon. There was a medieval warming period after all.' Well yeah there was. There it is right there," he says, pointing to a hockey stick graph on the screen.

"There are one there and two others. But compared to what is going on now, there is just no comparison. So if you look at a thousand years worth of temperature and compare it to a thousand years of CO2 you can see how closely they fit together."

I don't get it. Here's Gore explaining away the Medieval Warm Period (MWP) by saying it fits his carbon dioxide figures like a glove, and you'll recall he said this as well:

"When there is more carbon dioxide, the temperature gets warmer, because it traps more heat from the sun inside," he intones on screen, pointing to an ice core graph. "Carbon dioxide having never gone above 300 PPM, here is where CO2 is now [380]. We give off where it has never been as far back as this record will measure."

Well, if temperatures are only getting warmer because of CO2 emissions, and the evidence of the past 1,000 years fits like a glove, and it's true that Earth hasn't seen CO2 levels as high as they are now for 650,000 years, then how does he explain Harvard University's discovery that the Medieval Warm Period was warmer than the 20th century?

He doesn't. Gore is relying on the claims of hockey-stick graph guru Michael Mann, another of the global warming religion's high priests. Back in the late 1990s, and again in senate testimony in 2003,[94] Mann claimed the current warming period (CWP) was "unprecedented over at least the past millennium". The UN IPCC (Intergovernmental Panel on Climate Change), picking up on "the Mann-tra", told the world's media that the warmth of the 20th century was "*unprecedented over the entire past millennium*".

Mann has also stated on a Canadian TV broadcast[95] that "there is a 95 to 99% certainty that 1998 was the hottest year in the last one thousand years."

Mann used his dodgy data to construct an even dodgier graph, the hockey stick, purporting to show declining temperatures over the past millennium until the industrial revolution of the 20th century. This was the infamous hockey stick, it was utterly wrong, and it was the graph Gore displayed in his movie.

And yet here is a joint Harvard-Smithsonian investigation blowing Al Gore, Michael Mann and the UN IPCC out of the swamp.

So how much warmer was the MWP? One study published in

94 Testimony of Michael Mann to US Senate Committee on Environment and Public Works, 29 July 2003 http://epw.senate.gov/108th/Mann_072903.pdf

95 http://motls.blogspot.com/2007/04/cbc-global-warming-doomsday-called-off.html

Nature reveals the temperature by 1400AD in New Zealand was 0.75°C higher than the average temperature there now.[96]

Another study from Italy reveals winter temperatures around 1000AD that were 0.9°C higher on average than they are today.[97]

A Swedish study puts temperatures well above that:

"Between AD 900 and 1000, summer temperature anomalies were as much as 1.5°C warmer than the 1961-1990 base period."[98]

In Iceland, which provides clues about the Arctic Circle climate, a study last year[99] of sea temperatures found "that the peak warmth of the Medieval Warm Period was about 1°C *higher* than the peak warmth of the Current Warm Period."

In other words, the North Atlantic was a full one degree warmer than it currently is.

Sediment cores from a Portuguese estuary reveal the Atlantic Ocean was also much warmer off the Iberian Peninsula a thousand years ago.[100]

"The MWP was identified as occurring between AD 550 and 1300, during which time interval mean sea surface temperatures were between 1.5 and 2°C higher than the mean value of the past century, while peak MWP warmth was about 0.9°C greater than late 20th-century peak warmth."

Greenland may be thawing slightly, but temperatures there now are still a full one degree below where they were in the Medieval Warm Period, according to an ice core sample from the summit of the Greenland ice cap.[101]

Further south, in Mexico, a 2007 study reveals MWP temperatures 1.5°C higher than they are today.[102]

---

96 Wilson, A.T., Hendy, C.H. and Reynolds, C.P. 1979. Short-term climate change and New Zealand temperatures during the last millennium. *Nature* 279: 315-317.

97 Giraudi, C. 2005. Middle to Late Holocene glacial variations, periglacial processes and alluvial sedimentation on the higher Apennine massifs (Italy). *Quaternary Research* 64: 176-184

98 Linderholm, H.W. and Gunnarson, B.E. 2005. Summer temperature variability in central Scandinavia during the last 3600 years. *Geografiska Annaler* 87A: 231-241.

99 Sicre, M.-A., Jacob, J., Ezat, U., Rousse, S., Kissel, C., Yiou, P., Eiriksson, J., Knudsen, K.L., Jansen, E. and Turon, J.-L. 2008. Decadal variability of sea surface temperatures off North Iceland over the last 2000 years. *Earth and Planetary Science Letters* 268: 137-142.

100 Abrantes, F., Lebreiro, S., Rodrigues, T., Gil, I., Bartels-Jónsdóttir, H., Oliveira, P., Kissel, C. and Grimalt, J.O. 2005. Shallow-marine sediment cores record climate variability and earthquake activity off Lisbon (Portugal) for the last 2000 years. *Quaternary Science Reviews* 24: 2477-2494.

101 Johnsen, S.J., Dahl-Jensen, D., Gundestrup, N., Steffensen, J.P., Clausen, H.B., Miller, H., Masson-Delmotte, V., Sveinbjörnsdottir, A.E. and White, J. 2001. Oxygen isotope and palaeotemperature records from six Greenland ice-core stations: Camp Century, Dye-3, GRIP, GISP2, Renland and NorthGRIP. *Journal of Quaternary Science* 16: 299-307.

102 Richey, J.N., Poore, R.Z., Flower, B.P. and Quinn, T.M. 2007. 1400 yr multiproxy record of climate variability from the northern Gulf of Mexico. *Geology* 35: 423-426.

In Antarctica, the poster child for global warming believers, sediment samples from the Antarctic Peninsula reveal the MWP temperature was substantially higher than the current warm period, even allowing for recent warm spells there.[103]

Not content with ice core samples, American Antarctic researchers went looking for the remains of old seal breeding sites and radiocarbon dated the remains they found. What they discovered was that elephant seal breeding colonies were well south (further inland) from their current locations because the Medieval Warm Period had substantially retreated the ice sheets inland.

This extended "summer of seals" began around 600 BC and ended at 1400 AD, "broadly contemporaneous with the onset of Little Ice Age climatic conditions in the Northern Hemisphere and with glacier advance near [Victoria Land's] Terra Nova Bay."[104]

Their study indicated "warmer-than-present climate conditions" over a long period of time, and they said that "if, as proposed in the literature, the [Ross] ice shelf survived this period, it would have been exposed to environments substantially warmer than present."[105]

I'm sure you're starting to get the picture. Synopses of these and many other studies can be found on the website *CO2Science*,[106] and they make interesting reading in light of the claims that appear to dominate the news media.

There is little denial that the planet is warming up at the moment, but there is little evidence the warming is anthropogenic; human-caused.

Yet Al Gore had the hopeful audacity, despite many of these studies being published well before his film was made, to brush off the Medieval Warming Period with one glib reference: "Compared to what is going on now, there is just no comparison."

He's right, just not the way he intended. Study after study shows it was substantially warmer back then, and ice sheets substantially smaller. We've had a half degree Celsius rise in temperature over the past 150 years, but a thousand years ago it rose up to six times that

---

103 Khim, B.-K., Yoon, H.I., Kang, C.Y. and Bahk, J.J. 2002. Unstable climate oscillations during the Late Holocene in the Eastern Bransfield Basin, Antarctic Peninsula. *Quaternary Research* 58: 234-245.

104 Hall, B.L., Hoelzel, A.R., Baroni, C., Denton, G.H., Le Boeuf, B.J., Overturf, B. and Topf, A.L. 2006. Holocene elephant seal distribution implies warmer-than-present climate in the Ross Sea. *Proceedings of the National Academy of Sciences USA* 103: 10,213-10,217.

105 This, of course, casts doubt on the claims that Antarctica's big sea ice shelves date back 10,000 years, or even 2,000 years – particularly in the warmer West Antarctic region. They may be much younger, frequently receding and growing.

106 http://www.co2science.org/index.php

from its lows. So yes, Gore is correct when he says today's "global warming" doesn't hold a candle to the five hundred year warm spell a thousand years ago.

So that's where we stand in relation to recent global warming in modern history. Not only did it happen across the entire globe, but humanity and life in general thrived. Even the polar bears lived through it.

However, the full context of 20th century warming is set against the back drop of the Little Ice Age, which began around 1400 AD[107] and lasted through to about 1900. Although most of the blame for the LIA is suspected to lie with a reduction in the sun's intensity, known as the "Maunder Minimum", it triggered climate feedbacks and sent the civilised world reeling.

"The cold began on January 6, 1709, and lasted in all its rigour until the twenty-fourth," wrote one French priest in his journal. "The crops that had been sewn [sic] were all completely destroyed.... Most of the hens had died of cold, as had the beasts in the stables. When any poultry did survive the cold, their combs were seen to freeze and fall off. Many birds, ducks, partridges, woodcock, and blackbirds died and were found on the roads and on the thick ice and frequent snow. Oaks, ashes, and other valley trees split with cold. Two thirds of the vines died.... No grape harvest was gathered at all in Anjou.... I myself did not get enough wine from my vineyard to fill a nutshell."[108]

During the 1600s, France was hit by repeated famines caused by crop failures, which in turn led to disease outbreaks because of poor nutrition and the cold.

"Famine was widespread during this period, from the food shortages arising from the activities of the League in the 1590s, through subsistence crises of 1630, 1649, 1652, 1661, and 1694, to the great grainless winter of 1709 (Le Roy Ladurie 1987:272). The immediate triggers of these crises were periods of very cold or very damp winters and unusually wet summers. For example, the famines of 1630 and 1661 occurred during peace time and were entirely due to bad weather," writes ecology professor Peter Turchin.[109]

---

107 The dates of the MWP and Little Ice Age vary between the northern and southern hemispheres, due to built-in lag caused by sea currents, air flows and climate oscillations. Thus, the MWP began slightly later in the south and was still peaking around the time the north was already plunging into coldness.

108 *Times of Feast, Times of Famine*, Emmanuel Le Roy Ladurie, Doubleday, 1971

109 http://www.eeb.uconn.edu/people/turchin/PDF/5Valois.pdf

France suffered heavy population losses as it struggled to balance the needs of the people against the needs of the state.

"Historians proposed various explanations for [the population decline], of which two appear most probable: the significant worsening of the climate and the great demands placed on the French peasantry by the aggressive external policy of Louis XIV."[110]

Turchin's analysis found real wages in France peaked at the end of the "golden age" – corresponding to the end of the Medieval Warm Period, and were slashed to one sixth of their values by the Little Ice Age, recovering to a maximum level of half what they had been.

Across the English Channel, Britain wasn't faring much better. In the winter of 1683/84, the River Thames froze solid for two months, and in fact the English Channel itself froze up, Arctic style, between Dover and Calais. It wasn't just icebergs in the channel, you could almost walk on water from England to France.

As to the Thames, it had become quite partial to freezing during the Little Ice Age, and became the scene of "frost fairs" when it did:

"On the 20th of December, 1688, a very violent frost began, which lasted to the 6th of February, in so great extremity, that the pools were frozen 18 inches thick at least, and the Thames was so frozen that a great street from the Temple to Southwark was built with shops, and all manner of things sold. Hackney coaches plied there as in the streets. There were also bull-baiting, and a great many shows and tricks to be seen. This day the frost broke up. In the morning I saw a coach and six horses driven from Whitehall almost to the bridge (London Bridge) yet by three o'clock that day, February the 6th, next to Southwark the ice was gone, so as boats did row to and fro, and the next day all the frost was gone. On Candlemas Day I went to Croydon market, and led my horse over the ice to the Horseferry from Westminster to Lambeth; as I came back I led him from Lambeth upon the middle of the Thames to Whitefriars' stairs, and so led him up by them. And this day an ox was roasted whole, over against Whitehall. King Charles and the Queen ate part of it."[111]

And you can see from the records how the Little Ice Age started off gradually, bit hard, and then eased off. The Thames froze 26

110 Ibid

111 Original eyewitness account taken from *The Beauties of England and Wales*, reprinted in *The Mirror of Literature, Amusement, and Instruction,* Vol. 13, No. 355., Saturday, February 7, 1829, republished by The Gutenberg Project http://www.archive.org/stream/themirrorofliter10950gut/10950.txt

times during the LIA. Twice in the 15th century, four times in the 16th century, ten times in the 17th century, eight times in the 18th century (including 1709) and one final occasion in the 19th century (1814). That's the last time the Thames has ever frozen over at London, although it was reportedly solid enough on that occasion that an elephant was led over it.[112]

Thomas Wright's epic *History of France*[113] records of the winter of 1709 on the European mainland:

"While Louis's kingdom was still overwhelmed with the alarm caused by the disasters of the year 1708, it was struck with a new calamity. A frost, unexampled for its severity, set in on the 9th of January 1709, and continued without any relaxation for more than a month.

"Even the impetuous Rhone was frozen over; and spirituous liquors became congealed by the side of the fire...Olives, vines, fruit-trees in general, corn in the ground, all perished: so intense was the cold, that the solid trunks of great trees are said to have burst asunder as though they had been blown to pieces by gunpowder; and stones were split.

"With a prospect of the entire loss of the crops, the prices of provisions rose to an extravagant height, and the population of the country was exposed to actual starvation. Whole families were found frozen to death in their houses."

Think about that mental image for a moment. Tens of millions died of cold, disease and hunger in Europe (where the events are best-documented because of the civilisation level and literacy of inhabitants). Al Gore tries to counter this – without ever actually directly acknowledging in his movie anything you've just read – by talking about the much smaller death toll in recent heat waves:

"We have already seen some of the heat waves scientists are saying are going to be a lot more common. A couple of years ago in Europe they had that massive heat wave that killed 35,000 people,"

112 Global warming believers have tried to minimize the relevance of the Thames freezing over, by pointing out that it didn't in 1963's very cold winter, even though it did further upstream. They point to a change in the river's banks in the 1800s that made it flow faster through London, and suggest that's why it no longer freezes in the city. The implication is that it still would, because winters remain cold. However, this excuse fails to dial in the five or six degree Celsius benefit of the urban heat island effect over London now. Nor does it address the reality of the English Channel freezing over in 1684. It also clashes with their own logic: either the earth is warmer now, which is why the Arctic and the Thames don't ice up as much, or it isn't really any warmer and it's only human intervention preventing a Thames freeze-up.

113 *History of France*, Vol 2, by Thomas Wright, p228, published c1859 by The London Printing & Publishing Company

said Gore. "India didn't get as much attention, but the same year the temperature there went to 122 degrees Fahrenheit."[114]

It hit the same level this southern summer just gone in Melbourne, where temperatures of 45°C were endured, although that event was not global warming related.

As a proportion of human population, however, i.e., per capita, far fewer people die of climate related causes now than they did two centuries ago. Gore presents a population graph that shows human birth rates shot up from the late 1800s onwards. There's a reason for that: the Little Ice Age's effects kept mortality rates high and made famine common – hardly conducive to raising a family. Yet as the globe begins to warm and crops begin to thrive, so too the population grows.

If you look back a few hundred years, mothers frequently had eight to ten children. They had to, just to ensure three or four survived long enough to have kids of their own. So, low population rates in the Middle Ages were not from lack of sex, but from a climate that was hostile to crops, people and wildlife.

Grimm's Fairy Tales, and others from that period, frequently talk of ravenous wolves and hard times. Humans were facing more predation from wolves because the climate was hurting everyone. The imagery in this tale from Hans Christian Andersen speaks volumes about life in the Little Ice Age:

"In the winter, when the fields were covered with snow, and the water filled with large blocks of ice which I had blown up to the coast," continued the Wind, "great flocks of crows and ravens, dark and black as they usually are, came and alighted on the lonely, deserted ship. Then they croaked in harsh accents of the forest that now existed no more, of the many pretty birds' nests destroyed and the little ones left without a home; and all for the sake of that great bit of lumber, that proud ship, that never sailed forth. I made the snowflakes whirl till the snow lay like a great lake round the ship, and drifted over it. I let it hear my voice, that it might know what the storm has to say…"[115]

"I moaned through the broken window-panes, and the yawning clefts in the walls; I blew into the chests and drawers belonging to his daughters, wherein lay the clothes that had become faded and

114 http://forumpolitics.com/blogs/2007/03/17/an-inconvient-truth-transcript

115 Hans Christian Andersen, "The Story of the Wind", http://hca.gilead.org.il/the_wind.html

threadbare, from being worn over and over again. Such a song had not been sung, at the children's cradle as I sung now. The lordly life had changed to a life of penury. I was the only one who rejoiced aloud in that castle," said the Wind. "At last I snowed them up, and they say snow keeps people warm. It was good for them, for they had no wood, and the forest, from which they might have obtained it, had been cut down. The frost was very bitter, and I rushed through loop-holes and passages, over gables and roofs with keen and cutting swiftness. The three high-born daughters were lying in bed because of the cold, and their father crouching beneath his leather coverlet. Nothing to eat, nothing to burn, no fire on the hearth! Here was a life for high-born people! 'Give it up, give it up!' But my Lord Daa would not do that. 'After winter, spring will come,' he said, 'after want, good times. We must not lose patience, we must learn to wait."

By the time the "good times" arrived, Hans Christian Andersen (who was 11 years old when the northern hemisphere experienced "the year without a summer" in 1816[116]) and the fictional characters in his story had long been laid to rest.

Some of humanity's darkest literature dates from these times of endless winter – John Polidori's *The Vampyre* from 1819, or *Frankenstein* from 1818, for example.

This, then, is the fine balance facing today's citizens – the choice between cold and warm. The frost of 1709 that left whole families "frozen to death in their houses" happened 300 years ago this January just gone. The year without a summer took place less than 200 years ago. These events are some of the Little Ice Age low points from which the climate is only now climbing out of. They were entirely natural in origin, part of global cooling and warming cycles that can be severe and which have swung hotter and colder in the past thousand years than anything today.

One of the things global warming believers hammer constantly is the "runaway" effect, whereby slow warming can speed up because it triggers other climate deviances to come into play. For example,

---

116 Officially, this coincided with a drop in sunspot activity on the sun's surface known as 'the Dalton Minimum'. Added to the mix however was dust thrown into the atmosphere by a large volcanic eruption in Indonesia, so 1816 was a double whammy year. Nonetheless, astrophysicists point out that during the Maunder and Dalton Minimums, the sun physically shifted its position in the solar system, a stunt it apparently pulls every 178 to 180 years, known as 'inertial solar motion'. It is not yet known whether ISM has any added climate impact.

thawing permafrost releases methane and CO_2 into the atmosphere, thus increasing the effects of global warming because it boosts greenhouse gas levels.

Using that argument, however, provides a very simple explanation for why greenhouse gas levels are rising now. The natural warming phase that pulled Earth out of the Little Ice Age a hundred years ago began to melt Arctic ice (we know this from news reports from that era) and tundra, and of course other sensitive regions on the planet. This thawing has unlocked other greenhouse gases stored in the oceans, soil and frozen ground, leading to the big increases in atmospheric methane and CO_2 since 1850.

This process was set in train at some stage from the late 1700s onward, as the Little Ice Age began to loosen its grip, and had accelerated enough by the early 20th century to cause climatologists and journalists to remark on how warm it was becoming.

It is not a process to be feared, it is merely Earth returning some balance to the climate after the cold half-millennium. We have not yet returned to the warmth of the Medieval Warm Period, which was an entirely natural global cycle and which caused humans, animal and vegetable life to thrive.

Because humanity's industrial revolution coincided with the end of the big freeze and the start of the big thaw, we have understandably but mistakenly assumed the warmth is caused by us, even though on the best scientific data available to the IPCC and others, the amount of carbon and methane chucked up by humans since 1850 has contributed about 2.5% of the rise in global temperatures, or just over 0.01°C in 150 years.

Are we seeing *global* warming though? Not necessarily. Although temperatures have clearly gone up in some areas, they haven't in others. Overall, the average is an increase, but averages can hide a multitude of non-events. The question of whether warming is global or regional also depends on the quality of climate data currently available, and as you're about to see there are some very big questionmarks over that one.

Chapter Seven

What About The Sun?

"Global warming – at least the modern nightmare vision – is a myth. I am sure of it and so are a growing number of scientists. But what is really worrying is that the world's politicians and policy makers are not."

– David Bellamy, Daily Mail, July 9, 2004

Before we even begin to tackle the sun's role in global warming, let's see first how global warming believers try and explain it away. The sun, after all, is our only serious source of heat, and in real terms it is 30% more powerful than it was back when the Earth was young. Your average seven year old could probably do the math based on that data and work out that if the sun is getting gruntier long term, earth will probably heat up, long term.

Not, however, according to global warming believers.

At one of their worship centres, the "Scientists for Global Responsibility" organisation, they're in denial when it comes to solar energy having any significant role, listing it as their "No. 2 myth":[117]

2. *Current climate change is caused by the Sun*

• the Sun is obviously a very important factor in Earth's climate, eg variations in solar energy reaching the Earth are a major factor in moving in and out of ice ages (due to long-term changes in the Earth's orbit around the Sun)

• some sceptics argue that historic variations in global temperature correlate well with changes in incoming solar energy and that this solar energy is now at a level higher than for several centuries

• *however, these recent variations are so small that they do not adequately explain the size of the current temperature changes*

117 http://www.sgr.org.uk/climate/CCampMythNotes_Aug06.html

Contrast that generalised statement from Scientists for Global Responsibility, with some actual specific studies, and work out for yourself if the variations are really "so small".

A 2003 study by scientists from Germany's Max Planck Institute estimates that the sun could have been responsible for as much as 50% of global warming since 1970, although they estimate the more likely figure to be just under 30%.[118]

Those same scientists, when they studied new data in 2004, came back even more strongly, according to a report in London's *Telegraph*:[119]

"Global warming has finally been explained: the Earth is getting hotter because the sun is burning more brightly than at any time during the past 1,000 years, according to new research.

"A study by Swiss and German scientists suggests that increasing radiation from the sun is responsible for recent global climate changes.

"Dr Sami Solanki, the director of the renowned Max Planck Institute for Solar System Research in Gottingen, Germany, who led the research, said: 'The Sun has been at its strongest over the past 60 years and may now be affecting global temperatures. The Sun is in a changed state. It is brighter than it was a few hundred years ago and this brightening started relatively recently – in the last 100 to 150 years'."

The clue to discovering the sun is getting brighter came from recent ice core samples from Greenland. The cores showed the late 20th century had the lowest recorded levels of beryllium 10 since the Medieval Warm Period. This particular isotope is created by cosmic rays, but levels in the atmosphere drop if the sun's magnetic energy increases.

Solanki conceded that the brightly burning sun was not directly solely responsible for global warming since 1970, but that it could be causing unknown global warming effects in the upper atmosphere not yet identified by scientists, impacts possibly affecting the planet more than the direct sunlight itself.

Far from solar variations being minor, Solanki says the sun appears to be going through a unique phase:

"The unusually high number of sunspots during the past century suggests that we currently may be seeing a state of the solar dynamo that is uncharacteristic of the Sun at middle age. Also, the higher

118 "Can solar variability explain global warming since 1970?", S Solanki and N Krivova, *Journal of Geophysical Research*, vol 108, 2003

119 "The truth about global warming – it's the Sun that's to blame", *The Telegraph*, 19 July 2004

activity level implies more coronal mass ejections and more solar energetic particles hitting the Earth. Thus we expect that the late 20th century has been particularly rich in phenomena like geomagnetic storms and aurorae."

Solanki says further evidence of this solar flaring impacting Earth comes from tracking sunspots from 1850 through 1900. After the end of the Little Ice Age, the sun became more active, but towards 1900 that activity dropped away, and on Earth temperatures fell again.

"Both [sunspot numbers] and temperature show a slow decreasing trend just prior to 1900, followed by a steep rise that is unprecedented during the last millennium; (2) great minima in the SN [sunspot number] data are accompanied by cool periods while the generally higher levels of solar activity between about 1100 and 1300 correspond to a relatively higher temperature (the medieval warm period)."[120]

Now, again, can someone explain how Scientists for Global Responsibility can seriously claim "these recent variations are so small that they do not adequately explain the size of the current temperature changes"?

That bastion of global warming worship, *New Scientist* magazine, admits through clenched teeth that the sun has a "huge influence" on climate cycles, and also admits the obvious:[121]

"Solar forcing may have been largely responsible for warming in the late 19th and early 20th century, levelling off during the mid-century cooling."

But *New Scientist* magazine, then putting its credibility on the line, tries to convince global warming believers to hold to their faith, despite these annoying studies, by assuring them:

"For the period for which we have direct, reliable records, the Earth has warmed dramatically even though there has been no corresponding rise in any kind of solar activity."

Let's put that little fabrication to bed right now, just so you can

---

120 "Millennium-Scale Sunspot Number Reconstruction: Evidence for an Unusually Active Sun since the 1940s", I Usoskin, S Solanki et al, *Physical Review Letters*, vol 91, no. 21, 21 November 2003, http://cc.oulu.fi/~usoskin/personal/Sola2-PRL_published.pdf

121 Climate myths: Global warming is down to the Sun, not humans, by Fred Pearce, *New Scientist*, 16 May 2007 http://www.newscientist.com/article/dn11650 . To be fair to Pearce, he does try to play down the beryllium 10 findings, quoting James Hansen's GISS team. GISS tried to explain away Be10 by running new computer modelling that minimised its significance, but then again this is the same clique of climatologists trying to defend Michael Mann's dodgy hockey-stick graph, and the same GISS whose leader, James Hansen, has suggested a sea level rise of up to five metres this century, even though this would have required Greenland's glaciers to melt at a rate 70 times faster than they currently are, starting last year. So in other words, treat any study by GISS, Hansen, Mann and their friends with a well-deserved raised eyebrow.

judge for yourself whether *New Scientist* can still be trusted on climate change issues. The magazine baldly states "there has been no corresponding rise in *any* kind of solar activity".

They could have looked at a real science journal, *Nature*, which in 1999 published a study, headed "A Doubling of the Sun's Coronal Magnetic Field during the Last 100 Years."[122]

In fact, the study discloses a 230% increase in the magnetic field, "which may influence global climate change", since 1901, with the last 140% of that rise happening since 1964, when global temperatures started to rise more significantly.

Work on the sun's magnetic field is increasing, amid dawning realisation it may have some of the biggest impact of all[123], but even if you stick to traditional measures of Total Solar Irradiance it appears clear that TSI has been increasing too, according to the latest data:

"This finding has evident repercussions for climate change and solar physics. Increasing TSI between 1980 and 2000 could have contributed significantly to global warming during the last three decades [Scafetta and West, 2007, 2008]. Current climate models [Intergovernmental Panel on Climate Change, 2007] have assumed that the TSI did not vary significantly during the last 30 years and have therefore underestimated the solar contribution and overestimated the anthropogenic contribution to global warming."[124]

So *New Scientist's* outrageous fibs about the influence of the sun can be seen as bogus. Yet, inexplicably, whilst the magazine isn't prepared to accept a solar role in anything post 1950 it does admit the sun was "largely responsible" for global warming in the 19th century and first half of the 20th century. So if they accept the sun caused global warming heading into the modern age, why are they in denial that such warming may (indeed, according to their own computer models it *must*) have caused other climate feedbacks, like increased CO_2 and methane, thus enhancing the warming of the late 20th century?

Their denial just doesn't make sense.

New Scientist appeals to authority, in the form of the 2007 IPCC

---

122 Study by M Lockwood, R Stamper and M N Wild, *Nature*, Vol 399, 3 June 1999, pages 437-439

123 "Hansen knocks down straw man", by J Blethen, 1 September 2008, http://heliogenic.blogspot.com/2008/09/hansen-knocks-down-straw-man.html

124 Scafetta N., R. C. Willson (2009), "ACRIM-gap and TSI trend issue resolved using a surface magnetic flux TSI proxy model," *Geophys. Res. Lett.*, 36, L05701, doi:10.1029/2008GL036307

report, which it says downgraded the influence of the sun as a possible factor over the past 250 years from 40% to 20%. The "authority" however was not that reliable.

"It is possible to infer, therefore, that the IPCC's reduction in its estimate of the solar influence on terrestrial temperature changes may have been unsoundly based. It did not command the support of most solar physicists: indeed, it was authored by a very junior solar specialist," writes Britain's Viscount Christopher Monckton, a global warming critic, in response to the *New Scientist* claim.[125]

"The Symposium of the International Astronomical Union in 2004 concluded with a communiqué to the effect that most of the warming of the past half century was caused by the Sun; that the Sun's activity would soon diminish (it has indeed done so); and that global cooling would set in as a result (this, too, has begun to occur).

"It is not credible, therefore, to dismiss the Sun as a significant cause – and perhaps even the primary cause – of the warming which began 300 years ago and continued at a near-linear rate until 1998, when it ceased and has now been replaced by seven years [now ten] of cooling," warns Monckton.

But Monckton makes another point reinforcing what we've already covered, and it bears repeating. The global warming believers are adamant that warming causes "climate feedbacks" which in turn accelerate the warming process. Most scientists are agreed that this occurs, although the level of runaway caused by feedbacks is open to debate because the planet has very successfully self-regulated its climate up to now.

However, global warming believers may actually be covering up the role of the sun, according to Monckton, specifically to brainwash people into believing it must be CO2.[126]

"The UN dated its list of 'forcings' (influences on temperature) from 1750, when the sun, and consequently air temperature, was almost as warm as now. But its start-date for the increase in world temperature was 1900, when the sun, and temperature, were much cooler.

"Every 'forcing' produces 'climate feedbacks' making temperature rise faster. For instance, as temperature rises in response to a forcing, the air carries more water vapour, the most important greenhouse

---

125 Ibid, see comments posted on *New Scientist* website after article.

126 "The sun is warmer now than for the past 11,400 years", Christopher Monckton, *The Telegraph*, 5 November 2006 http://www.telegraph.co.uk/news/uknews/1533312/The-sun-is-warmer-now-than-for-the-past-11,400-years.html

gas; and polar ice melts, increasing heat absorption. Up goes the temperature again. The UN more than doubled the base forcings from greenhouse gases to allow for climate feedbacks. It didn't do the same for the base solar forcing.

"Sami Solanki, a solar physicist, says that in the past half-century the sun has been warmer, for longer, than at any time in at least the past 11,400 years, contributing a base forcing equivalent to a quarter of the past century's warming. That's before adding climate feedbacks," says Monckton.

"The UN expresses its heat-energy forcings in watts per square metre per second. It estimates that the sun caused just 0.3 watts of forcing since 1750. Begin in 1900 to match the temperature start-date, and the base solar forcing more than doubles to 0.7 watts. Multiply by 2.7, which the Royal Society suggests is the UN's current factor for climate feedbacks, and you get 1.9 watts – more than six times the UN's figure.

"The entire 20th-century warming from all sources was below 2 watts. The sun could have caused just about all of it."

Much of it hinges on sunspots. Climate science is not clean and linear, but has a range of different ingredients to it. In the case of the medieval warm period, it's believed that higher than usual activity on the sun caused the global warming back then, possibly by cranking up the North Atlantic Oscillation. Conversely, we know from historical records since the telescope was invented that sunspots during the Little Ice Age were very infrequent – the sun was operating on reduced power if you like, apart from a blip in the mid 1700s.

In the 1850s we see sunspot activity beginning to increase, and the warmth since then is, of course, history. The sunspots, however, were not working alone.

When examining the sun's role in climate change, there's also the impact of longer term solar cycles, and long term response times in the environment. These are more subtle, yet powerful trends that can have climate impacts well down the track. Common wisdom says, for example, the CO_2 levels rise, which then causes temperatures to rise. But scientists studying ocean currents have found evidence that the oceans warmed first, then CO_2 rose, not the other way around, as this report from a 2007 study shows:

"Deep-sea temperatures warmed by ~2°C between 19 and 17 thousand years before the present (ky B.P.), leading the rise in atmospheric

CO_2 and tropical–surface-ocean warming by ~1000 years. The cause of this deglacial deep-water warming does not lie within the tropics, nor can its early onset between 19 and 17 ky B.P. be attributed to CO_2 forcing. Increasing austral-spring insolation [higher seasonal solar radiation in the Southern hemisphere] combined with sea-ice albedo [heat reflectivity] feedbacks appear to be the key factors responsible for this warming."[127]

All very technical, but what this study found was that solar heat in the southern hemisphere warmed the oceans enough that ice melted and CO_2 was released, but that it took up to a thousand years for the warmth to trigger CO_2 release in any major way. In other words, far from CO_2 being the "forcer" or instigator of warming, it was a result of warming that had begun a millennium earlier deep within the sea.

This idea that CO_2 is released *after* warming has already begun is backed up by another study, which examined ice cores from Vostok in Antarctica going back 240,000 years. Those scientists found "the CO_2 increase lagged Antarctic deglacial warming by 800 (±200) years and preceded the Northern Hemisphere deglaciation."[128]

Again, in simple terms, Antarctica warmed up, slowly releasing CO_2 trapped in the ice and nearby ocean, and that extra CO_2 didn't reach significant levels for around 800 years. The warming cycle, perhaps then aided by CO_2, then helped trigger melt in the Northern Hemisphere.

The 800 year time lag between warming and then release of CO_2 was seen as entirely reasonable:

"A delay of about 800 years seems to be a reasonable time period to transform an initial Antarctic temperature increase into a CO_2 atmospheric increase through oceanic processes. Indeed, it is not clear whether the link between the southern ocean climate and CO_2 is the result of a physical mechanism, such as a change in the vertical ocean mixing or sea-ice cover changes, or a biological mechanism, such as atmospheric dust flux and ocean productivity. The 800-year lag cannot really rule out any of these mechanisms

127 "Southern Hemisphere and Deep-Sea Warming Led Deglacial Atmospheric CO_2 Rise and Tropical Warming", L Stott, A Timmerman, R Thunell, *Science* 19 October 2007: Vol. 318. no. 5849, pp. 435 – 438 DOI: 10.1126/science.1143791

128 "Timing of Atmospheric CO2 and Antarctic Temperature Changes Across Termination III", Caillon et al, *Science* 14 March 2003, vol 299, page 1728 http://icebubbles.ucsd.edu/Publications/CaillonTermIII.pdf

as having sole control. Any of these mechanisms might plausibly require a finite amount of warming before CO_2 outgassing becomes significant," says the study.

So much for the popular theory that CO_2 levels increase first, *then* warming. All of this is vital in identifying the culprit behind the current warm period. If these two studies are to be believed, then it's conceivable that Earth is currently heating as a direct result of the sun's warmth during the Medieval Warm Period in AD 1000. Jump ahead 800 years, and the planet started to warm up around 1850, a warming trend that continues into the present. Based on these studies, that ancient heat would have caused carbon levels to rise starting in the 1800s, which they have. Whilst that carbon may have a secondary effect in helping warm the atmosphere now, it wasn't the initial cause of the warming, nor is the human contribution (about 1.7%) significant in comparison to the 98.3% of CO_2 emitted naturally.

Any extra oomph caused by greater sunspot activity was icing on the cake if it coincided with these longer term warming cycles.

If you've been to the beach and watched the waves coming in, you'll know what I mean. You can get a small wave approaching the shoreline, trickling in, when a larger wave catches it up and the combined effect is the addition of both waves together. Likewise, as a wave is surging back out away from the beach, it can take the sting out of an incoming wave.

Solar variations and other climate deviations are like that. They can boost each other, or partially cancel each other out, depending on timing. Depending on how many wave cycles are lining up at a given moment in time, the effect can be devastating, or minor. Our current computer models are not sophisticated enough to factor all these possibilities in.

So, if the current CO_2 release worldwide is the result of warming a thousand years ago coinciding with extra CO_2 released by human industry, its effects are, or have been, enhanced by the growth in solar activity during the 20th century.

Frankly, this makes far more sense than being asked to believe that a gas forming only 0.038% of the atmosphere, and which experts acknowledge the human contribution is less than 1/40th of that already tiny amount, is the *main* cause of current warming.

Putting it another way, humanity's contribution to atmospheric carbon dioxide is believed to be around 0.00095% of the total atmosphere by volume. And to put *that* another way, imagine having a 10,000 litre water tank (about the size of a household rain-collection tank), and adding one litre of lukewarm water to it. That's the scale of influence we're talking about.

In our water tank scenario, you don't actually add the whole human contribution (one litre) all at once. You add it at the rate of about eight ml, or 1.5 tsp, a year. Smart readers will have guessed that even that isn't quite correct, because we churn out CO_2 every day, not just once a year. You're right: imagine a tap dripping so slowly with lukewarm water that it took a year to fill up one and a half teaspoons.

The real rate of adding warm water to the big 10,000L cold water tank outside is *one drop of water every three days.*

Do you really believe this would have a catastrophic effect on the heat of the water in your tank, even if the drips held their heat?

It is even less credible to believe that cancelling human CO_2 emissions tomorrow (turning off our leaky tap at the rate of one drip every three days), is going to have a measurable effect on global temperatures fifty years from now.

Which brings us back to solar activity.

World temperatures in the past 40 years peaked in 1998, a major El Niño year and also approaching the highpoint in sunspot solar cycle 23. Satellites out in space, their view unobscured by clouds, have been directly measuring the energy from the sun as it hits Earth's atmosphere.

"The estimated increase [in energy output between 1986 and 1996] of 0.04% would induce appreciable climate change if it persists for a sufficient number of solar cycles and if the climate system feedbacks reached their full equilibrium response to the forcing," records the official website for one of those satellite monitoring projects.[129]

Interestingly, the peak of solar activity broke at the end of the 1990s, with a brief blip in March 2002,[130] then slumped significantly to almost no sunspot activity, lessening the sun's grip on our climate and heralding the start of a cooling trend from 1999 onwards that is still continuing, El Niño fluctuations notwithstanding.

By 2008, misleadingly named "the tenth warmest year on record", sunspot activity had dwindled to just 6 spots for the entire year,

129 http://lasp.colorado.edu/sorce/data/tsi_data.htm

130 As further proof of the sunspot link, the years 1999, 2000 and 2001 don't appear in the top 5 warmest years list from NASA, but 2002 ranks right up there behind 1998.

down from nearly 175 a year at the start of the decade. Officially, the sun was "blank" for 266 days of 2008, making it the quietest year on the sun since 1913.[131]

That means the sun has been less active in 2008 than in any of the preceding 95 years. Has it run out of fizz? Are we heading for a new little ice age? Not necessarily. It is too early to make predictions. Astronomers have no choice but to watch and wait for sunspots in cycle 24 to emerge, and see whether the sun revs up again or whether it is indeed slowing down.

Back in 2006, however, NASA was predicting that the upcoming solar cycle 24 would be a biggie:

"Solar cycle 24, due to peak in 2010 or 2011 'looks like its going to be one of the most intense cycles since record-keeping began almost 400 years ago,' says solar physicist David Hathaway of the Marshall Space Flight Center. He and colleague Robert Wilson presented this conclusion last week at the American Geophysical Union meeting in San Francisco."[132]

However, in March 2009, with the sun as blank as the proverbial millpond, NASA's David Hathaway issued another news release, explaining that the predicted peak was being shifted out to early 2013.[133] So far, solar cycle 24 is one of the quietest on record, which based on past experience will lead to colder weather.

But there are some other indications of irregularity in the sun's behaviour. At Colorado University's Laboratory for Atmospheric and Space Physics, Associate Director Tom Woods has told *Popular Science* magazine that the sun's magnetic field is very weak this year, 40% lower in strength at the poles than it usually hits at its lowest point.[134]

"If you look at the last time the polar field was that weak," says Woods, "it goes back to the early 1800s during a time called the Dalton Minimum [the infamous "year without a summer"], when we had a low solar activity cycle."

Now, you'll recall that study in *Nature* I mentioned where the sun's magnetic field had doubled in strength by 1999? Now it has dropped

131 Minnesota Public Radio, 5 January 2009 http://minnesota.publicradio.org/collections/special/columns/updraft/archive/2009/01/solar_minimum.shtml

132 "Scientists Predict Big Solar Cycle," NASA news release, 21 December 2006

133 http://wattsupwiththat.com/2009/03/08/more-revisions-to-the-nasa-solar-cycle-prediction/#more-6087

134 "What's happening to the sun: could its unusual behavior herald a new ice age?" *Popular Science*, January 2009

again as the Sun goes quiet, and suddenly we're cooling down.[135]

In correlation with the current 95 year low in solar activity, the winter of 2008/2009 in the northern hemisphere was the harshest in a hundred years, by some descriptions.

"At 7:51 a.m. today it was the coldest it's ever been in Cedar Rapids, at least on record," reported the *Des Moines Register*.[136] "Temperatures in Cedar Rapids dipped to 29 degrees F below zero (-34°C) this morning, setting an all-time record low, according to the National Weather Service.

"Cedar Rapids' previous lows of 28 degrees below zero were recorded on Dec. 28, 1924 and Jan. 12, 1974."

"January 2009 enters the record books as the city's 10th coldest" noted the *Chicago Tribune* this year.[137] "Its average temperature of 15.9 degrees F (-9°C) makes it the coldest January since 1994. Saturday's high of 38 degrees (3.3°C) was the month's peak reading, making it the first January since 1985 and only the ninth since 1871 with so low a maximum temperature."

Even in tropical Hawaii they were complaining.[138]

"The high temperature yesterday in Kahului, Maui, was just 72 degrees (22°C), tying the record for coolest high temperature for the date last set in 1975."

In New York state, Utica recorded a low of minus 3°F (-19.5°C) mid January,[139] and subsequent reports confirmed January 2009 was the coldest in western New York since 1977, and among the 18 coldest winters of the past 139 years.[140]

"Meteorologist Tom Niziol...working with National Weather Service statistical guru Dave Sage, used each day's average daily temperature – the average of the day's high and low temperatures – and then took the monthly average of those daily numbers.

"The coldest January on record? The blizzard year, 1977, when the average daily January temperature was a frigid 13.8 degrees (-10°C).

---

135 We could be in for a very significant cooling. Not only is solar cycle 24 extremely quiet, but NASA was already on record predicting that solar cycle 25, allegedly due to peak in 2022, could be "one of the weakest in centuries" according to David Hathaway (http://science.nasa.gov/headlines/y2006/10may_longrange.htm) This is because the solar 'conveyor belt' that moves around the sun has significantly slowed down. "We've never seen speeds so low," said Hathaway. All of this adds up to much less solar activity, and a colder Earth.

136 "Cedar Rapids reports -29 degrees: all-time coldest reading", *Des Moines Register,* 15 January 2009

137 "January was among coldest months on record", *Chicago Tribune*, 31 January 2009

138 "Record-low high temperatures set on Kauai, Maui", *Honolulu Advertiser,* 29 January 2009

139 "Many people forced to brave record-low temperatures", WKTV News, 14 January 2009

140 http://www.buffalonews.com/cityregion/story/561186.html

The warmest was 1932, when the average January temperature was 37.2 degrees (2.9⁰C).

"The top five coldest Januarys in the Buffalo area all date back more than 30 years, and all but 1977 date back more than 60 years.

"Does that say anything about global warming?

"Those few statistics are not enough to make an objective statement about whether this has anything to say about global warming," Niziol said. "But those are fascinating statistics."

Over in Europe, Bulgaria reported deep cold:

"Bulgaria is still in the grip of an unusual cold spell. The lowest temperature of –19 degrees Celsius was measured in the town of Kneja, Northern Bulgaria," reported Radio Bulgaria.

Bulgaria, however, had it easy:

"Slovenia registered the lowest temperatures ever. At the Bohin resort, a half frozen weatherman standing outside, reported minus 49°C (-56⁰F). Slovenian media have reported recommendations of the meteorological institute of Germany, which alarms [sic] over the risks of having piercings – the metal earrings on people's body could cause dangerous freezing.[141]

"No metal objects attached to the body should be worn, warns the media, for people who must venture outside. For everyone else, Slovenian media urges its citizens to stay in their homes."

In Canada, records also fell: "How cold is cold? So cold.. that at least six record lows were set across Manitoba Tuesday."[142]

Britain's temperatures hovered 1 to 1.5 degrees Celsius below average. "The last time winter months were as cold was in 1995-1996, when the lowest national temperature on record of minus 27.2⁰C was equalled in the Scottish village of Altnaharra," reported one newspaper.

It's worth pointing out that 1995/96 was the last solar minimum, or sunspot low point.

"London colder than Antarctica," bristled a headline in *The Telegraph*. "The capital reached a high of 3⁰C yesterday, which was the same as Nuuk, the capital of Greenland.[143]

"At Antarctica's Jubany Scientific Station, which is home to 60 people, 16,000 penguins and 650 sea lions, a relatively balmy high of 5⁰C was recorded."

141 "Slovenia with record low temperature -49", *Macedonia International News Agency*, 10 January 2009
142 http://www.cjob.com/News/Local/Story.aspx?ID=1053168
143 "London colder than Antarctica", *The Telegraph*, 5 January 2009

The paper noted overnight lows in London were hitting -10°C.

"New Zealand cooler in 2008 than 141 years ago," bellowed a statement from climate sceptics who'd crunched the numbers.

"New Zealand's national average temperature of 12.9 degrees C during 2008, described by NIWA [National Institute of Water and Atmospheric Research] as 'milder than normal' was in fact cooler than it was 141 years ago."[144]

The scientists made the point that carbon dioxide levels of 385ppm in 2008 are "a third higher than the 1867 level of 290ppm when the annual national temperature was officially recorded as 13.1 degrees C (55.7°F), 0.2 degrees warmer than this year."

The 1867 temperature readings came from the Royal Society of New Zealand's Transactions & Proceedings, which documented data provided by the "Inspector of Meteorological Stations" in 1867.

So why is the climate being thrown around so much? There appears to be nothing "global" about global warming. Rather, larger variations in different regions are impacting on the overall, and artificial, average.

So harsh has the northern winter of 08/09 been that Arctic wildlife is turning up as far south as New Jersey to escape the cold.

"Snowy owls, who normally breed in the Arctic and winter in northern Canada [have headed] south in search of food. Snowy owls have camped out in the Meadowlands this month, dive-bombing rats, mice and ducks," reported the *Hackensack Record*.[145]

Another Arctic bird to fly south is the rough-legged hawk.

"Raptors that usually stay well north of here are showing up. They are being driven down here because the ice and snow back home are so deep that the rodents they feed on have become difficult to find. About a dozen rough-legged hawks have made the Meadowlands home this January."

New Jersey's regular birdlife, meanwhile, has decamped to the warmer climes of South Carolina.

Despite a cooler 2008 and cold start to 2009 the British Met Office, one of the key conspirators in the global warming belief system, nonetheless took the opportunity to predict a really warm 2009, prompting scorn from *Telegraph* columnist Christopher Booker:[146]

"Last week, as Britain shivered in sub-zero temperatures, the Met

144 "NZ cooler in 2008 than 141 years ago", NZ Climate Science Coalition, 14 January 2009
145 "Really cold winter has birds befuddled" *The Record*, 1 February 2009
146 "More hot air from Met Office", Christopher Booker, *The Telegraph*, 3 January 2009

Office predicted that 2009 would be one of 'the five warmest years on record'. This statement entertained various US climate experts, such as Dr Roger Pielke Sr of Colorado University, who recalled how last September the Met Office forecast that this winter in the UK would be 'milder than average', just before we enjoyed the coldest autumn and winter for decades.

"Dr Pielke also recalled the Met Office's prediction two years ago that 2007 would be globally 'the warmest year on record'...Even as they made that prediction, temperatures began their steepest plunge since they toppled off that 1998 highpoint, dropping by nearly 0.7°C – equivalent to the entire net warming of the 20th century.

"The Met Office, which played a key part in setting up the IPCC, has long since abandoned any pretence that it is an impartial scientific body when it comes to promoting its favourite cause of man-made climate change. It stages seminars to equip 'professionals in Government and the public sector' to 'dispel scepticism about climate change in your organisation'.

"It's just a pity that our Met Office's comically consistent inability to predict weather even a few weeks ahead (let alone a century hence) is beginning to make it an international laughing stock," said Booker.

Mind you, this is the same UK Met Office that spent 33 million quid on a Cray supercomputer[147] to improve climate change forecasting, only to be forced to admit the machine's carbon footprint was a whacking great 14,400 tonnes of carbon emissions a year[148] – equivalent to the carbon footprint of 2,400 homes.

Before you laugh at the Met's misfortune, bear in mind that doing two Google searches from your own home computer allegedly wastes as much CO2 as boiling a kettle, because of the number of computer servers involved. Google, probably wanting to avoid paying extra carbon credits on those kinds of estimates, has tried to downplay the revelation.[149]

---

147 "Met Office forecasts a supercomputer embarrassment", *The Times of London*, 17 January 2009

148 I'm frankly a bit sceptical about the whole "carbon emissions" calculation game. Based on the above figures, a supercomputer in a room is responsible for nearly 40 tonnes of atmospheric carbon dioxide a day. Really? To put this in perspective, monitors attached to car exhaust pipes find emissions of around 150 grams per kilometre. If you drive your car a thousand kilometres, your car will pump out 150kg of CO2. Ten thousand kilometers (about six months' usage for an average motorist) will generate 1.5 tons of CO2.

149 "Google responds to search carbon cost claims", *The Independent*, 12 January 2009 http://www.independent.co.uk/life-style/gadgets-and-tech/news/google-responds-to-search-carbon-cost-claims-1324968.html

One of the aspects of the sun attracting a lot of scientific attention at present is the spectrum of solar radiation that the average person can't see: its magnetic field. As we've just seen, the sun's magnetic field is currently 40% weaker than its usual lows, which could be a bad sign for the weather.

When the field is strong, Earth's magnetic field strengthens in response, shielding the planet from a greater range of incoming cosmic rays and particles. But when the Sun's magnetic field is weak, so is ours and our upper atmosphere gets bombarded by radiation from elsewhere in space.

New studies have shown those cosmic ray particles can interact with high level clouds, and essentially help cool the planet down. But they might also explain why Australia has been taking a pounding with record heat and fires this past summer.[150]

"The sun's magnetic field may have a significant impact on weather and climatic parameters in Australia and other countries in the northern and southern hemispheres," reported *Science Daily* late in 2008.

"According to a study in *Geographical Research*, the droughts are related to the solar magnetic phases and not the greenhouse effect."

Essentially, the study links magnetic field fluctuations to "the incidence of ultraviolet radiation over the tropical Pacific, and changes in sea surface temperatures with cloud cover." This turns out to provide more accurate explanations for rainfall, or lack of it.

The experience with the desertification of the Sahara is a warning of what could happen to Australia if it continues to cook because of solar fluctuations. The essence of global warming, of course, is the greenhouse effect, where alleged triggers like rising CO_2 set a chain of events in motion that causes more water vapour to deposit in the atmosphere and cloak the planet in warm humid conditions. Australia may be many things, but it's a *lack* of moisture the "lucky country" is suffering from, and a *lack* of cloud cover.

Late in 2008, NASA announced another disturbing change in the sun, unprecedented since records began (to borrow a phrase used by global warming believers):[151]

"Data from the Ulysses spacecraft, a joint NASA-European Space

150 More on the fires later in the book

151 "Ulysses Reveals Global Solar Wind Plasma Output At 50-Year Low", NASA news release, 23 September 2008, http://www.nasa.gov/home/hqnews/2008/sep/HQ_08241_Ulysses.html

Agency mission, show the sun has reduced its output of solar wind to the lowest levels since accurate readings became available. The sun's current state could reduce the natural shielding that envelops our solar system.

" 'The sun's million mile-per-hour solar wind inflates a protective bubble, or heliosphere, around the solar system. It influences how things work here on Earth and even out at the boundary of our solar system where it meets the galaxy,' said Dave McComas, Ulysses' solar wind instrument principal investigator and senior executive director at the Southwest Research Institute In San Antonio. 'Ulysses data indicate the solar wind's global pressure is the lowest we have seen since the beginning of the space age.'

"The sun's solar wind plasma is a stream of charged particles ejected from the sun's upper atmosphere. The solar wind interacts with every planet in our solar system. It also defines the border between our solar system and interstellar space.

"This border, called the heliopause, surrounds our solar system where the solar wind's strength is no longer great enough to push back the wind of other stars. The region around the heliopause also acts as a shield for our solar system, warding off a significant portion of the cosmic rays outside the galaxy.

" 'Galactic cosmic rays carry with them radiation from other parts of our galaxy,' said Ed Smith, NASA's Ulysses project scientist at the Jet Propulsion Laboratory in Pasadena, Calif. 'With the solar wind at an all-time low, there is an excellent chance the heliosphere will diminish in size and strength. If that occurs, more galactic cosmic rays will make it into the inner part of our solar system.'

"In 2007, Ulysses made its third rapid scan of the solar wind and magnetic field from the sun's south to north pole. When the results were compared with observations from the previous solar cycle, the strength of the solar wind pressure and the magnetic field embedded in the solar wind were found to have decreased by 20 percent. The field strength near the spacecraft has decreased by 36 percent.

" 'The sun cycles between periods of great activity and lesser activity,' Smith said. 'Right now, we are in a period of minimal activity that has stretched on longer than anyone anticipated'."

The UN IPCC *Summary for Policymakers*, released late 2007 and being used by governments worldwide to draft responses to global warming, claims solar influence is "negligible".

However, a new scientific study in 2008 calls the IPCC claim into serious doubt. Nicola Scafetta and Bruce West looked at data the IPCC and many climatologists had ignored as just "background noise", and found that the "noise" matched near perfectly with known solar activity for the past four centuries at least.

On the garbage in/garbage out principle, they found existing computer modelling done by the IPCC and others was not producing accurate information because they had not entered key data on solar flares and other activity into the models.

"The nonequilibrium thermodynamic models we used suggest that the Sun is influencing climate significantly more than the IPCC report claims," Scafetta and West reported in the journal *Physics Today*. "If climate is as sensitive to solar changes as the...findings suggest, the current anthropogenic contribution to global warming is significantly overestimated.

"We estimate that the Sun could account for as much as 69% of the increase in Earth's average temperature," they added.

"Furthermore, if the Sun does cool off, as some solar forecasts predict will happen over the next few decades, that cooling could stabilize Earth's climate and avoid the catastrophic consequences predicted in the IPCC report."[152, 153]

All of which lends incredible weight to the suspicion that the real driver of climate change is the sun, and things about it that we're only discovering now. For example, the quiet periods on the sun in the past (Maunder Minimum, Dalton Minimum) have caused quite rapid cooling of the planet, and increased glaciation. Ironically, most of the melt we've recently seen on the world's glaciers may have nothing to do with current CO_2 emissions, and everything to do with solar activity hundreds of years ago.

Al Gore and others have pointed to retreating glaciers as proof of the perils of human-caused global warming on sensitive ecosystems. But according to glaciologists, we may be reading too much into glacial changes.

"To make a case that glaciers are retreating, and that the problem is global warming, is very hard to do," said Keith Echelmeyer of the Uni-

---

152 Scafetta, N. and West, B.J. 2008. Is climate sensitive to solar variability? *Physics Today* 61 (3): 50-51

153 See also Scafetta's follow-up paper: Scafetta N., R. C. Willson (2009), ACRIM-gap and TSI trend issue resolved using a surface magnetic flux TSI proxy model, *Geophys. Res. Lett.*, 36, L05701, doi:10.1029/2008GL036307

versity of Alaska's Geophysical Institute back in 1997.[154] At the time, Vice President Gore was holding news conferences about retreating glaciers, but Echelmeyer warned that "the physics are very complex. There is much more involved than just the climate response."

The US National Science Foundation, in a 1999 essay now only available via the Web Archive,[155] said the size and location of glaciers plays a huge role in how quickly they respond to modern climate change.

"How changes in the environment are 'felt' by the glacier depends on the 'security' of the glacial system. This means the size and setting of the ice. Remember that there are temperate glacial settings such as Chile and New Zealand, and polar glacial settings such as Antarctica. A climatic change, even a small one, will be felt by smaller glaciers in the warmer settings more than larger glaciers in the polar settings. Antarctica is well below freezing. Raising the temperature 5°C (9°F) permanently will not directly melt most of the Antarctic Ice Sheet. Very cold is still very cold! Environmental influences such as enhanced movement across a debris layer and rising sea level are more likely to directly influence glaciers in polar as well as other temperature settings."

Significantly, the NSF said a large polar ice sheet would take between 10,000 and 100,000 years to respond to global warming occurring now; a large mountain glacier between 1,000 and 10,000 years, and a small mountain glacier between 100 and 1,000 years to respond to warming happening now.

In other words, if glaciers are suddenly retreating or growing, they may be reacting to warming events that happened hundreds or even thousands of years ago.

"How quickly an ice body responds to an environmental change is called the response time.

"In general, the smaller the ice mass, the faster the response to a change in its environment. Valley glaciers will shrink much faster than a continental ice sheet if the climate warms and all other factors are equal. In fact, a large ice sheet might not react at all to temperature or precipitation [rainfall, snowfall] changes that happen over the period of a year or even a decade. In effect, they do not "feel" the small-scale climate changes."

---

154 http://www.sepp.org/Archive/Publications/pressrel/goreglac.html

155 http://web.archive.org/web/19991009085822/http://www.glacier.rice.edu/land/5_glaciersandtheir2.html

So again, if the National Science Foundation is correct, then movements of smaller glaciers may be related to the warming that first began in the late 1700s and became more apparent in the 1850s, while larger glaciers may be responding to the Medieval Warm Period. And if you are unaware of all these little fishhooks, you might just look at melting glaciers as proof of catastrophic global warming – just like many TV journalists do.

For what it's worth, Alaska's glaciers are growing again for the first time in 250 years. Michael Asher's *Daily Tech* website notes that the expansion followed record levels of winter snowfall which, because of the cold, stayed on the ground longer in the winter of 2006/07.

"In mid June [2008] I was surprised to see snow still at sea level in Prince William Sound," US Geological Survey glaciologist Bruce Molnia is quoted. "In general, the weather this summer was the worst I have seen in at least 20 years. On the Juneau Icefield, there was still 20 feet (6m) of new snow on the surface in late July. At Bering Glacier, a landslide I am studying [did] not become snow free until early August."

Written records of glacier extents in the area date back to the mid 1700s.

Meanwhile, Asher also notes the re-growth of Scandinavian glaciers last year.

"After years of decline, glaciers in Norway are growing again, reports the Norwegian Water Resources and Energy Directorate." The growth apparently began in 2007.

New Zealand's 50 South Island glaciers include some of the most accessible in the world for visitors – the Fox Glacier and Franz Josef. Despite global warming, the massive Fox terminates just 250 metres above sea level. And according to the company running daily tours on the Fox, it's been growing since the mid 80s.[156]

"Glaciers constantly advance and retreat, held in delicate balance by the accumulation of snow gained in the upper glacier and ice melting in the lower part. An increase in snowfall at the nevé [snow collection points at the top of the glacier] will result in the glacier advancing. Correspondingly, a faster melt will result in the glacier retreating. Overall Fox Glacier has been advancing since 1985."

It's the same story for the rest of them:[157]

156 http://www.foxguides.co.nz/facts.asp
157 http://www.niwa.cri.nz/news/mr/2005/2005-08-30-1

"Glaciers in New Zealand's Southern Alps gained ice mass again in the past year. Fifty glaciers are monitored annually by the National Institute of Water & Atmospheric Research (NIWA).

"NIWA's most recent survey of the glaciers was undertaken in March this year. Dr Jim Salinger of NIWA said today that analysis of the aerial photographs shows the glaciers had gained much more ice than they had lost during the past glacier year (March 2004 – February 2005).

"This year's gains are due to more snow in the Southern Alps, particularly from late winter to early summer 2004. During this five month period, more depressions ('lows') to the southeast of the Chatham Islands brought frequent episodes of strong cold southwesterly winds, and temperatures were 0.6°C below average, producing more snow."

Most of the NZ glaciers are like Fox and Franz Josef and respond very swiftly to climate, within five to 20 years, but some of the larger ones, like the lumbering Tasman glacier, are in what the National Science Foundation would call the 100 to 1000 year response time category.

While a 2008 news release[158] carried by every major news website around the world misleadingly suggested the Tasman is now melting in response to modern global warming, a briefing page on NIWA's website admits the Tasman is only now giving up the ground it gained during the Little Ice Age 300 years ago:[159]

"Their long response times have meant that they have simply absorbed any snow gains into their shrinking masses," wrote NIWA's Jim Salinger, "maintaining their areas while their surfaces have lowered like sinking lids. These glaciers have kept their LIA areas for all but the last couple of decades. Most of the relatively thick ice filling their over-deepened valleys was a remnant from the LIA, insulated by mantles of rocky moraine. They lost ice only by slow surface melt, with little or no terminus retreat. Rapid glacial lake expansion and glacier calving is now changing this situation."

In his movie, Al Gore spends a bit of time on glaciers, wailing about the impact of failing to ratify Kyoto.

"And now we're beginning to see the impact in the real world.

158 "Scientist: New Zealand glacier shrinking fast – Country's biggest glacier said to be melting at fastest rate in recent history", *Associated Press*, 24 April 2008 http://www.msnbc.msn.com/id/24287887

159 http://www.niwa.cri.nz/pubs/wa/ma/16-3/glacier

This is Mount Kilimanjaro more than 30 years ago, and more recently. And a friend of mine just came back from Kilimanjaro with a picture he took a couple of months ago. Another friend of mine Lonnie Thompson studies glaciers. Here's Lonnie with a sliver of a once mighty glacier. Within the decade there will be no more snows of Kilimanjaro."

The bad news for Gore is that while Kilimanjaro made pretty pictures, the melt of the mountain's snow and ice has nothing to do with modern global warming.

"Kilimanjaro's ice has been melting away for more than a century, and most of that melt occurred before 1953, prior to the period where science begins to be conclusive about atmospheric warming in that region," reported *LiveScience* on the findings of a new study on Kilimanjaro's plight.[160]

The problem on that mountain is deforestation on the lower slopes by farmers and landowners. Those forests have been providing moisture into the atmosphere which condenses as snow above the freezing level to feed the glacier and summit. Without the trees, moisture is down, meaning less snowfall and shrinking glacier.

Additionally, because Kilimanjaro's glacier is entirely well above freezing level (unlike the New Zealand glaciers), its shrinkage isn't actually caused by melting because it's too cold to melt. Instead, the drier air sucks moisture out of the ice by a process known as "sublimation". Sublimation skips the ice to water phase, and turns ice directly to water vapour at temperatures below freezing.

So while Kilimanjaro is losing its clothes thanks to human intervention, it is not suffering from higher global temperatures caused by CO_2, which is what Gore was trying to claim. That hasn't stopped Gore from misleading millions about glacier melt regardless.

"This is happening in [Alaska's] Glacier National Park," he says in the movie. "I climbed to the top of this in 1998 with one of my daughters. Within 15 years this will be the park formerly known as Glacier.

"Here is what has been happening year by year to the Columbia Glacier. It just retreats more and more every year. And it is a shame because these glaciers are so beautiful. People who go up to see them, here is what they are seeing every day now."

Not one mention from Gore about the reality the glaciers are

160 "Global Warming Not Behind Kilimanjaro Meltdown", *LiveScience*, 11 June 2007, http://www.livescience.com/environment/070611_gw_kilimanjaro.html

reacting to much earlier warm periods than our own. Another inconvenient fact to be avoided.

Indeed, as climatologist Doug Hoyt, the author of nearly a hundred scientific papers on global warming, notes,[161] many of the glaciers receding worldwide since the end of the Little Ice Age have not retreated back to where they were in the medieval warm period and are still in what could be called 'occupied territory'.

"As examples, the Aletsch and Grindelwald glaciers (Switzerland) were much smaller than today between 800 and 1000 AD. In 1588, the Grindelwald glacier broke through its end moraine and it is still larger than it was in 1588 and earlier years. In Iceland today, the outlet glaciers of Drangajökull and Vatnajökull are far advanced over what they were in the Middle Ages and farms remain buried beneath the ice."

Both the Medieval Warm Period and the effects of deforestation on Kilimanjaro provide clues to help solve the mystery over what's caused the increase in CO_2 levels and how that fits with global warming.

Here's something to think about. The Earth has clearly enjoyed warmer periods than now, with lower CO_2 levels. Those warmer periods *must* have created the same melts, with release of CO_2 and methane from tundra and the oceans that the climate models predict now.

But all that extra CO_2 doesn't show up in the records from those times. If warming releases CO_2 (and we know it does), why didn't that extra CO_2 show up in atmospheric samples trapped in the ice cores?

Presumably it was reabsorbed by the planet, perhaps because of greater forest growth (CO_2 being great plant food). This indicates the modern planet has an ecosystem capable of regulating emissions levels. But if that's true (and I believe it is), then why haven't human emissions been absorbed by the planet as well over the past 150 years?

To an extent they have. Only around half of our CO_2 emissions remain in the atmosphere, the rest is soaked up. But I would point the finger at deforestation as the reason for the surplus. Normally, rising CO_2 would result in bigger forests – that's what keeps the system in balance – but industrialization and pasturalization in the past 200 years saw many forests – carbon sinks – chopped down. This would explain the sudden increase in CO_2 levels beginning last century. At a time not just of human development but also natural warming as we emerged from the Little Ice Age, we nobbled one

161 http://www.warwickhughes.com/hoyt/scorecard.htm

of the planet's major systems for re-absorbing CO_2. Like a global game of musical chairs, we took away one of the chairs and left a portion of the increased emissions with nowhere to go.

This coincidence of factors, where we saw increased CO_2 and warming, led scientists to falsely assume the rise in CO_2 *caused* the warming. Yet as you've seen, the scientific data suggests the warming comes first, not rising CO_2.

The real answer is that if we hadn't cut down the great jungles and forests, we probably would not have detected a rise in CO_2 over this past century, just as we didn't detect a major CO_2 rise for the MWP. So the warmth, during the MWP and the 20th century, was caused by the sun, not gases. The only reason we're *seeing* the gases and panicking is because in the past they were re-absorbed.

Correlation, as they say, does not necessarily equal causation. Essentially that means just because CO_2 levels jumped, doesn't mean they've caused the warmth. They jumped because we cut down forests, at the same time as the sun was going through a hot phase. Coincidence, not cause and effect.

I would lay good odds that, left to its own devices, Planet Earth would quickly soak up the excess atmospheric CO_2 by re-growing its big forests, if we allowed it to.

There is some evidence that's exactly what's happening.

"CHILIBRE, Panama: The land where Marta Ortega de Wing raised hundreds of pigs until 10 years ago is being overtaken by galloping jungle – palms, lizards and ants," reported the *New York Times* this year.[162]

"Instead of farming, she now shops at the supermarket and her grown children and grandchildren live in places like Panama City and New York. Here, and in other tropical countries around the world, small holdings like Ms Ortega de Wing's – and much larger swaths of farmland – are reverting to nature, as people abandon their land and move to the cities in search of better livings.

"These new 'secondary' forests are emerging in Latin America, Asia and other tropical regions at such a fast pace that the trend has set off a serious debate about whether saving primeval rainforest – an iconic environmental cause – may be less urgent than once thought.

"By one estimate," reported the *New York Times*, "for every acre

162 "New jungles prompt a debate on rain forests", *New York Times*, 29 January 2009, http://www.nytimes.com/2009/01/30/science/earth/30forest.html?_r=3&hp

of rainforest cut down each year, more than 50 acres of new forest are growing in the tropics on land that was once farmed, logged or ravaged by natural disaster."

That's a staggering figure they don't teach in schools – the replenishment of rainforests naturally at a rate of 50 new acres for every one cut down. Plants grow faster and bigger in greenhouses with a CO_2 concentration of 700 ppm/v. It makes sense that higher CO_2 levels in our modern atmosphere are stimulating growth of Earth's lungs – the forests – to soak it up.

With some cunning planning, countries that produce large volumes of food efficiently for export markets could be encouraged to become the breadbaskets of the world, taking the pressure off rainforest environments and allowing small land holders to let the land return to bush. There may be costs in subsidizing Third World countries to achieve this, but it would be far cheaper than spending massive amounts of money on carbon-trading schemes.

What has become apparent to me, writing this book, is that the debate between both sides has become corrupted. The websites of global warming critics tend to be heavily science-based, quoting studies, assembling data, trying to win the intellectual debate; whilst the websites of global warming believers, and their cheerleaders in the media, are almost dumbed down, ignoring specifics in many cases to argue heart issues (cute polar bears), generalities and theories aimed at school students and other vulnerable members of the public who might be fooled by the aura of credibility.

Perhaps that's because the believers are so convinced of the truth of their new religion, that they feel the evidence will eventually support it but saving the planet requires immediate mobilisation. Research scientist and believer, Stephen Schneider, admitted as much in a *Discover* magazine interview 20 years ago.[163]

"On the one hand, as scientists we are ethically bound to the scientific method, in effect promising to tell the truth, the whole truth, and nothing but – which means that we must include all the doubts, the caveats, the ifs, ands, and buts. On the other hand, we are not just scientists but human beings as well. And like most people we'd like to see the world a better place, which in this context translates into our working to reduce the risk of potentially disastrous climatic

163 *Discover*, pp. 45–48, Oct. 1989

change. To do that we need to get some broadbased support, to capture the public's imagination. That, of course, entails getting loads of media coverage. So we have to offer up scary scenarios, make simplified, dramatic statements, and make little mention of any doubts we might have. This 'double ethical bind' we frequently find ourselves in cannot be solved by any formula. Each of us has to decide what the right balance is between being effective and being honest. I hope that means being both."

One thing's for certain, global warming belief websites and movies are fully of "scary scenarios" and "simplified, dramatic statements", and evidence that doesn't fit is glossed over or ignored.

Keep it in mind when you see or read scare stories in the news media or from political leaders being advised by global warming believers.

So, if we now know that the UN Intergovernmental Panel on Climate Change has used a fake hockey stick graph, and deliberately excluded data confirming the sun's influence in global warming, can we be confident they've actually been measuring global temperatures accurately, or have they been overinflated as well? What you're about to read is another dirty little secret of the global warming debate.

Chapter Eight

Urban Heat Islands: What If We've Been Reading The Wrong Data?

"The current anthropogenic contribution to global warming is significantly overestimated"

– ***Scafetta and West, Physics Today***

One of the most controversial aspects of the global warming debate is quantifying how much, if any, the planet itself is actually heating up. As you've seen over the last few chapters, there's clear physical evidence of a warming trend manifesting itself in a number of areas, and those hotspots are big enough to push global temperature "averages" up and thus justify use of the phrase "global warming".

However, averages can easily be misleading. If you hold a garage sale where the prices of 100 items are $5 each, and you sell all of them, your average sale price is $5. If you sell 99 items for $5 and one item for $1,000, your average sales price rockets to $14.95, which doesn't really reflect the value of most of the items on sale. It's the same with global temperature averages, where hot and cold spots can distort the results.

To be truly meaningful, current temperature readings for each area must be compared with historic temperature readings from the same location. That can identify a trend.

It can also, however, identify a major weakness in temperature data collection, and that's exactly what has happened in the global warming community. Virtually all of our data on the "global" warming trend since 1850 has come from historic daily temperature readings at weather stations around the world. I use the phrase "weather station" loosely. These can range from a professional meteorological bureau in a big city, down to Farmer Jones' thermometer on the back porch – if Farmer Jones has historically been in the habit of filing his weather readings back to a central collection point then chances are his data is part of the global warming database.

Now, you might think that it's backyard weather jockeys like Farmer Jones that we have to be wary of. In fact, it's the reverse. Most weather stations are close to, or in the middle of, towns and cities, and far away from farms. That's because in the 1870s no one saw the point in riding 60 miles to read a thermometer out in the wop-wops. Weather stations were only handy as long as they were nearby.

In the 1870s, black asphalt was not yet the order of the day on city streets, which tended to be dirt and gravel. Dirt doesn't get as hot as black asphalt. If you've seen the shimmer hovering over roads and footpaths in summer, you've probably twigged that our streets and buildings (particularly those with darker roofs) are capable of throwing an incredible amount of heat back into the atmosphere.

This phenomenon is known as the "Urban Heat Island Effect", or UHIE (you-ee). Weather stations that once collected the 3pm temperature in a small town a hundred and forty years ago now find themselves surrounded by buildings and busy streets; the small town now a large city.

Britain's Met Office explains how this has impacted London.[164]

"Chandler (1965) found that, under clear skies and light winds, temperatures in central London during the spring reached a minimum of 11°C, whereas in the suburbs they dropped to 5°C. Indeed, the term urban heat island is used to describe the dome of warm air that frequently builds up over towns and cities."

The more populated an area is, reports the Met Office, the greater the UHIE, and these islands of warmth can cause spectacular weather patterns to dump on surrounding areas.

"These can enhance convectional uplift, and the strong thermals that are generated during the summer months may serve to generate or intensify thunderstorms over or downwind of urban areas. Storms cells passing over cities can be 'refuelled' by contact with the warm surfaces and the addition of hygroscopic particles. Both can lead to enhanced rainfall, but this usually occurs downwind of the urban area."

Many of those great weather pictures that roll out on the nightly

---

164 http://www.metoffice.gov.uk/education/secondary/students/microclimates.html

news as "proof" of climate change have more to do with urban heat islands than global warming.

Britain's BBC, in an analysis of the problem admits that city weather is now way off kilter with the rest of the countryside.[165]

"The effect is an increase in temperature and that often means a massive difference between London and its surrounding areas. On some days an increase of around ten degrees has been known. This is illustrated on infrared satellite images where in the summer in particular, the major cities are seen as much darker areas in comparison with the rest of the country.

"Another reason why large towns and cities are generally warmer is due to decreased amounts of evaporation. In the countryside there is more water. As the water evaporates the process of changing from a liquid to a gas uses latent heat, which cools the surroundings.

"Roads and pavements do not retain moisture as fields and vegetation do, adding to the warming effect. You can easily feel this by standing on a black tarmac road on a hot summer's day, and then walking onto some parkland. You will immediately feel cooler.

"In America, researchers at NASA have been studying the effect and they say that some cities are so hot they can even generate their own weather, such as causing violent thunder and lightning storms.

"Certainly in London we see less rainfall and snowfall as well as higher temperatures."

Sadly, "knowing" something and remembering to factor it in are two different things. London's UHI effect has been well documented above, but it didn't prevent the media from attributing a reduction in foggy days in London this year to global warming.[166]

"Global warming may...have contributed to the drop in fog," reported *Bloomberg* news agency, "because cooler days favour its formation."

The logic being that warmer days mean less fog. That makes sense, but with a UHIE of up to six degrees in extra warmth, we can explain the drop in London's fogginess without invoking global warming at all. Yet it was a UK Met Office official who gave *Bloomberg* the "global warming" explanation. Do these people not read the scientific studies?

The story continued with a revelation that the UK Met Office

165 http://www.bbc.co.uk/weather/features/understanding/urban_heat_islands.shtml
166 "London fog clears in phenomenon that adds to warming", *Bloomberg*, 19 January 2009

had recorded an average temperature in Central England for the 30 years ending in 2008, of 9.95 degrees Celsius. That's up nearly half a degree from the 9.49 average of the 30 years ending in 1978. But did the data correct for Urban Heat Island Effect? I was unable to find any proof on the UK Met's Hadley Centre site that it did.

Stephen McIntyre on the *ClimateAudit* website fired similar bullets.

"Neither CRU [UK Met Office] nor NOAA have archived any source code for their calculations, so it is impossible to know for sure exactly what they do. However, I am unaware of any published documents by either of these agencies that indicate that they "correct" their temperature index for UHI effect," wrote McIntyre early 2009.[167]

Given that scientists know all about UHIE and the massive impact on temperatures, you would assume it had been factored into temperature readings being used for global warming calculations. The IPCC and others assure us that they've worked these details into their computer models and compensated for them.

Likewise, NASA's Gavin Schmidt assured readers that "The UHI effect is real enough, but it is corrected for".[168] However, there's no proof of that – no data has been offered to back up his claim.

The closest the UK Met gets is a statement that says they factor in 0.1 degrees Celsius per century as a general [not specific and cross-checked] allowance for UHI in Britain – but the figure is plucked from thin air and bears no resemblance to the documented UHI effect of up to 6.0 degrees Celsius.

The US domestic weather station network monitored by NASA's GISS [Goddard Institute of Space Studies] has access to some high quality rural weather stations which are not affected by urbanisation, but "outside the US, there is no corresponding network," writes Steve McIntyre.[169]

"A lot of the stations are in cities and virtually all of the recent data (post-1990) is from airports. GISS uses hopelessly obsolete population meta-data to supposedly identify 'rural' stations, but GISS 'rural' is all too often small city (or even large city). Unlike the US, GISS methods don't find sure ground and thus their adjustments end up being essentially random, mostly reflecting random site relocations and having nothing to do with UHI adjustment. They may say that they adjust for UHI, but this cannot be demonstrated in their actual adjustment."

167 "Real Climate and disinformation on UHI", Steve McIntyre, *ClimateAudit*, 20 January 2009, http://www.climateaudit.org/?p=4901
168 Ibid
169 Ibid

Testing the UHIE theory, scientists established weather stations in America's "northernmost settlement", the small town of Barrow, Alaska,[170] installing 54 temperature sensors in 2001, half within the urban area and the remainder scattered over 150km2 of tundra outside of town. They found the effect of buildings and 4600 residents enabled an average temperature difference in winter of 3.2⁰C between town (warmer) and tundra (colder).[171] But under the right conditions the difference could be as high as 9⁰C![172]

And that's not a city with skyscrapers and vast amounts of traffic. It's just an Alaskan village.

Studies of Mexico's big city temperatures have shown an average increase in temperature of 0.57⁰C per decade (remember, the IPCC reckons global temperatures have risen that much in a century), which suggests Mexico's largest cities are now, on average, nearly 6⁰C hotter than they were a hundred years ago, thanks to UHIE.[173] That underlines another point: you can't rule out UHIE impact by saying that it only biases the initial crossover from rural to urban. Some have tried to argue that once a weather station is in a UHIE area, you can attribute any further rises to global warming because the stats already factor in a UHIE baseline. Not true. What the Mexico study and others show is that as city populations grow, so does the UHIE in the city (more vehicles, more people, more air conditioning pumping warm air out of buildings), which is why Mexico's recording nearly 0.6⁰C increase per decade.[174]

Others, likewise, are far from convinced that climate scientists are telling the truth about UHIE's impact on temperature records. Viscount Monckton takes particular issue with the UN's decision to ignore a "fundamental" law of physics when calculating the amount of temperature rise the Earth might suffer from global warming.

"Even a 0.6⁰C temperature rise [per century] wasn't enough. So the

---

170 Hinkel, K.M., Nelson, F.E., Klene, A.E. and Bell, J.H. 2003. "The urban heat island in winter at Barrow, Alaska.", *International Journal of Climatology* 23: 1889-1905.

171 Hinkel, K.M. and Nelson, F.E. 2007. "Anthropogenic heat island at Barrow, Alaska, during winter: 2001-2005." *Journal of Geophysical Research* 112: 10.1029/2006JD007837

172 Temperature (Urbanization Effects – North America) – Summary", http://www.co2science.org/subject/u/summaries/uhinorthamerica.php

173 Jáuregui, E. 2005. "Possible impact of urbanization on the thermal climate of some large cities in Mexico." *Atmosfera* 18: 249-252.

174 In late March, a further devastating blow to urban temperature data was confirmed when the Hadley Centre's Phil Jones published a study of UHIE in China: "Urban-related warming over China is shown to be about 0.1°C decade^{-1} over the period 1951–2004". That's 0.5°C in just 50 years. See Jones, P. D., D. H. Lister, and Q. Li (2008), "Urbanization effects in large-scale temperature records, with an emphasis on China," *J. Geophys. Res.*, 113, D16122, doi:10.1029/2008JD009916.

UN repealed a fundamental physical law. Buried in a sub-chapter in its 2001 report is a short but revealing section discussing "lambda": the crucial factor converting forcings to temperature. The UN said its climate models had found lambda near-invariant at 0.5°C per watt of forcing.

"You don't need computer models to "find" lambda. Its value is given by a century-old law, derived experimentally by a Slovenian professor and proved by his Austrian student (who later committed suicide when his scientific compatriots refused to believe in atoms). The Stefan-Boltzmann law, not mentioned once in the UN's 2001 report, is as central to the thermodynamics of climate as Einstein's later equation is to astrophysics. Like Einstein's, it relates energy to the square of the speed of light, but by reference to temperature rather than mass."

The *Encyclopaedia Britannica* describes the Stefan-Boltzmann law as, "a fundamental constant of physics occurring in nearly every statistical formulation of both classical and quantum physics."

It would be one thing if this law was airy-fairy speculation yet to be proven. In such circumstances the UN IPCC might be within its rights to ignore it. But it isn't. It's a proven law of physics, in the same way that the law of gravity dictates apples do not fall up.

But ignore it they did. And they declared they had "found" lambda to be a 0.5°C temperature increase per watt of forcing.

"The bigger the value of lambda," points out Christopher Monckton,[175] "the bigger the temperature increase the UN could predict. Using poor Ludwig Boltzmann's law, lambda's true value is just 0.22-0.3°C per watt. In 2001, the UN effectively repealed the law, doubling lambda to 0.5°C per watt. A recent paper by James Hansen says lambda should be 0.67, 0.75 or 1°C: take your pick. Sir John Houghton, who chaired the UN's scientific assessment working group until recently, tells me it now puts lambda at 0.8°C: that's 3°C for a 3.7-watt doubling of airborne CO_2. Most of the UN's computer models have used 1°C. Stern implies 1.9°C.

"On the UN's figures, the entire greenhouse-gas forcing in the 20th century was 2 watts. Multiplying by the correct value of lambda gives a temperature increase of 0.44 to 0.6°C, in line with observation. But using Stern's 1.9°C per watt gives 3.8°C. Where did 85 per

175 "The sun is warmer now than for the past 11,400 years", Christopher Monckton, *Sunday Telegraph*, 5 November 2006, http://www.telegraph.co.uk/news/uknews/1533312/The-sun-is-warmer-now-than-for-the-past-11,400-years.html

cent of his imagined 20th-century warming go? As Professor Dick Lindzen of MIT pointed out in *The Sunday Telegraph* last week, the UK's Hadley Centre had the same problem, and solved it by dividing its modelled output by three to 'predict' 20th-century temperature correctly," says Christopher Monckton.

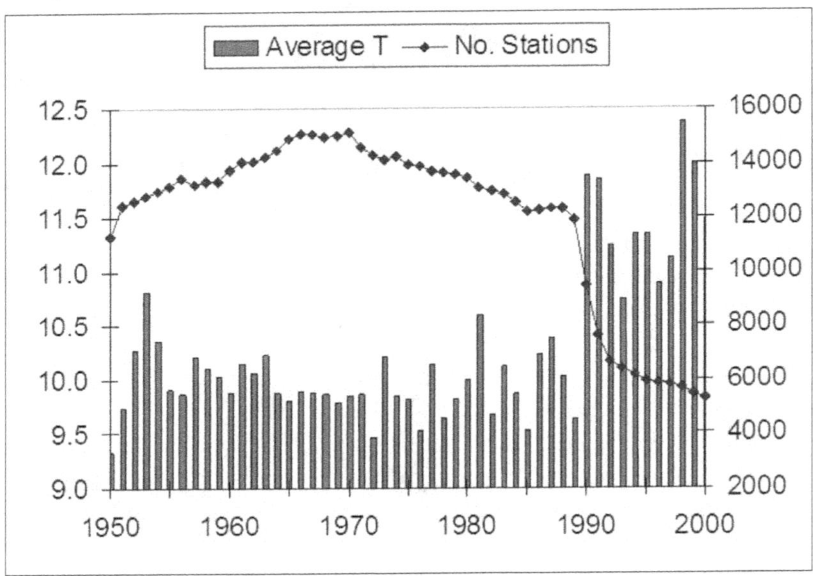

So if climatologists can't do maths and have to invent figures to make their calculations work, and the Senate's Expert Panel revealed they can't do statistics either, should we trust them to honestly collect temperature data?

Not much, by all accounts.

You see, there's another 'hockey stick' that hasn't been talked about much: the disappearance in 1990 of two thirds of the world's temperature stations from the global database.

Up until 1989, data was being compiled from a network of thousands of temperature stations around the world. All well and good, except in 1990 there was a 'cull' of weather stations, mostly in the Soviet Union (Russia), China, Africa and South America. You can see from the accompanying graph[176] that when the switches were flicked on for the remainder of the weather stations in 1990, aver-

176 Graph courtesy Ross McKitrick, see note 177

age temperatures mysteriously soared, eclipsing the slight cooling of the previous decades.

It's hard not to suspect there was a major bias in the remaining temperature stations on the data network, perhaps they were primarily urban centres rather than rural (less costly to monitor and maintain), and prone to the urban heat island effect.

Canadian Ross McKitrick, an associate professor of environmental economics, was sufficiently puzzled by this mysterious temperature surge that accompanied the big drop in reporting stations, to double check the data with NOAA and NASA'S GISS, then graph it above.[177]

He says, and readers will probably agree, that before the world accepts at face value that the 1990s were "the warmest decade", it should first be proven that that reduction to a much smaller network of temperature stations (with vastly reduced global coverage) isn't the cause.

US meteorologist Joe D'Aleo is another who's studied the problem.

"At the end of 1989 more than half of the stations in the USSR were shut down. The bulk of the difference between the satellite and surface trends occurs at the time and location where this network discontinuity occurs. Hoyt estimates that this "major flaw in the surface network" accounts for half of the observed warming, bringing the satellite and surface trends much closer into agreement."

Douglas Hoyt, a solar physicist and climatologist with nearly a hundred published scientific papers to his name,[178] is a senior player in debunking many of the extravagant claims surrounding global warming.

If the planet is really heating up as much as claimed, he says, the data (proxies) should prove it: tree rings, biological indicators, chemical changes – all the things you would expect a warmer climate to have an impact on. But the hard evidence isn't backing up the UN IPCC claims of increased global warming.[179]

"The IPCC theoretical predictions for surface warming range from 0.09 to 0.27°C/decade at the surface and greater in the mid-troposphere (about 0.11 to 0.32°C/decade).

"In support of the theoretical claim, it is often pointed out that

177 http://www.uoguelph.ca/~rmckitri/research/nvst.html

178 http://www.warwickhughes.com/hoyt/bio.htm

179 "Urban heat islands and land use changes", Douglas Hoyt, http://www.warwickhughes.com/hoyt/uhi.htm

the surface is warming by about 0.18°C/decade[180], based upon using surface thermometers.

"Given this claim of strong surface warming in the last 25 years, one would expect to see a very strong signal in the proxy measurements of temperature, but it is definitely not present in the tree ring data, the ice core data, or the Mg/Ca isotope data. Some of the proxy series show strong increases up to 1980 but then have strong reverses in the 1980s (e.g., some of the northern tree line series.) Tree ring densities have also declined consistent with cooling rather than warming. From looking at proxy data, one would never know that 1998 was 'the warmest year of the millennium'. It is likely temperatures have not risen as much as claimed," explains Hoyt.

Smart call, really. A warmer climate should leave genuine fingerprints, as it has in the past. But not this time. Which again turns suspicion back on the network of thermometers: are they telling us about real temperature increases, or imaginary ones caused by "garbage in/garbage out" computer modelling?

Is the "global warming" we think we feel, nothing more than our cities getting substantially hotter thanks to the UHI effect. It feels hotter because it is – in cities. Our temperature readings are getting hotter because our weather stations are in those cities.

A classic example of just this kind of weakness in the data comes from New Zealand's Auckland city over this past southern summer. At issue, allegedly the city's hottest day:

"Auckland sweltered yesterday in the hottest temperature in more than 130 years – straight after a record-warmest night," reported the *Herald* newspaper. "Met Service records show the mercury hit 32.4°C at Whenuapai Airbase about 3.30pm, equalling the previous hottest day, recorded at the Auckland Domain in February 1872."[181]

Now, what's wrong with that report? To an ordinary citizen, probably nothing. But to someone aware of the weaknesses in climate change theory, everything. Firstly, the temperature reading comes from an airbase surrounded by hot tarmac and jet engines. Unfortunately, as Steve McIntyre pointed out a few moments ago, much of the modern surface temperature data is taken from airports, making the temperature readings suspect right from the outset.

180 Jones, P. D. and A. Moberg, 2003. "Hemispheric and Large-Scale Air Temperature Variations: An Extensive Revision and Update to 2001." *Journal of Climate*, 16, 206-223

181 "Aucklanders swelter as south shivers," *NZ Herald*, 13 February 2009, http://www.nzherald.co.nz/nz/news/article.cfm?c_id=1&objectid=10556519

Secondly, it turns out to have been equally hot 130 years ago, but on that occasion the temperature was taken in a large parkland, Auckland Domain. So, if pre-industrial Auckland in 1872 could manage a surface air temperature above grassland of 32.4°C, and in February 2009 matched that temperature over hot tarmac and with an Urban Heat Island Effect now operating, doesn't that really mean the latest temperature reading is low in real terms? Arguably, with at least two or three degrees of UHI built in, Auckland would have needed to reach at least 35°C at Whenuapai Air Base in order to match the 130 year old figure, relatively speaking.

Thirdly, they were comparing different weather stations about 20km apart, one at sea level on tarmac and the other about 100 metres above sea level on grass. Apart from all that, the comparison was ideal.

But this is the problem. News stories about "record temperatures" are based on raw temperature readings from city weather stations on the day. No attempt is made to calculate or factor out the UHIE,[182] which as you've seen may be as high as 10°C in some cities at some times. The margin of error on modern city temperature readings is therefore so high that any claim about "record temperatures" is meaningless.

As proof of UHI in action, we downloaded weather data for several weather stations in the Auckland region for 2 March 2009. Two inner city suburbs, Penrose and Glendowie, both recorded highs for the day of 30.1°C. In contrast, rural stations reported highs of only 21.3°C (South Kaipara) and 22.5°C (Waiuku Sandspit Road). Albany City on Auckland's North Shore recorded 28.3°C, while Whenuapai hit 25.7°C.

Narrowing it down further, the Waiuku weather station (rural) recorded 29.9°C on February 12, the day that Whenuapai recorded 32.4°C.

Of course average temperatures in cities around the world have risen. They've risen significantly. But we know why, and it has nothing to do with global warming and carbon dioxide.

182 I can categorically state that, as of March 2009, none of the New Zealand temperature data fed to NOAA has been adjusted for UHIE based on known conditions in NZ cities. A query to the National Institute of Water and Atmospheric Research (NIWA) confirmed that "this phenomenon has not been studied in great detail in New Zealand", and their database contains only one study of UHIE in New Zealand, dating back to 1981: "Modelling the Winter Urban Heat Island over Christchurch", Tapper et al, *Journal of Applied Meteorology*, April 1981, Vol 20, pp 365-376. I have however since located (late 2009) an explosive paper by NZ climate scientist Jim Hessell completed in 1980, which revealed most of NZ's temperature stations were corrupted by UHIE. I recommend it as a 'must read' because the discoveries are relevant to temp stations worldwide. You can find it here http://www.investigatemagazine.com/hessell1980.pdf

Climatologist Doug Hoyt, looking for answers, lays out just how bad it currently looks for the IPCC's predictions, based on proxies that should show signs of real warming:

"These proxies provide no support for the assertion that the last 10 years are warmer than any years in the last millennium which is often claimed. Another proxy often used to assert temperatures are rising is the recession of glaciers, but glaciers respond to precipitation, evaporation, sunlight, cloud cover, aerosol deposition, and many other factors besides temperature so they are not a good proxy for temperature alone. For example, Kilimanjaro in Africa is often cited as evidence for rising temperatures, but actually it is receding due to local land use changes and lack of precipitation that have nothing to do with greenhouse gases.[183]

"The proxies do not support the strong warming seen in the surface thermometer network. The warming at the surface is not supported by balloon borne pressure transducers[184], by balloon borne thermistors[185], or by satellite observations[186]. These three independent measurements of temperature in the mid-troposphere give a warming about 0.07°C/decade, or about one third the IPCC warming trend. The balloon borne pressure sensors allow the lower 75 millibars of the atmosphere to be sampled and these are consistent with very modest warming in the mid-troposphere and strongly indicate there are problems with the surface thermometer measurements. Since the satellite observations are validated by the balloon observations, they are the best temperature measurements available and their results are inconsistent with the IPCC theoretical predictions."

In simple terms, the network of surface thermometers monitored by NASA, NOAA, the UK Met Office and other government weather bureaux around the world is throwing up numbers that don't match the latest biological indicators, chemical indicators, ice core data or even the low level weather balloon samplings and satellite samplings.

---

183 Molg, T., D. R. Hardy, and G. Kaser, 2003. "Solar-radiation-maintained glacier recession on Kilimanjaro drawn from combined ice-radiation geometry modeling," *J. Geophys. Res.*, 108, 4731

184 Pielke, R. A., Sr., J. Eastman, T. N. Chase, J. Knaff, and T. G. F. Kittel, 1998. "The 1973-1996 trends in depth-averaged tropospheric temperature," *J. Geophys. Res.*, 103, 16,927-16,933

185 Ibid, and Christy, J. R., R. W. Spencer, W. B. Norris, W. D. Braswell and D. E. Parker, 2003. "Error estimates of Version 5.0 of MSU/AMSU bulk atmospheric temperatures." *Journal of Atmospheric and Oceanic Technology*, 20, 613-629

186 Christy, J. R. And R. W. Spencer, 2004. "25 years of satellite data show 'global warming' of only 0.34C". Press release, University of Alabama, Huntsville, AL (http://uahnews.uah.edu/scienceread.asp?newsID=196).

Given that it's the surface thermometers whose data is trumpeted by the news media to give us our annual "warmest year on record" story, do we trust that data any more, if it is not matching all the other checks and balances in the system?

The short answer is, 'No', and journalists need to start treating these claims more sceptically and put them to the test, instead of treating the pronouncements like it's Moses coming down the mountain with a couple of stone tablets.

The entire "the planet is heating up catastrophically" argument is being run and co-ordinated from the NASA unit run by global warming's grand poobah, James Hansen (more on whom, shortly), whose data and public statements appear increasingly at odds with actual hard evidence. Perhaps this is the IPCC equivalent of 'Stockholm Syndrome', where otherwise intelligent scientists have been trapped too long in dark rooms with their computer models and they've lost touch.

Doug Hoyt believes history will show this is where the global warming believers of the late 20th century ran off the tracks,

"There are many things wrong with the surface network such as poor locations of thermometers[187], poor geographical coverage with less than 30% of the globe having sensors[188], urban heat island contamination (e.g.,[189, 190]), land use changes[191, 192], soot on snow causing warming[193], and so forth.

"These problems with the surface thermometers lead to measurements of greater warming than is actually occurring compared to a situation where we had perfect measurements. For example Kalnay estimates 40% of the surface warming is coming from land use

187 Davey, C. A. and R. A. Pielke Sr., 2004: "Microclimate exposures of surface-based weather stations – implications for the assessment of long-term temperature trends." *Bull. Amer. Meteor. Soc.*, submitted

188 Santer, B. D., et al., 2000. "Interpreting differential temperature trends at the surface and in the lower troposphere". *Science*, 287, 1227-1232.

189 Torok, S. J., Morris, C. J. G., Skinner, C. and Plummer, N. 2001. "Urban heat island features of southeast Australian towns". *Australian Meteorological Magazine*, 50, 1-13

190 Oke, T.R. 1973. "City size and the urban heat island". *Atmospheric Environment*, 7 769-779

191 Marland, G., R. A. Pielke, Sr., M. Apps, R. Avissar, R.A. Betts, K.J. Davis, P.C. Frumhoff, S.T. Jackson, L. Joyce, P. Kauppi, J. Katzenberger, K.G. MacDicken, R. Neilson, J.O. Niles, D. dutta S. Niyogi, R.J. Norby, N. Pena, N. Sampson, and Y. Xue, 2003. "The climatic impacts of land surface change and carbon management, and the implications for climate-change mitigation policy." *Climate Policy 3*, 2003, 149-157, doi:10.1016/S1469-3062(03)00028-7

192 Kalnay, E. and M. Cai, 2003. "Impact of urbanization and land-use change on climate." *Nature* 423, 528 – 531

193 Hansen, J., and L. Nazarenko 2003. "Soot climate forcing via snow and ice albedos". *Proc. Natl. Acad. Sci.*, 101, 423-428

changes (i.e., 0.07°C/decade). Marland et al. (2003) say more than half the warming may be coming from these land use changes (i.e., >0.09°C/decade). Hansen says 25% of the twentieth century warming may come from increased soot on snow which is a land use change (i.e., 0.015°C/decade).

"Urban heat islands and poor siting of the thermometers may account for yet more spurious warming. Subtracting these spurious signals from the claimed warming of 0.18°C/decade reduces it to less than 0.08°C/decade. Based on climate models, the surface warming from carbon dioxide must be less than the mid-tropospheric warming of 0.07°C/decade and correcting for these identified problems in the surface network brings the surface observations into near agreement with the satellite observations.

"In summary," concludes Hoyt, "the warming of the globe is only about 0.07°C/decade and even if all of it were attributed to greenhouse gases, it is far less than the warming predicted by the IPCC models. The models are not validated by the magnitude of the temperature trends."[194]

As if to demonstrate just how dodgy the surface data fed to the news media (and therefore the public) really is, news broke just before Christmas 2008 that NASA's GISS had goofed significantly in some of its temperature claims.

Labelling it "a surreal scientific blunder", the *Telegraph* in London reported that NASA had got it horribly wrong when it announced a week earlier that October 2008 was "the hottest October on record".[195]

"This was startling," continued Christopher Booker in the *Telegraph*.

"Across the world there were reports of unseasonal snow and plummeting temperatures last month, from the American Great Plains to China, and from the Alps to New Zealand. China's official news agency reported that Tibet had suffered its "worst snowstorm ever". In the US, the National Oceanic and Atmospheric Administration registered 63 local snowfall records and 115 lowest-ever temperatures for the month, and ranked it as only the 70th-warmest October in 114 years.

---

194 For a more detailed analysis of what the computer models predicted versus what actually happened, I highly recommend Hoyt's "scorecard": http://www.warwickhughes.com/hoyt/scorecard.htm

195 "The world has never seen such freezing heat", Christopher Booker, *The Telegraph*, 18 December 2008, http://www.telegraph.co.uk/comment/columnists/christopherbooker/3563532/The-world-has-never-seen-such-freezing-heat.html

"So what explained the anomaly? GISS's computerised temperature maps seemed to show readings across a large part of Russia had been up to 10 degrees higher than normal. But when expert readers of the two leading warming-sceptic blogs, *WattsUpWithThat*[196] and *ClimateAudit*[197], began detailed analysis of the GISS data they made an astonishing discovery. The reason for the freak figures was that scores of temperature records from Russia and elsewhere were not based on October readings at all. Figures from the previous month had simply been carried over and repeated two months running," said Booker.

This in itself was bad enough, but it was a case of human error. What followed, however, is enough to give world leaders serious cause for concern about the quality of the mushroom food they're being force-fed by global warming believers:

"GISS began hastily revising its figures," Booker continued. "This only made the confusion worse because, to compensate for the lowered temperatures in Russia, GISS claimed to have discovered a new "hotspot" in the Arctic – in a month when satellite images were showing Arctic sea-ice recovering so fast from its summer melt that three weeks ago it was 30 per cent more extensive than at the same time last year.

"A GISS spokesman lamely explained that *the reason for the error in the Russian figures was that they were obtained from another body, and that GISS did not have resources to exercise proper quality control over the data it was supplied with.* [My emphasis] This is an astonishing admission: the figures published by Dr Hansen's institute are not only one of the four data sets that the UN's Intergovernmental Panel on Climate Change (IPCC) relies on to promote its case for global warming, but they are the most widely quoted, since they consistently show higher temperatures than the others," wrote an outraged Booker in the *Telegraph*.

Given the admission that there's no quality control check on the incoming temperature data every day, there's now actually no rational reason left to trust NASA, NOAA or the UK Met Office announcements about record surface temperature readings.

I'll repeat the point because it's really important: NASA is admitting temperature data provided by the rest of the world's temperature stations is not checked for accuracy.

---

196 http://WattsUpWithThat.com
197 http://www.climateaudit.org

To underscore just how far the global warming industry is prepared to go to present fake data, no lesser personage than IPCC chairman Rajendra Pachauri gave a speech in Sydney last October where he pulled a Gore – presenting a fake graph of temperature increases over the past decade – which left *Sydney Morning Herald* columnist Michael Duffy shocked:[198]

"As this was shown on the screen, Pachauri told his large audience: 'We're at a stage where warming is taking place at a much faster rate [than before]'.

"Now, this is completely wrong. For most of the past seven years, those temperatures have actually been on a plateau. For the past year, there's been a sharp cooling. These are facts, not opinion: the major sources of these figures, such as the Hadley Centre in Britain, agree on what has happened, and you can check for yourself by going to their websites. Sure, interpretations of the significance of this halt in global warming vary greatly, but the facts are clear.

"So it's disturbing that Rajendra Pachauri's presentation was so erroneous, and would have misled everyone in the audience unaware of the real situation. This was particularly so because he was giving the talk on the occasion of receiving an honorary science degree from the university.

"Later that night, on ABC TV's *Lateline* program, Pachauri claimed that those who disagree with his own views on global warming are 'flat-earthers' who deny 'the overwhelming weight of scientific evidence'. But what evidence could be more important than the temperature record, which Pachauri himself had fudged only a few hours earlier?"

It's just as well Pachauri's science degree was "honorary". Who knows what he'd do with a real one.

Someone actually posted quotes and a link to the critical article on Pachauri's own blog,[199] but the former railway engineer deigned not to answer any questions from mere mortals or taxpayers. Embarrassingly for the UN IPCC boss, he'd actually been fronted on the misleading warming stats back in January 2008 by Britain's

198 "Truly inconvenient truths about climate change being ignored", Michael Duffy, *Sydney Morning Herald*, 8 November 2008, http://www.smh.com.au/news/opinion/michael-duffy/truly-inconvenient-truths-about-climate-change-being-ignored/2008/11/07/1225561134617.html?page=fullpage#contentSwap1

199 http://blog.rkpachauri.org/blog/10/1225876187.htm

The Guardian,[200] and told them he "would look into the apparent temperature plateau so far this century".

By his appearance in Sydney late last year, it was evident Pachauri had failed to look and was ignoring the science in favour of spin.

Has Earth actually warmed at all in the past decade or two? Or are we trusting our personal taxes, and the money we'll be spending on higher fuel and food prices, and a reduced standard of living, to people whose idea of forecasting is licking a finger and sticking it out the window to test the wind, and a United Nations executive team eager to become some kind of defacto world government?

Have all of your efforts to reduce your carbon footprint been an absolute waste of your time, energy and hard-earned wages?[201]

Heading towards 2010's December meeting to construct a new "Kyoto", the credibility of the global warming believers is in shreds, but sadly the politicians are still listening to them and still ready to make you pay.

---

200 The original *Guardian* story appears to have been pulled, but it ran internationally and the following links to a version carried by Reuters on 11 January 2008: http://www.reuters.com/article/environmentNews/idUSL1171501720080111?feedType=RSS&feedName=environmentNews&rpc=22&sp=true

201 If you sat in a darkened house for 'Earth Hour', but even one member of your household broke wind, you cancelled any gains from switching off the lights.

Chapter Nine

Fancy A Cold Bear?

"The global warming scenario is pretty grim. I'm not sure I like the idea of polar bears under a palm tree."

– ***Lenny Henry, comedian***

Possibly the ultimate pin-up of global warming, the Polar Bear extinction myth is entirely false. Firstly, the iconic photo of polar bears trapped on an ice floe was actually taken just a short distance away from much larger ice floes, and misused by Environment Canada to push global warming fears.

Al Gore incorporated the photo into his global warming travelling show:

"Their habitat is melting... beautiful animals, literally being forced off the planet," Gore is quoted as saying, glancing back at the screen for effect. "They're in trouble, got nowhere else to go." (Audience members let out gasps of sympathy...)[202]

Newspapers around the world carried the image – it made the front page of the *New York Times*. But it wasn't until Australia's ABC TV *Media Watch* programme did a little digging that we found out what the photo was really illustrating.

Amanda Byrd, an Australian marine biology student, snapped the shot one Arctic summer, August 2004, for much the same reason that Ed Hillary climbed Everest: because it was there. She wasn't trying to make a point about stranded polar bears, because they weren't actually stranded.

"They did not appear to be in danger," she confirmed to *Media Watch*.[203]

Her photo, however, was stored electronically and retrieved by another on the vessel, Dan Crosbie.

202 http://newsbusters.org/node/11879

203 http://www.abc.net.au/mediawatch/transcripts/0706_byrd.pdf

"Dan Crosbie gave the image to the Canadian Ice Service, who gave the image to Environment Canada, who distributed the image to seven media agencies including AP."

Which accounts for how a teddy bears' picnic on ice came to star in Al Gore's slideshow.

Associated Press gave the photo a bit of marketing oomph on the day the UN released a major climate change report and, hey presto, instant polar bear scandal, just add water (no ice).

"The photograph represents polar bears standing on ice that's melting," conceded the editor of Sydney's *Sunday Telegraph*, Neil Breen, after being questioned about his use of the shot. "Now obviously there's a disputed account of when that was taken now, and maybe it was taken in the Alaskan Summer when you would naturally expect ice to melt but at the time it was sent to us, *Associated Press* in their caption to us told us that the picture was taken of melting ice caps and to do with global warming and that it was sent to them by a Canadian ice authority and we had no reason to question it."

No, the media didn't. If we had to independently verify the bona fides of every photograph or news story supplied to us by wire services and photo agencies, no work would actually get done. At some level, journalists have to trust they're not being played for suckers by their suppliers.

At least one official in Environment Canada, the organisation partly responsible for releasing the photo, doesn't think the bears in the picture were stranded either.

"You have to keep in mind that the bears are not in danger at all," EC's Denis Simard told Canada's *National Post* newspaper.[204] "This is a perfect picture for climate change...you have the impression they are in the middle of the ocean and they are going to die...But they were not that far from the coast, and it was possible for them to swim...They are still alive and having fun."

So, the photo was misused to push global warming. It's not the first time polar bears have been misused in the services of the warming myth.

The WWF has made the bizarre prediction that polar bears will be unable to breed within five years:

"Assuming the current rate of ice shrinkage and accompanying

204 "Gore pays for photo after Canada didn't," *National Post*, 23 March 2007

weight loss in the Hudson Bay region, female bears could become so thin by 2012 they may no longer be able to reproduce, " Lara Hansen, chief scientist for the World Wildlife Fund, told journalists.[205]

Meanwhile, other lobbyists are predicting global warming will eradicate two thirds of all polar bears.

"I don't think there is any question polar bears are in danger from global warming," biology professor and World Conservation Union spokesman Andrew Derocher told Britain's *Telegraph* newspaper. "People who deny that have a clear interest in hunting bears."[206]

"They are declining due to global warming, and changes in when the ice freezes and melts in Hudson's Bay," said Derocher, describing the problem as "habitat loss".

The idea that polar bears must have ice or they die resonates with some scientists:

"Polar bears depend entirely on sea ice for survival. In recent years, a warming climate has caused major changes in the Arctic sea ice environment, leading to concerns regarding the status of polar bear populations,[207]" begins one US Geological Survey study.

And yet polar bears – a variant of the brown bear – can often be found on land, playing and occasionally hunting. While their main diet consists of ringed seal, walrus or fish, polar bears have been known to eat reindeer, caribou and even berries, as well as raiding human rubbish dumps.

They survived the Medieval Warm Period, and the Roman Warm Period and presumably other warm periods over the previous 100-200,000 years that they've existed as a subgroup.

Intriguingly, a joint Canadian/Russian study found the Russian Arctic was sufficiently warm only a few thousand years ago that great forests extended all the way to the water's edge in the Arctic.

"This implies that over much of northern Eurasia summers may have been 2.5° to 7.0°C warmer than today during the period; 9000 to 4000 yr Before Present." [208]

---

205 http://www.associatedcontent.com/article/213986/climate_change_affecting_arctic_species.html?cat=47

206 "Polar bears 'thriving as the Arctic warms up', *The Telegraph*, 9 March 2007, http://www.telegraph.co.uk/news/worldnews/1545036/Polar-bears-'thriving-as-the-Arctic-warms-up'.html

207 "Polar Bear Population Status in the Southern Beaufort Sea" By Eric V. Regehr and Steven C. Amstrup, U.S. Geological Survey; and Ian Stirling, Canadian Wildlife Service, U.S. GEOLOGICAL SURVEY Open-File Report 2006-1337, http://pubs.usgs.gov/of/2006/1337/

208 "Holocene Treeline History and Climate Change Across Northern Eurasia", *Quaternary Research* 53, 302–311 (2000), doi:10.1006/qres.1999.2123 http://www.sscnet.ucla.edu/geog/downloads/634/269.pdf

We know, as a matter of fact, that polar bears were around in their current form in 2000 BC, and clearly they're still here now.

A US Fish and Wildlife Service (FWS) report on the Alaskan bears reveals that too much ice, such as that recently experienced only 20 years ago, can cause the bears to go hungry, ironically.

"In the southern Beaufort Sea heavy ice conditions in the mid-1970s and mid-1980s caused significant declines in productivity of ringed seals (Stirling 2002). Each event lasted approximately three years and caused similar declines in the natality of polar bears and survival of subadults, after which reproductive success and survival of both species increased again.

"In the Viscount Melville Sound area," notes the report, "ringed seals occurred at lower densities than in most other areas of polar bear habitat from Alaska east to West Greenland, possibly because there is greater proportion of multi-year ice in this area, which is less preferred by ringed seals."

The good news is, according to the FWS report, is that the bears can adapt to what nature throws at them:

"The fluctuating sea-ice condition in regions like the Beaufort Sea or Baffin Bay, however, may require modifications of foraging strategy from month to month or even day to day during break-up, freeze-up, or periods of strong winds (Ferguson et al. 2001). Polar bears are adaptable enough to modify their foraging patterns for the extreme range of sea-ice scenarios (Ferguson)."

None of which prevented National Public Radio from running a beat-up story on the plight of the polar bear.[209]

"Cubs in northern Alaska aren't surviving at nearly the rate of recent decades, and more bears are spending summers on land – even denning there. Land isn't the best place for polar bears because they're cut off from their main food source," reported NPR.

But that's not what the Fish and Wildlife Service has found; cubs were dying because of too much sea ice which lowered seal birth rates and affected food supplies for those years. Nor is denning on land unusual for polar bears, it's virtually compulsory and has been since the time of Methuselah.

"Throughout their range, most pregnant female polar bears excavate dens in snow located on land in the fall- early winter period

209 "Polar bear population struggles as sea ice melts", National Public Radio, 21 January 2008, http://www.npr.org/templates/story/story.php?storyId=18121378

(Harington 1968, Lentfer and Hensel 1980, Ramsay and Stirling 1990, Amstrup and Gardner 1994). The only known exceptions are in Western and Southern Hudson Bay where polar bears excavate earthen dens and later reposition into adjacent snow drifts (Jonkel et al 1972, Richardson et al. 2005b), and in the southern Beaufort Sea where a portion of the population dens in snow caves located on pack and shorefast ice."[210]

As to claims that their primary food sources will disappear, studies in the early 1990s found both polar bears and walruses adapted to retreating ice:[211]

"Kochnev (2006) reports that in the autumn seasons of 1990, 1991, 1993, 1995, 1996, and 1997 the ice edge retreated 80-380 km to the north and to the west of Wrangel Island. During these years walruses occupied coastal haulout sites in substantial numbers for protracted periods of time. Walrus carcasses on the beaches became a food-source for polar bears and was the main factor attracting bears to these locations (Kochnev 2001). Following a walrus mortality event such as a stampede, the number of bears increased and usually reached a peak in the second half of October.

"The relationship between number of bears present and walrus carcasses continued to exist until the freezing of the sea," says the US FWS report.

Once upon a time, it was open season on polar bears. Inuit hunted them, and furriers hunted them. By the 1970s, total bear populations are believed to have been as low as 5,000 throughout the world. Today, thanks to bans on uncontrolled hunting, there are now an estimated 25,000 polar bears in the north.

One polar bear expert who doesn't believe for a second the creatures are endangered is Canadian scientist Mitch Taylor, who's spent more than 20 years studying Canada's polar bears up close.

"There aren't just a few more bears," he told the *Telegraph*. "There are a hell of a lot more bears."[212]

As examples, he cites the Davis Strait bears, one of 20 known polar bear populations in the Arctic. In the mid-80s they counted

210 Range-wide status review of the polar bear, US Fish & Wildlife Service, December 2006 http://alaska.fws.gov/fisheries/mmm/polarbear/pdf/Polar_Bear_%20Status_Assessment.pdf

211 "Polar Bear Population Status in the Southern Beaufort Sea" By Eric V. Regehr and Steven C. Amstrup, U.S. Geological Survey; and Ian Stirling, Canadian Wildlife Service, U.S. GEOLOGICAL SURVEY Open-File Report 2006-1337, http://pubs.usgs.gov/of/2006/1337/

212 "Polar bears 'thriving as the Arctic warms up', *The Telegraph*, 9 March 2007

850 bears in the Davis. Today there are 2,100. Whilst one population on the West Hudson has fallen 22%, from 1,200 bears in 1987 down to 934 currently, what the doomsayers fail to mention is that the Hudson population once numbered only 500 bears, so is still running at double what it used to be.

Over in the Norwegian arctic is the tourist destination of Longyearbyen, where it is mandatory for tourists to carry a gun each when leaving town, because of the risk of polar bear attacks. Whilst the tourist handbook for new arrivals expressly warns that polar bears are a protected species, the next clause points out they're also very common, and very dangerous. The local authority runs a firearms training course for those who haven't used a gun before, and guns are available for daily hire.

"Only 1500 people live in Longyearbyen – the capital of the territory. Another thousand is spread out in two mining communities and two scientific stations. That's all humans. There is probably bigger number of polar bears," notes one blogger who travelled there recently.[213]

When last year's 'Australian of the Year', Tim Flannery, waxed lyrical about the impending doom for polar bears because of global warming, Mitch Taylor jumped on his computer in Nunavut, in the Canadian arctic, and fired off a missive:[214]

> Tim Flannery is one of Australia's best-known scientists and authors. That doesn't mean what he says is correct or accurate. That was clearly demonstrated when he recently ventured into the subject of climate change and polar bears. Climate change is threatening to drive polar bears into extinction within 25 years, according to Flannery. That is a startling conclusion and certainly is a surprising revelation to the polar bear researchers who work here and to the people who live here. We really had no idea.
>
> The evidence for climate change effects on polar bears described by Flannery is incorrect. He says polar bears typically gave birth to triplets, but now they usually have just one cub. That is wrong.
>
> All research and traditional knowledge shows that triplets, though they do occur, are very infrequent and are by no means typical. Polar bears generally have two cubs – sometimes three and sometimes one. He says the bears'

213 http://www.globosapiens.net/travel-information/Longyearbyen-391.html?page=2
214 "Last stand of our wild polar bears", Dr Mitchell Taylor, 1 May 2006, http://meteo.lcd.lu/globalwarming/Taylor/last_stand_of_our_wild_polar_bears.html

weaning time has risen to 18 months from 12. That is wrong. The weaning period has not changed. Polar bears worldwide have a three-year reproduction cycle, except for one part of Hudson Bay for a period in the mid-1980s when the cycle was shorter.

One polar bear population (western Hudson Bay) has declined since the 1980s and the reproductive success of females in that area seems to have decreased. We are not certain why, but it appears that ecological conditions in the mid-1980s were exceptionally good.

Climate change is having an effect on the west Hudson population of polar bears, but really, there is no need to panic. Of the 13 populations of polar bears in Canada, 11 are stable or increasing in number. They are not going extinct, or even appear to be affected at present.

It is noteworthy that the neighbouring population of southern Hudson Bay does not appear to have declined, and another southern population (Davis Strait) may actually be over-abundant.

I understand that people who do not live in the north generally have difficulty grasping the concept of too many polar bears in an area. People who live here have a pretty good grasp of what that is like to have too many polar bears around.

This complexity is why so many people find the truth less entertaining than a good story. It is entirely appropriate to be concerned about climate change, but it is just silly to predict the demise of polar bears in 25 years based on media-assisted hysteria.

Dr. Mitchell Taylor,
Polar Bear Biologist, Department of the Environment

The now less-sceptical environmentalist, Bjorn Lomborg himself, waded into this debate late last year calling for an end to the polar bear scare stories. For a start, he says, of the 20 polar bear communities, only two are declining, and those "populations are in areas that have gotten colder over the past 50 years."[215]

In contrast he says, the two bear populations doing the best are at polar opposites of the hungry cold bears:

"The habitats of the two thriving groups have actually become warmer. If polar bears are today's 'canaries in the coalmine', then the coalmine does not appear half as fearsome as some claim," says Lomborg.

One polar bear scare story doing the rounds, and tied to the mis-

---

215 "The not so disappearing polar bear", Bjorn Lomborg, *The Telegraph*, 10 November 2008 http://www.telegraph.co.uk/earth/earthcomment/3310555/The-not-so-disappearing-polar-bear.html

chievous use of Amanda Byrd's iconic photo, is the claim that polar bears are "drowning" trying to swim away from drifting icebergs.

"Studies show that bears have drowned because the shrinking ice cover means they have to swim long distances," exclaimed the earnest National Public Radio broadcast on polar bears.[216]

Well, actually, "studies" never showed that because there aren't any "studies". This is a global warming myth that has grown in the telling.

"Al Gore," says Bjorn Lomborg, "bases his claim of 'drowning' bears on a single sighting of four dead bears the day after an abrupt windstorm. The sighting occurred in an area where polar bear numbers are increasing."

Lomborg says that even if you take the worst performing polar bear community, at West Hudson Bay, and extrapolate their losses out, "we are losing 15 bears a year to climate change."

If we can save 15 bears a year, we beat the climate change impact he says.

National Public Radio helpfully gave listeners suggestions on how they could save polar bears:[217]

"Jim Gessler has been visiting polar bears here since he was a child. Concern about their fate has pushed him to do what he can about climate change.

" 'I'm turning off lights when I leave the room. I don't have a car anymore,' Gessler said.

"His daughter, Ann Gessler, 22, has given up meat.

" 'How much energy it takes to produce a hamburger is really distressing,' she said."

She would probably do more for hungry polar bears if she simply posted her unwanted Big Mac north, but apparently she has other ideas.

"If you just make small changes in your lifestyle, that's a lot more beneficial."

You can almost sense Bjorn Lomborg banging his head on the desk in frustration.[218]

---

216 "Polar bear population struggles as sea ice melts", National Public Radio, 21 January 2008, http://www.npr.org/templates/story/story.php?storyId=18121378

217 "Polar bear population struggles as sea ice melts", National Public Radio, 21 January 2008, http://www.npr.org/templates/story/story.php?storyId=18121378

218 "The not so disappearing polar bear", Bjorn Lomborg, *The Telegraph*, 10 November 2008 http://www.telegraph.co.uk/earth/earthcomment/3310555/The-not-so-disappearing-polar-bear.html

"The Kyoto Protocol will cost [the UK] $180 billion, yet do almost no good: it would save just 0.06 polar bear each year.

"There are dramatically smarter options available if we care about polar bears. Hunters shoot 49 bears from Western Hudson Bay each year. For each bear we save with climate change policies, we could save 260 by revoking hunting rights and clamping down on poachers."

He's right, the solution to the problem is smart: stop shooting the bears. But it's not as catchy as "Turn off the lights, save a bear!"

Nonetheless, in May last year, and against the advice of many who work with polar bears in the Arctic, the creatures were listed as "endangered" by the US Fish and Wildlife Service, on the grounds of global warming. They made their decision on the strength of the FWS study footnoted in this chapter, and future projections of global warming and disappearing sea ice.

The decision has been made, but it's worth noting what a scientific audit team – University of Pennsylvania's J Scott Armstrong, New Zealand's Kesten Green of Monash University, and the Harvard-Smithsonian Centre of Astrophysics' Willie Soon – had to say about the quality of the advice that US FWS based their ruling on:[219]

"Calls to list polar bears as a threatened species under the U.S. Endangered Species Act are based on forecasts of substantial long-term declines in their population. Nine government reports were prepared to inform the U.S. Fish and Wildlife Service decision on whether or not to list polar bears as threatened under the Endangered Species Act.

"We assessed these reports in light of evidence-based (scientific) forecasting principles. None of the reports referred to works on scientific forecasting methodology. Of the nine, Amstrup, Marcot and Douglas (2007) and Hunter et al. (2007) were the most relevant to the listing decision. Their forecasts were products of complex sets of assumptions.

"The first in both cases was the erroneous assumption that General Circulation Models [also known as Global Climate Models, computer modelling of global warming] provide valid forecasts of summer sea ice in the regions inhabited by polar bears. We never-

219 "Polar Bear Population Forecasts: A Public-Policy Forecasting Audit," Armstrong, Green, Soon, *Interfaces* Vol. 38, No. 5, September–October 2008, pp. 382–405, http://kestencgreen.com/polarbears.pdf

Urban Heat Island Effect: Many of the world's temperature recordings are taken at airports or other built-up areas, and suffer from environmental heat pollution. This photo at Rome Airport shows one reason why Rome might be recording high temperatures. BELOW: The weather data for Detroit Lakes (collected by the unit atop the white pole) is contaminated by the heatwash from an air-con unit. Photos: *WattsUpWithThat*

A full-colour version of this photo insert is available for download from www.investigatemagazine.com/airconphotos.pdf

By far the bulk of US surface stations surveyed so far reveal uncorrected UHIE errors in excess of two degrees Celsius. And the US network is the highest quality in the world. Claims of rising temperatures are based on this data. Photo: *Surfacestations.org*

CRN Rating key	1	2	3	4	5	
Estimated Error in °C (per NOAA)	Error ≤1°C	Error ≤1°C	Error ≥1°C	Error ≥2°C	Error ≥5°C	Unrated
Quality	Best	Good	Fair	Poor	Worst	Closed

ABOVE: Yet another US MMTS weather station baking in reflected heat off the tarmac, hot car engines and from air conditioners. BELOW: If Antarctic warming is as serious as the journal *Nature* claimed this year, why is Antarctic sea ice growing well above average levels? Photo: *Surfacestations.org; Graphic: NSIDC*

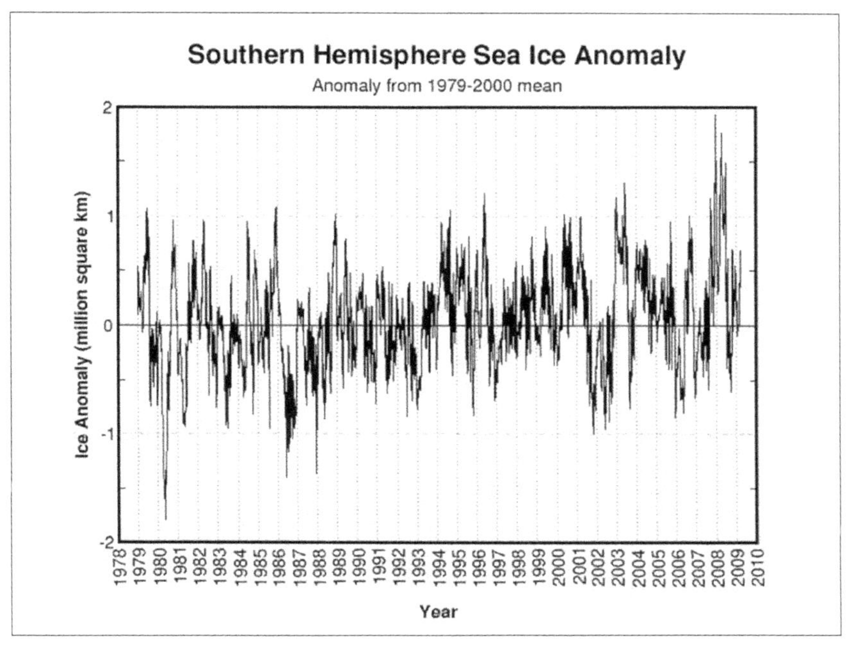

The top image measures temperatures taken by satellites, and shows an overall cooling of Antarctica between 1981 and 2004. BELOW: The more controversial "Antarctica is warming" graphic based on a mix of surface station and later satellite data between 1957 and 2006. Photos: *NASA*

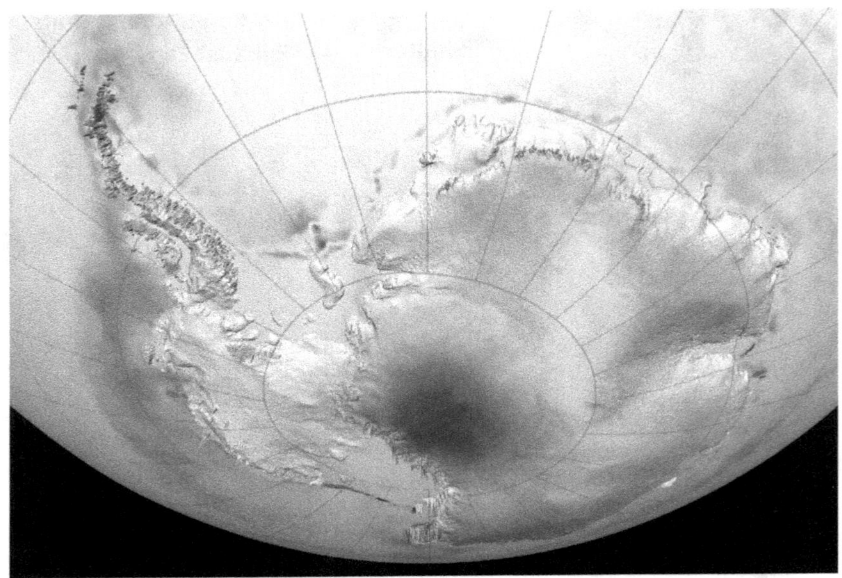

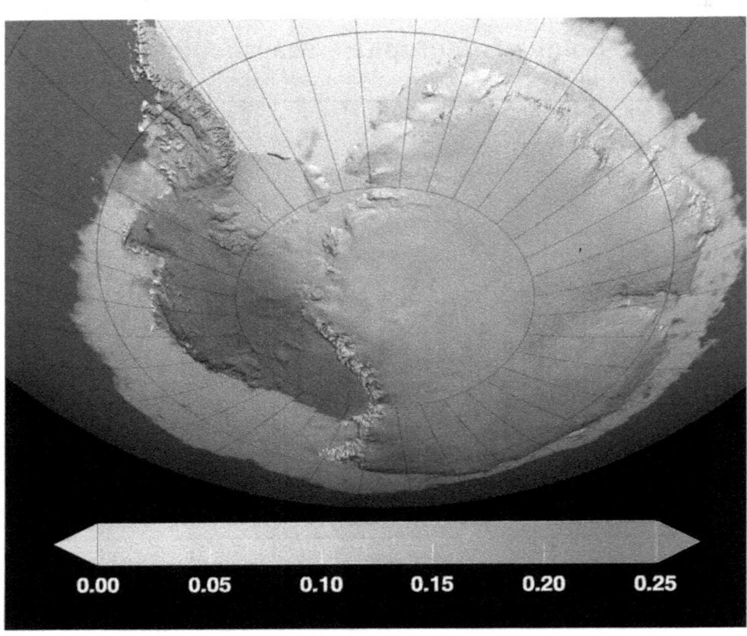

Not so wild about Harry: The top image shows "Harry", one of the Antarctic weather stations that showed the highest warming trend in the recent Nature paper, buried in snow in 2006. Although it may seem counterintuitive, snow actually traps heat, which is why Inuit people found igloos could be cosy, so the station and others across Antarctica could have been reading the warmth of their own electronics. BELOW: "Harry" in happier times. Photos: *Surfacestations.org*

In 2002, the journal *Nature* published a study by the US National Science Foundation's Peter Doran and others that continues to spike the guns of global warming believers:

"Antarctica overall has cooled measurably during the last 35 years – despite a global average increase in air temperature of 0.6 degrees Celsius during the 20th century.

"Our 14-year continuous weather station record from the shore of Lake Hoare reveals that seasonally averaged surface air temperature has *decreased* by 0.7 degrees Celsius per decade," they write. "The temperature decrease is most pronounced in summer and autumn. Continental cooling, especially the seasonality of cooling, poses challenges to models of climate and ecosystem change."

Whilst Doran et al acknowledged other studies had found some warming, he reminded *Nature* readers that the warming was largely confined to "between 1958 and 1978".

This is likely to be a key reason the 1981 to 2004 satellite readings on the previous page show overall cooling, but the 1957 to 2006 surface data published early 2009 shows slight warming – by including the earlier warming the average was bumped up.

Despite hating the fact his 2002 paper has been cited by sceptics, Doran gave an interview to *Seed* magazine in late 2006 admitting that "[the] cooling...is still going on today", and adding, "We also went out and looked at what's been going on continent-wide. We teamed up with John Walsh, now at the University of Alaska Fairbanks, who does more large-scale climate measurements, and we showed that the entire continent – about 60 percent of it – has, over the previous 30 years, been cooling."

Doran wrote in 2002 that too much data was being collected from the warm Antarctic peninsula, and not enough from colder and larger East Antarctica: "The Peninsula itself is warming dramatically, the authors note, and there are many more weather stations on the Peninsula than elsewhere on the continent..Our approach shows that if you remove the Peninsula from the dataset, and look at the spatial trend. The majority of the continent is cooling," said Doran. Photo: *Peter Doran/NSF*

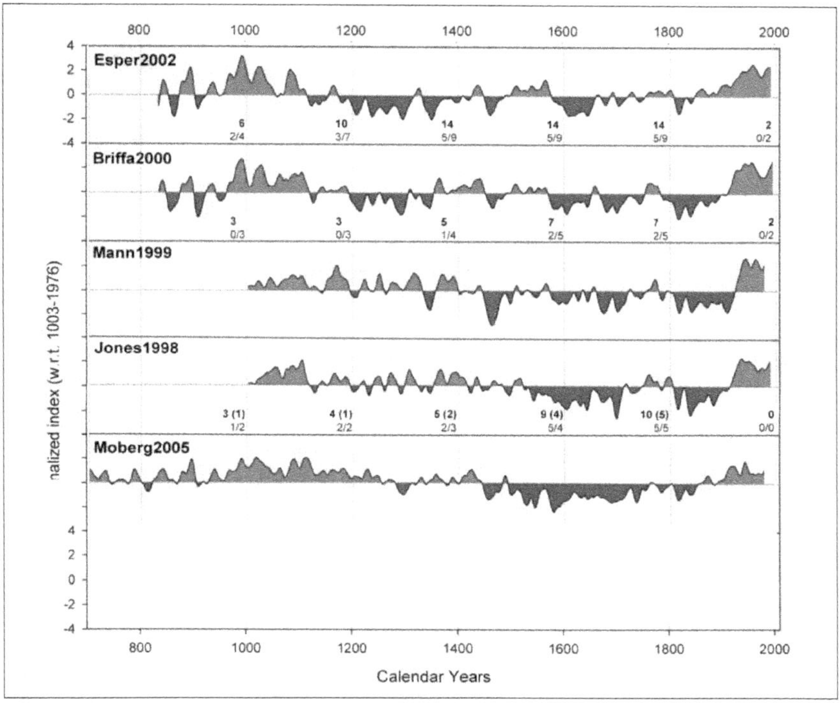

Esper2002

Briffa2000

Mann1999

Jones1998

Moberg2005

nalized index (w.r.t. 1003-1976)

Calendar Years

How to make the Medieval Warm Period disappear. Note both Moberg and Esper's studies show MWP was more significant than today, while Mann's clearly implies today's warm period is hotter and bigger. Mann's 'hockey stick' below shows little trace of the MWP. Graphics: *Wegman Report to the US Senate*

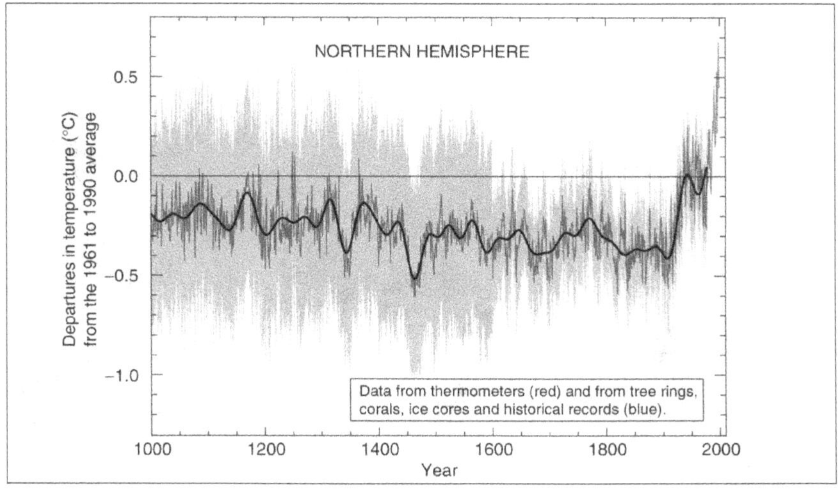

Global warming's 'rogue's gallery': LEFT: NASA GISS boss James Hansen. RIGHT: UN IPCC chair Rajendra Pachauri. BELOW: The man who left his palatial garden and letterbox lights on during 'Earth Hour' according to media reports, Al Gore. Photos: *Lizette Kabré* (Hansen and Pachauri), *WENN* (Gore)

theless audited their conditional forecasts of what would happen to the polar bear population assuming, as the authors did, that the extent of summer sea ice would decrease substantially over the coming decades.

"We found that Amstrup et al. properly applied only 15% of relevant forecasting principles and Hunter et al. only 10%, while 46% were clearly contravened and 23% were apparently contravened. As a consequence their forecasts are unscientific and of no consequence to decision makers," the audit team reported.

Never let the facts get in the way of a good ~~scare~~ bear story.

Chapter Ten

How Good Are The Computer Models Anyway?

"If you asked me to name the three scariest threats facing the human race, I would give the same answer that most people would: nuclear war, global warming and Windows."

– Dave Barry, columnist

So much of the "proof" of global warming is not found in the environment, but in software packages. At the centre of this whole debate are various "computer models" that purport to predict the future based on existing known data.

As we all know, computers are only as good as the information programmed into them to start with, and if you miss out the important bits the data your machine prints out may be utterly useless and inaccurate in real terms.

And one of the big problems is that much of the climate change we're seeing is regional, rather than global.

"What particularly interests me," says British glaciologist Dr David Vaughan on a blog post,[220] "is that the rate of climate change differs so radically from one region to another as exemplified not just in the Antarctic Peninsula but also in Alaska and on the Siberian Plateau. This is the reality, some areas will warm more rapidly, some areas will be cooler. Some areas will change differently. Regional climate change will be more profound for people than global mean warming. Yet we have no mechanism for predicting regional climate change. The output from the models we use do not produce accurate regional forecasts. I would say we are ten years away from the measurements we need in order to produce the information necessary to adapt to climate change."

220 http://climatex.org/articles/climate-change-info/larsen-ice-shelf-b-antarctic-2002/

So global warming is a valid description for regional warming? Better not tell global warming believer Michael le Page whose unintentionally hilarious attack on what he termed sceptics' "climate myths" in *New Scientist* magazine tries to dismiss events like the Little Ice Age on the grounds that they might be regional:[221]

"The term "Little Ice Age" is somewhat questionable, because there was no single, well-defined period of prolonged cold around the entire planet. After 1600, there are records of average winter temperatures in Europe and North America that were as much as 2°C lower than present (although the third coldest winter in England since 1659 was in 1963).

"Comparisons of temperature indicators such as tree-ring records from around the northern hemisphere suggest there were several widespread cold intervals between 1580 and 1850.

"Yet while there is some evidence of cold intervals in parts of the southern hemisphere during this time, they do not appear to coincide with those in the northern hemisphere. Such findings suggest the Little Ice Age may have been more of a regional phenomenon than a global one."

This "it's regional so it doesn't count" line was also used by global warming believer Jean Lynch-Stieglitz at the Georgia Institute of Technology in Atlanta.[222]

Lynch-Stieglitz and her team discovered the Gulf Stream, which circulates warm currents through the Atlantic, was 10 percent weaker during the Little Ice Age, "implying that diminished oceanic heat transport may have contributed to Little Ice Age cooling in the North Atlantic."[223]

From her study, Lynch-Stieglitz suggested the LIA was "confined mainly to the northern hemisphere", allowing hockey-stick guru Michael Mann to argue that this is why the LIA did not appear on his global warming graph, because it was only *regional*.[224] Of course, it turns out to have been global after all, but that's another story as you'll see in a moment.

"We're seeing a rearrangement of heat around the globe – so globally overall it's not colder," says Lynch-Stieglitz.

---

221 "Climate Myths", *New Scientist*, 16 May 2007 http://www.newscientist.com/article/dn11645-climate-myths-we-are-simply-recovering-from-the-little-ice-age.html

222 http://www.newscientist.com/article/mg19225804.300

223 "Gulf Stream density structure and transport during the past millennium", Steiglitz et al, *Nature* 444, 601-604 (30 November 2006) | doi:10.1038/nature05277;

224 "Climate: The great hockey stick debate" by Fred Pearce, *New Scientist*, 18 March 2006 http://www.newscientist.com/article/mg18925431.400-climate-the-great-hockey-stick-debate.html

Which, if we take Lynch-Stieglitz at her word, means we can't attribute Arctic and Greenland icemelt to global warming either, because it's balanced by expanding ice mass in Antarctica and therefore not relevant at a global level.

Perhaps sensing growing public frustration, Al Gore successfully tried to bolster belief in the computer models:

"So the temperature increases are taking place all over the world, including in the ocean. This is the natural range of variability for temperature in the ocean. You know people say, 'Aw, it just naturally go up and down, so don't worry about it.' This is the range that would be expected over the last 60 years. But the scientists that specialize in global warming have computer models that long ago predicted this range of temperature increase."

But scientists doing computer modelling haven't found it that easy at all.

Another scientist poking borax at "global" warming in the Arctic is Syun-Ichi Akasofu, who looked at James Hansen's 2006 historical study showing massive warming in the Arctic, much higher than the rest of the planet, yet simultaneous cooling over Greenland and eastern Siberia, and pondered.[225]

Could the computer models predict, in hindsight, he wondered, the Arctic warming at much higher levels than anywhere else, but also provide for the cool spots that appeared over the North Atlantic and Siberia – the actual real-time data? Given what the Global Climate Models (GCMs, or computer models) have accurately achieved plotting CO_2 emissions, Akasofu was hopeful.

"Thus, we asked the IPCC arctic group (consisting of 14 subgroups headed by V. Kattsov) to 'hindcast' geographic distribution of the temperature change during the last half of the last century. To 'hindcast' means to ask whether a model can produce results that match the known observations of the past; if a model can do this at least qualitatively, we can be much more confident that the model is reliable for predicting future conditions."

However, despite running the real data, with its inconvenient cold spots but massively overheated warm spots nearby, through 14 of the leading computer models used by the IPCC, not one of the models was able to accurately predict what actually had hap-

225 "Is the Earth still recovering from the "Little Ice Age"?", Syun-Ichi Akasofu, 7 May 2007, http://www.iarc.uaf.edu/highlights/2007/akasofu_3_07/Earth_recovering_from_LIA_R.pdf

pened. Rather, their CO_2 forecasts picked a warm Arctic across the board, but nowhere near as hot as it managed to reach, with no cold areas either.

Which, as we know, didn't happen.

"It took a week or so before we began to realize another possibility of this discrepancy: If 14 GCMs cannot reproduce prominent warming in the continental arctic, perhaps much of this warming is not caused by the greenhouse effect at all," wrote Akasofu.

"Therefore, our conclusion at the present time is that much of the prominent continental arctic warming and cooling in Greenland during the last half of the last century is due to natural changes, perhaps to multi-decadal oscillations like Arctic Oscillation, the Pacific Decadal Oscillation, and the El Niño."[226]

In other words, "global" warming in the Arctic now appears to be "regional warming".

And what was it *New Scientist* tried to say to dismiss sceptics who quoted the Little Ice Age?

"[It's] more of a regional phenomenon than a global one."

I'm enjoying these attempts by global warming believers to try and discredit scientific studies or events they don't like. The problem for them is, it cuts both ways.

CO2Science, meanwhile, have hit back at claims the Little Ice Age was confined to the Northern Hemisphere, citing a study of the ocean near Indonesia.[227]

"In contradiction of the climate-alarmist claim that the Medieval Warm Period and Little Ice Age were merely *regional* phenomena confined to countries surrounding the North Atlantic Ocean, Newton *et al.*[228] say their data from the Makassar Strait of Indonesia clearly indicate that "climate changes during the Medieval Warm Period and Little Ice Age were *not* confined to the high latitudes," nor, we would add, to countries bordering the North Atlantic Ocean."

While *New Scientist* magazine and Michael Mann[229] have tried

226 In case you're wondering, Akasofu is no slouch. A prominent geophysicist, he's the founding director of the International Arctic Research Centre at University of Alaska, Fairbanks, and has served previously at the Geophysical Institute and as an associate editor on several scientific journals. He is one of the world's most-cited scientists. http://en.wikipedia.org/wiki/Syun-Ichi_Akasofu

227 http://www.co2science.org//articles/V10/N6/C3.php

228 Newton, A., Thunell, R. and Stott, L. 2006. Climate and hydrographic variability in the Indo-Pacific Warm Pool during the last millennium. *Geophysical Research Letters* 33: 10.1029/2006GL027234.

229 "The great hockey stick debate", *New Scientist*, http://www.newscientist.com/article/mg18925431.400-climate-the-great-hockey-stick-debate.html

to claim the Little Ice Age was "regional" and confined largely to the northern hemisphere, down south at New Zealand's National Institute of Water and Atmospheric Research (NIWA), climate scientist Jim Salinger clearly and explicitly blames the Little Ice Age for New Zealand's extensive glaciation: [230]

"Their long response times have meant that they have simply absorbed any snow gains into their shrinking masses," wrote NIWA's Jim Salinger, "maintaining their areas while their surfaces have lowered like sinking lids. These glaciers have kept their LIA areas for all but the last couple of decades. Most of the relatively thick ice filling their over-deepened valleys was a remnant from the LIA."

So, who do we believe? Mann or Salinger? Are New Zealand's glacial advances since the 1700s imaginary?

One of the things that's become apparent is that despite alleged warming, Antarctica and even the Arctic have benefitted from significantly higher snowfalls, such that overall ice mass may actually be growing worldwide, not shrinking. While the computer models predicted increased precipitation as a result of sea water evaporating in hotter weather, they didn't anticipate the amount of moisture that would fall back to earth as snow.

According to a study in *Nature*[231] based on satellite measurements, we're seeing a 6.5% increase in precipitation for every 1°C increase in temperature. While most of the computer climate models were picking somewhere between 1 and 3%.

Not only does extra snow rebuild ice fields, but the energy involved in evaporating and then condensing water soaks up a chunk of heat as well. There is good reason to believe the higher than expected precipitation rates are a planetary feedback that helps regulate against runaway warming.

Unfortunately, the computer models have not been sophisticated enough to cope with surprises like this.

In March this year, a controversial new scientific study was published that may explain this, and in doing so set the cool cat among the warming pigeons.[232]

---

230 http://www.niwa.cri.nz/pubs/wa/ma/16-3/glacier

231 "Satellite observations suggest climate models are wrong on rainfall", H Leifert, *Nature*, 31 May 2007 http://www.nature.com/news/2007/070528/full/news070528-9.html

232 "A peak behind the curtain," *ClimateAudit*, 4 March 2009, http://www.climateaudit.org/?p=5416#more-5416

The study was rejected by one climate journal last year, not because the scientific team were nobodies, but because their research has chilling implications for the global warming industry if true.

What was so radical? The team analysed radiosonde data from the mid to upper troposphere, which is a fancy way of saying weather balloon data collected where most of the high clouds form and up to about the level that passenger jets cruise at.

Surprisingly, the researchers found humidity has decreased at those high levels, which isn't what existing computer models are predicting. That means that although extra moisture evaporates from the oceans and land, it's tending to stay at low altitude levels, meaning we're not getting the "greenhouse" blanket effect in the upper atmosphere that scientists feared. In essence lower humidity higher up has more of a negative (cooling) effect rather than positive (warming). The extra moisture close to the surface is simply recycled as rain or snow for the most part.

"The bottom line is that, if (repeat 'if') one could believe the NCEP [US National Centers for Environmental Prediction] data 'as is', water vapour feedback over the last 35 years has been negative. And if the pattern were to continue into the future, one would expect water vapour feedback in the climate system to halve rather than double the temperature rise due to increasing CO_2," explained scientist Garth Paltridge who led the team.

This could indeed explain why we're seeing more snow [which creates more ice] and rain, and a failure by the earth to catch fire as the global warming computer models predicted – that water vapour in the atmosphere is helping to cool the planet down.

"The only object I can see for this paper," hissed one peer reviewer for *The Journal of Climate*, "is for the authors to get something in the peer-reviewed literature which the ignorant can cite as supporting lower climate sensitivity than the standard IPCC range".

So, on the advice of its peer reviewer, the *Journal* didn't publish the new study. Apparently it's now official: if you dare to speak heresy against global warming, the religious believers will have you excommunicated.

A few weeks after last year's rejection, the scientists involved were at a conference on water vapour and told the story of what happened, seeking views from the audience as to whether it was proper to publish such radical data.

"The audience was split as to whether the existence of the NCEP trends in humidity should be reported in the literature. Those 'against' (among them a number of people from NASA GISS) simply said that the radiosonde data were too 'iffy' to report the trends publicly in a political climate where there are horrible people who might make sinful use of them. Those 'for' simply said that scientific reportage shouldn't be constrained by the politically correct. The matter was dropped. I found after the event that the journal editor had come (I think specifically) to hear the talk. He didn't bother to introduce himself.

"I guess the story doesn't amount to much. Perhaps it is significant only in that it shows how naïve we were to imagine that climate scientists might welcome the challenge to examine properly and in detail even the smell of a possibility that global warming might not be as bad as it is made out to be. Silly us."

Now, admittedly weather balloon data at those altitudes is sometimes unreliable, but as Paltridge points out, no more so than the current fad of reporting surface temperatures as gospel:

"The attempt would be similar in principle to the current efforts at abstracting a believable global warming signal from the networks of surface-temperature observations."

"After some kerfuffle, the paper was accepted by *Theoretical and Applied Climatology* and appeared on February 26 on the journal's web site.[233] We presume it will be ignored. Being paranoiac from way back, we wonder at the happy chance by which a one-page general-interest article appeared in *Science* on February 20. With some self-referencing, it extolled the virtue of the latest modelling research, and of new(?) satellite observations of short-term, large amplitude, water vapour variability, which (say the authors) strongly support model predictions of long-term positive water vapour feedback. Well, maybe. It would be easy enough to argue against that conclusion. The paranoia arises because of another issue. We know that at least one of the authors is well aware of the contrary story told by the raw balloon data. But there is no mention of it in their article."[234]

233 "Trends in middle- and upper-level tropospheric humidity from NCEP reanalysis data", Paltridge et al, *Journal of Theoretical and Applied Climatology*, online version 26 February 2009, DOI 10.1007/s00704-009-0117-x

234 The *Science* article is a perspectives piece by Dessler & Sherwood which presents itself as a meta-analysis of earlier studies – all except the latest controversial one that indicates the earlier studies are

And that, folks, is the issue you should be worried about: global warming believers acting like the Spanish Inquisition to erase or marginalise data that doesn't suit the Green Bible.

But Paltridge's work has received corroboration he may not even be aware of. In another scientific study published at the end of 2007, the University of Alabama (Huntsville), one of the leading climate research centres, discovered the UN IPCC's computer models may be wildly overestimating global warming because of an improper understanding of the role of clouds.

"The widely accepted (albeit unproven) theory that manmade global warming will accelerate itself by creating more heat-trapping clouds is challenged this month in new research," reported *ScienceDaily*.[235]

"All leading climate models," explained lead researcher Roy Spencer, "forecast that as the atmosphere warms there should be an increase in high altitude cirrus clouds, which would amplify any warming caused by manmade greenhouse gases.

"That amplification is a positive feedback. What we found...was a strongly negative feedback. As the tropical atmosphere warms, cirrus clouds decrease. That allows more infrared heat to escape from the atmosphere to outer space.

"To give an idea of how strong this enhanced cooling mechanism is," said Spencer, "if it was operating on global warming it would reduce estimates of future warming by over 75%."

That's a staggering admission from one of the leading climate research universities: the UN IPCC reports may be overestimating global warming this century by up to 75% because their computer models have not properly understood how clouds work!

"The role of clouds in global warming is widely agreed to be pretty uncertain," Spencer told *ScienceDaily*. "Right now, all climate models predict that clouds will amplify warming. I'm betting that if the climate models' 'clouds' were made to behave the way we see these clouds behave in nature, it would substantially reduce the amount

---

flawed, and which the global warming cartel felt the public should not be told about yet. See "A matter of humidity", A Dessler & S Sherwood, *Science*, 20 Feb 2009, http://www.sciencemag.org/cgi/content/summary/323/5917/1020. A more detailed overview of its content can be viewed here: http://climatechangepsychology.blogspot.com/2009/02/andrew-dessler-steven-sherwood-matter.html

235 "Cirrus Disappearance: warming might thin heat-trapping clouds", Roy Spencer, John Christy et al, University of Alabama at Huntsville and Lawrence Livermore National Laboratory, published in *Geophysical Research Letters*, reported via *ScienceDaily*, 5 November 2007, http://www.sciencedaily.com/releases/2007/11/071102152636.htm

of climate change the models predict for the coming decades."

NASA's James Hansen should have known about this, because a paper from his own Goddard Institute back in 2001 covered a similar theme,[236] explaining that the Pacific Ocean "may be able to open a 'vent' in its heat trapping cirrus cloud cover and release enough energy into space to significantly diminish the projected climate warming.

"This newly-discovered effect – which is not seen in current climate prediction models – could significantly reduce estimates of future climate warming."

While the public faces of global warming belief attempt to write off the uncertainties, even the *Journal of Climate* has warned that clouds are one of the biggest grey areas in the global warming computer models because so little is known about them.

One scientist – Graeme Stephens at Colorado State University's Department of Atmospheric Science – has warned that computer models involve simplistic descriptions of cloud feedbacks on climate that contain "levels of empiricism and assumptions that are hard to evaluate with current global observations."[237]

"Out of necessity, most studies make ad hoc assumptions about the overriding importance of one process over all others generally ignoring other key processes, and notably the influence of atmospheric dynamics on cloudiness. Generally little discussion is offered as to what the system is let alone justification for the assumptions given. Much more detail on system and its assumptions are needed in order to judge the value of any study.

"Thus we are led to conclude that the diagnostic tools currently in use by the climate community to study feedback, at least as implemented, are problematic and immature and generally cannot be verified using observations."

And yet, these limited value computer model simulations are the ones that Dessler and Sherwood argued in *Science* magazine that we should have confidence in.[238]

236 "Heat vent in Pacific cloud cover could diminish greenhouse warming", Arthur Y. Hou et al, NASA Goddard and MIT, published in the *Bulletin of the American Meteorological Society*, March 2001, reported via *ScienceDaily*, 6 March 2001, http://www.sciencedaily.com/releases/2001/03/010301072351.htm

237 "Cloud Feedbacks in the Climate System: A Critical Review", Graeme Stephens, *Journal of Climate*, 15 January 2005, Vol 18, pp 237-273, http://www.gfdl.gov/~ih/jerusalem_papers/Stephens_JClim_2005.pdf

238 For a specific rebuttal of Dessler & Sherwood, see Roy Spencer's comments: http://www.drroyspencer.com/2009/02/what-about-the-clouds-andy/

The problem is that, behind the scenes, there is so much that the computer models don't accurately handle and, adding to the confusion, scientists are still trying to come to grips with where we're at on the Milankovitch cycles – the orbital variations and wobbles that can make it hotter or colder on earth over thousands of years.

A study by Yale's John Imbrie in 1980 established that the earth began heading into an ice age 6000 years ago.[239] Except, of course, we haven't. Which is where climatologist Stephen Vavrus and his team come in.

Vavrus – based at the University of Wisconsin-Madison's Centre for Climatic Research and the Nelson Institute for Environmental Studies – reckons we should be in an ice age, but greenhouse gases have held it off.[240]

The Centre for Climatic Research says the Antarctic ice core samples show human-induced climate change didn't begin a hundred and fifty years ago, but actually kicked off between 5,000 and 8,000 years ago when humans began clearing forests to cultivate the land.

"Between 5,000 and 8,000 years ago," said the Centre's John Kutzbach, "both methane and carbon dioxide started an upward trend, unlike during previous interglacial periods."

The Centre puts this down to "the introduction of large-scale rice agriculture in Asia, coupled with extensive deforestation in Europe [which] began to alter world climate by pumping significant amounts of greenhouse gases – methane from terraced rice paddies and carbon dioxide from burning forests – into the atmosphere."

Sounds great, again, in theory. Apart from the fact that it would be unusual for Earth to head into an Ice Age just a thousand or so years after coming out of one. And there's another little inconvenient truth: human population numbers. According to the Population Reference Bureau, the best estimate of human population 10,000 years ago (8,000 BC) was five million people worldwide.[241] Five million barely out of the Stone Age. They were not planting "large-scale" rice paddies, certainly not enough to rise above the methane emitted by vast herds of animals (now extinct) which once roamed prehistoric plains and forests.

---

239 J Imbrie, J Z Imbrie (1980). "Modeling the Climatic Response to Orbital Variations". *Science* 207 (1980/02/29): 943–953.

240 University of Wisconsin media release, 17 December 2008, http://www.eurekalert.org/pub_releases/2008-12/uow-sde121708.php

241 http://www.prb.org/Articles/2002/HowManyPeopleHaveEverLivedonEarth.aspx

In comparison to today's oil-pumping, beer-swilling, rainforest-burning world population of nearly 7 billion, those five million ancient tree-huggers of yesteryear were a non-event in terms of climate except, perhaps, in Australia, as we're about to discover.

Chapter Eleven

How Do We Sleep While Our Beds Are Burning?

"Don't worry about the world coming to an end today. It is already tomorrow in Australia."

– ***Charles M. Schulz, creator of Peanuts***

By all accounts, Australia has not always been desert. Fossils and evidence in the rocks show Australia once had lush tropical forests and a monsoon climate. But the continent was in a slightly different location relative to the tropics at that time, capable of supporting vast interior rainforests.

All that started to change around five million years ago, when changes in the Earth's climate (possibly caused by the seaway between North and South America closing up to form Panama and changing the equatorial current), combined with the drift of Australia towards its current location, placed enough stress on the interior jungles that they started dying back.

The more they died back, the less moisture they generated. And the less moisture they generated, the harder it became for rainclouds to form and survive above the centre of the continent.

Australia's deserts currently fall in an area of the world known as "the Horse Latitudes", a sub-tropical weather ridge where there's little rain or moisture. If you look along from Australia, similar dry zones have formed in southern Africa with the Kalahari and Namib deserts, and South America's Atacama Desert.

The northern hemisphere equivalents include the Sahara, the Arabian Desert and northern Mexico/Southwestern USA.

In other words, purely by virtue of where Australia sits on the planet, it is currently doomed to a hot dry interior.

By the time the first humans arrived, most (although not all) the great rainforests had already retreated from the interior closer to their current positions around the coasts. Human settlement of the lucky country began around 60,000 years ago,[242] but took around 10,000 years to have its first impact – extinction of large and scary monsters that roamed the outback at the time.

The aboriginal people felt none too safe at nights when giant eight metre long lizards were stalking their campsites. Not to mention sharp-fanged, meat-eating giant kangaroos and marsupial lions said to have been a more dangerous predator than modern African lions. But it was the regular bushfires that probably caused the most terror. After all, a man armed with a sharp stake can do much against a biological predator, but a man and family facing a raging conflagration can do little, as this account shows.

"Pastures had withered; creeks had become fissured clay-pans' water holes had disappeared; sheep and cattle had perished in great numbers, and the sunburnt plains were strewn with their bleached skeletons; the very leaves upon the trees crackled in the heat and appeared to be as inflammable as tinder...on the morning of February 6, the air which blew down from the north resembled the breath of a furnace.

"A fierce wind arose, gathering strength and velocity from hour to hour, until about noon it blew with the violence of a tornado. By some inexplicable means it wrapped the whole country in a sheet of flame – fierce, awful and irresistible..."

That could have been a narrative of this year's worst-recorded fire disaster in Victoria, except this year's happened on February 7, not February 6. The account above is from 1851,[243] long before atmospheric CO_2 became an issue.

According to contemporary reports of the 1851 blaze, temperatures on the day reached 47°C by 11am, comparable to the figure in 2009. The country had been locked in a long drought cycle, as it is now, and there had been record temperatures and heat for the preceding two months.

In other words, the 1851 bushfire context is almost an exact fit to the 2009 disaster, even down to the temperatures, and of course the blazes struck within 24 hours of each other, albeit 158 years apart.

There is one big difference, however. The 1851 blaze destroyed five

242 http://www.environment.gov.au/soe/2001/publications/report/settlements.html

243 Bushfires of Black Thursday, 1851: 'When the smoke turned day into night' http://www.chig.asn.au/black_thursday_bushfires_1851.htm

million hectares but only killed 12 people (that they know of). The 2009 bushfires by their peak had only hit half a million hectares, so their scale was significantly smaller, but they killed more than 170 people.

Why so deadly then? Because, ironically, environmentalists in local and state governments had enacted laws encouraging people to blend their homes into the Australian bush, and forbidding clearance of dry leaf and timber debris from the forest floor.

This meant that when a fire eventually hit that area, as statistically it was bound to do, there was enough fuel on the ground to quickly turn the blaze into a raging firestorm.

In one area, a homeowner who defied the Green restrictions, and cleared debris and trees several hundred metres back from his house, was fined more than $60,000 dollars for breaching environmental bylaws. But in a tragic lesson to Australian bureaucrats and environmentalists, his house was the only one left standing in his street, untouched by the fire.

One local council at Yarra had forbidden controlled burnoffs, with a 2007 council report stating that more information was needed about the impact of burnoffs on flora and fauna, and in particular the breeding habits of the Leadbeater's possum.[244] It's understood a substantially larger number of possums were roasted when the big one eventually hit.

Another warning from former government bushfire scientist David Packham in 2003, was ignored by local councillors.[245]

"The mix of fuel, unsafe roadsides and embedded houses, some with zero protection and no hope of survival, will all ensure that when a large fire impinges upon the area a major disaster will result," warned Packham to a deaf audience.

Tragically, the worst-hit areas in 2009 were in that council's zone. The council went even further and didn't just ban the removal of dry debris, but it actually required householders to plant trees up to their houses so as to increase bush planting.

A timber industry official, Scott Gentle, warned in July 2007:[246]

"Living in an area like Healesville, whether because of dumb luck or whatever, we have not experienced a fire...since...about 1963.

244 "Burnoffs following Victoria bushfires a 'threat to biodiversity'", *The Australian*, 12 February 2009
245 "Council ignored warning over trees before Victoria bushfires", *The Australian*, 11 February 2009
246 "Green ideas must take blame for deaths", Miranda Devine, *Sydney Morning Herald*, 12 February 2009

God help us if we ever do, because it will make Ash Wednesday look like a picnic."

Healesville, chillingly, was decimated by the 2009 fires, as predicted.

But Gentle also warned in 2007 of interference from greens in the Kinglake area several years ago, during attempts by forestry workers to make the area safer.

"The contractors were out working on the fire lines. They put in containment strategies and cleared off some of the fire trails. Two weeks later [a small 2007 fire] broke out, but unfortunately those trails had been blocked up again [by greens] to turn it back to its natural state… Instances like that are just too numerous to mention."

Melbourne meteorologist David Karoly, a climate change believer, posted an essay on Michael Mann and Gavin Schmidt's *RealClimate* website the week after the fires, blaming anthropogenic global warming for the severity of the disaster.[247]

Even so, Karoly was forced to admit it was speculative in many areas. For example, low humidity was a big contributing factor but, he conceded, "No specific studies have attributed reduced relative humidity in Australia to anthropogenic climate change".

Because of its hot central desert, Australia has a climate all its own that has nothing to do with global warming. The hot dry desert winds, when they are sucked out to the south eastern Victorian coast by weather systems like El Niño/La Niña, scorch everything in their path, stripping the air and vegetation of moisture.

And because the winds were coming from the central desert, Karoly concedes that this too had a natural explanation.

"Extreme fire danger events in south-east Australia are associated with very strong northerly winds bringing hot dry air from central Australia. The weather pattern and northerly winds on 7 February were similar to those on Ash Wednesday and Black Friday, and the very high winds do not appear to be exceptional or related to climate change."

Noting that Victoria was going through a near record drought, Karoly admits even that is a hard one to call:

"While south-east Australia is expected to have reduced rainfall and more droughts due to anthropogenic climate change, it is difficult to quantify the relative contributions of natural variability and climate change to the low rainfall at the start of 2009."

247 http://www.realclimate.org/index.php/archives/2009/02/bushfires-and-climate/

About the only aspect of the fires that Karoly is confident in sheeting home to human-caused global warming is the temperatures:

"A recent analysis of observed and modelled extremes in Australia finds a trend to warming of temperature extremes and a significant increase in the duration of heat waves from 1957 to 1999 (Alexander and Arblaster, 2009). Hence, anthropogenic climate change is likely an important contributing factor in the unprecedented maximum temperatures on 7 February 2009."

And yet, if the temperatures were virtually the same in the 1851 firestorm, but the size of the 2009 fire was ten times smaller, are we really suffering worse fires because of global warming, or is it simply environmental stupidity that built entire towns in tinder dry eucalypt forests?

The point of all this is to show that iconic events like bushfires do not a climate change case make. Unfortunately, as you've already seen in this book, the facts are of no consequence when those pushing human-induced global warming theory are involved. Who are these people, exactly?

One, in particular, deserves some very close scrutiny.

Chapter Twelve

The Emperor's Birthday Suit

"Effective execution of Agenda 21 will require a profound reorientation of all human society, unlike anything the world has ever experienced – a major shift in the priorities of both governments and individuals and an unprecedented redeployment of human and financial resources. This shift will demand that a concern for the environmental consequences of every human action be integrated into individual and collective decision making at every level"

– Agenda 21 document[248]

If you look at the global warming religion, you can peg it back to just a handful of high profile individuals, true believers if you like, in powerful positions. NASA's James Hansen is perhaps the ringleader.

Back in 1988, Hansen was one of the first to light the fuse on the global warming rocket, with testimony before a senate hearing.

"It's time to stop waffling so much and say the evidence is pretty strong that the greenhouse effect is here," said Hansen, pricking up the ears of journalists and environmentalists worldwide.

Now, as head of NASA's Goddard Institute of Space Studies, Hansen is the gatekeeper over much of the global warming data accumulated by weather stations and satellites. It is Hansen's claims that underpinned much of Al Gore's movie, and Hansen's philosophy is the hand up the puppet-holes of the IPCC panel and their ongoing "unprecedented" warnings about impending climate doom.

Asked by one interviewer three years ago whether we're reaching the point of no return, Hansen didn't pull any punches.[249]

---

248 *Agenda 21: The Earth Summit Strategy to Save The Planet* by Daniel Sitarz (Ed.), (EarthPress, 1993)
249 http://www.loe.org/shows/segments.htm?programID=06-P13-0005&segmentID=3

"I think that we are. I think that if we continue on the path of business as usual for another decade it will be impractical to keep the global warming less than one degree Celsius. And the reason that's an important level is that's the most warming that we've had in the last 700,000 years, and probably the last million years.[250] We just don't have the data for the full million years. But we know that the changes in that period, although they're significant, they're probably something we could adapt to.

"But if you start talking two or three degrees Celsius, then you're really talking about a different planet from the one we know. There would be no sea ice in the Arctic in the summer and fall. That means the species that live there now – polar bears, the seals that live on sea ice, and reindeer on the tundra – they would not be able to survive. So it's like the million-year flood; it's never happened in the past century. So, we're talking large regional changes."

But it was Hansen's 2006 claim that the Bush administration was trying to silence him that provoked serious media interest.

"Recently you told the *New York Times* that the Bush administration has been trying to silence you about your findings on climate change and the implications for public policy. What's going on?" quizzed interviewer Steve Curwood.

"Well, the public affairs office at NASA headquarters has put unusual restrictions on me with regard to speaking to the media, requiring that any request for interview be that I not respond to it, but rather just to send it to headquarters. And they would have the right of first refusal, which means someone there will actually do the interview rather than me. I have some objection to that because that policy has not been enforced on other people."

Then, speaking like a man who'd successfully been silenced, Hansen spilled the beans on what he felt was about to happen on Earth.

"If we continue on this path, which is business as usual, with the rate of emissions, of greenhouse gasses, continuing to increase at a couple of percent per year; then this century we would have warming of two and a half or three degrees Celsius. Once we get that kind of temperature we'll be having a sea level change at a rate of probably a few metres per century."

---

250 As we now know from the Harvard-Smithsonian study flagged earlier, this claim is simply untrue. The Medieval Warm Period was between one and two degrees Celsius warmer than the current warm period, according to an analysis of more than 200 scientific studies, and results from 6,000 boreholes drilled around the world. Additionally, data indicates temperatures 7,000 years ago were also substantially higher than today.

For Hansen, the gagging tipping point was not being able to release data to the public.

"It became clear that the new constraints on my communications were going to be a real impediment when I was forced to take down from our website our routine posting of updated global temperature analysis."[251]

The protestation is a little rich, given that Hansen and his colleague Michael Mann have been criticised for not releasing details on the GISS website and others of how they've reached some of their key conclusions, which in turn has prevented other scientists from checking the quality of their work.

The vetting of the website and Hansen's media comments might also have come about because of oversaucing the goose, like Michael Mann's claim that 1998 was almost 99% certain to have been "the hottest year in the last one thousand years".[252] Hansen's GISS had to yank data tables off their website in 2007 after sceptics proved the NASA climate team had made mistakes in drawing up the list of the hottest years on record.

It turns out that 1998 was not the hottest year on record, nor in fact was it the hottest year of the 20th century. Nor is the last decade dominant in the revised list of hottest years, which reads as follows:

1. 1934
2. 1998
3. 1921
4. 1931
5. 2006
6. 1999
7. 1953
8. 1990
9. 1938
10. 1939

It turns out the 20th century had two warm spells, the first of which is widely regarded, even by the IPCC and GISS, as too soon for human CO_2 emissions to have played any role at all. So much for the current warm period being the warmest in 700,000 years as

251 "NASA climate scientist 'gagged' by White House", *The Independent*, 20 March 2007, http://www.independent.co.uk/environment/climate-change/nasas-climate-scientists-gagged-by-white-house-440962.html

252 "1998 was warmest year of millennium, climate researchers report" – American Geophysical Union news release, 3 March 1999, http://www.agu.org/sci_soc/prrl/prrl9906.html

Hansen told interviewer Steve Curwood – it wasn't even the warmest in sixty years. Three of the top five spots, including number one, go to years earlier than 1935.

You may notice that only *one* year from the past decade, 2006, makes the top ten.

So what happened, how did NASA's list of the ten warmest years end up so wrong, and different from the versions that have dominated the news media?

Because Hansen's team refused to release full data, sceptics had been forced to poke around on their own in an attempt to cross-check. One of them, Canadian statistician Steve McIntyre, believed NASA's GISS database had a software glitch.

"Unfortunately, it was hard to prove," writes blogger Warren Meyer.[253] "Why? Well, that highlights one of the great travesties of climate science. Government scientists using taxpayer money to develop the GISS temperature database at taxpayer expense refuse to publicly release their temperature adjustment algorithms or software (in much the same way Michael Mann refused to release the details for scrutiny of his methodology behind the hockey stick).

"Using the data, though, McIntyre made a compelling case that the GISS database had systematic discontinuities that bore all the hallmarks of a software bug. Today [August 8, 2007], the GISS admitted that McIntyre was correct, and has started to republish its data with the bug fixed."

Of the top 20 warmest years since 1900, the majority are pre-1990.

The years 1998 and 1999 appear, I suggest, as a direct result of the massive El Niño which began in late 1997 and continued through 1998. El Niño also powered up in the southern summers of 02/03, 04/05 and 06/07, which would make this natural effect again the prime suspect in 2006's appearance in the top ten.

Far from being "the tenth warmest year on record", the latest NASA GISS data reveals 2008 came in at number 56 on the list.[254]

Global warming believers will be leaping to their feet about now, protesting that these revised top ten numbers apply only to the United States mainland. Yes, they do, and Hansen's GISS has so far refused to make any changes to its much publicised "World's warmest" list, which continues to highlight the last decade as the

---

253 http://www.coyoteblog.com/coyote_blog/2007/08/official-us-cli.html
254 http://data.giss.nasa.gov/gistemp/graphs/Fig.D.txt

hottest on record. However, there's good reason to believe the official world list is bogus too.

"The GISS today makes it clear that these adjustments only affect US data and do not change any of their conclusions about worldwide data," notes blogger Warren Meyer.[255]

"But consider this: For all of its faults, the US has the most robust historical climate network in the world. If we have these problems, what would we find in the data from, say, China? And the US and parts of Europe are the only major parts of the world that actually have 100 years of data at rural locations. No one was measuring temperature reliably in rural China, or Paraguay or the Congo in 1900.

"That means much of the world is relying on urban temperature measurement points that have substantial biases from urban heat."

The point Meyer makes is valid and flows on from the discussion earlier in the book: the global figures simply cannot be trusted because there is no proper correction for UHIE, but there's another aspect to this as well. North America, as one of the large landmasses on the planet, is more prone to global warming than, say, New Zealand or Australia in the southern hemisphere. That's because land absorbs heat more rapidly than seawater does. Most of the world's land is in the northern hemisphere, which by definition increases the likelihood that the north will warm faster than the south, and cool faster.

It's a point backed up by scientific data:

"Land-sea area changes may affect climate significantly in many ways: by changing the net radiation balance (since oceans have lower albedo [reflectivity of sunlight back into space] than land in general); by changing the geographical distribution of albedo; by changing the seasonality of the climate (even today the Southern Hemisphere, with its large ocean areas has a much smaller seasonal temperature amplitude (6°C) than the Northern Hemisphere (14°C) by enhancing the likelihood of polar glaciation, which is easier with a landmass at the pole; by changing the ocean circulation (see below); and so on," writes climate scientist Tom Wigley,[256] the man who later advised President Clinton that Kyoto was a waste of effort

255 http://www.coyoteblog.com/coyote_blog/2007/08/official-us-cli.html
256 "Climate and Paleoclimate: what we can learn about Solar Luminosity Variations", T M Wigley, Climatic Research Unit, University of East Anglia, *Solar Physics* 74 (1981) 435-471 http://www.springerlink.com/content/q2p4633u64082323/fulltext.pdf

because it would not actually reduce warming.

If the US continent turns out not to have been heating up much over the past two decades, as the figures now suggest, what does that say for the rest of the world where the data is not reliable? Are we really heating up globally, or have local science agencies jumped on the global warming bandwagon without properly checking their data?

It would help if the global temperature data was open to independent scrutiny, but it isn't.

"We believe if the global data set, which is not disclosed to enable peer review, were given equal scrutiny to the US set, global warming would either vanish or be barely detectable," writes one climate impact scientist.[257]

It's also important to remember that Hansen is on record admitting that data provided by international agencies outside the US is not checked for accuracy. All of this means that the world figures are likely to be far less reliable than the US data.

Before we leave this debate about the accuracy of the GISS top ten list, and James Hansen's claims about being 'gagged' by the Bush administration into staying quiet about global warming (which he hasn't), it would be remiss of me not to offer another possible explanation for NASA putting him on a short leash.

James Hansen endorsed Democrat John Kerry for president in 2005. There's also the inconvenient truth about his financial links to Kerry via the presidential candidate's wife, Teresa Heinz-Kerry.

"James Hansen, one of the leading climate scientists, he says it's crystal clear," said a CNN anchor to Republican senator James Inhofe. "What do you say?"[258]

"I say that that's James Hansen who was paid $250,000 by the Heinz Foundation. And I think he'd say almost anything you'd ask him to say," Inhofe retorted.

A third possible reason for the 'gagging' is that Hansen was a consultant to Al Gore's travelling global warming slideshow presentation. Add all this together, and Hansen is political in a whole range of ways.

But the fourth, and most likely reason, is allegedly that Hansen was simply an "embarrassment" to his NASA colleagues. In a let-

257 Dr John Everett, formerly of NOAA, http://www.climatechangefacts.info

258 http://www.conservative.org/pressroom/2007/speech_inhofe.asp

ter to a US senate committee released late January this year, one of Hansen's senior colleagues explains:[259]

"I was, in effect, Hansen's supervisor because I had to justify his funding, allocate his resources, and evaluate his results…Hansen was never muzzled even though he violated NASA's official agency position on climate forecasting (i.e., we did not know enough to forecast climate change or mankind's effect on it). Hansen thus embarrassed NASA by coming out with his claims of global warming in 1988 in his testimony before Congress," wrote Dr John Theon, NASA's former head of the Climate Processes Research Programme, and former chief of NASA's Atmospheric Dynamics and Radiation Branch.

Theon then shocked his readers, by declaring he no longer believed in anthropogenic [human-caused] global warming.

"I appreciate the opportunity to add my name to those who disagree that global warming is manmade."

Summing up the state of good research in 2009, Theon says existing global warming "climate models are useless."

"My own belief concerning anthropogenic climate change is that the models do not realistically simulate the climate system because there are many very important sub-grid scale processes that the models either replicate poorly or completely omit.

"Furthermore, some scientists have manipulated the observed data to justify their model results. In doing so, they neither explain what they have modified in the observations, nor explain how they did it. They have resisted making their work transparent so that it can be replicated independently by other scientists. This is clearly contrary to how science should be done. Thus there is no rational justification for using climate model forecasts to determine public policy.

"As Chief of several of NASA Headquarters' programs (1982-94), an SES position, I was responsible for all weather and climate research in the entire agency, including the research work by James Hansen, Roy Spencer, Joanne Simpson, and several hundred other scientists at NASA field centers, in academia, and in the private sector who worked on climate research," Theon wrote. "This required a thorough understanding of the state of the science. I have kept up with climate science since retiring by reading books and journal articles."

259 "James Hansen's former NASA supervisor declares himself a sceptic", news release from US Senate Environment & Public Works Ctte, minority office, 27 January 2009, http://epw.senate.gov/public/index.cfm?FuseAction=Minority.Blogs&ContentRecord_id=1a5e6e32-802a-23ad-40ed-ecd53Cd3d320

One of the huge problems in computer modelling is getting accurate historical temperature data to measure against. NASA and NOAA talk about average world temperatures, but it's only since 1978 that satellites have provided reliable telemetry on surface and air temperatures. Before that, we're relying on a sprinkling of weather stations on land. With oceans covering 70% of the earth's surface, big questions hang over the reliability of temperature readings taken on board ships in the past.

All said, lists of "the hottest years" can only be compared apples for apples, weather station for weather station, and any claim to know accurate global temperatures a century ago is sheer hubris.

As one cynical commenter on Warren Meyer's blog puts it:

"The earth has many temperatures, everyplace there is a thermometer there is a temperature, but what is the temperature between thermometers? … This is where the problem of averaging comes in…

"To pretend that the earth even has a temperature is voodoo. If it is 70 degrees here and 5 miles away it is 75 degrees then how do you measure an average temperature … assume that at 2.5 miles the temperature is 72.5°F?

"The recorded earth temperatures are only valid for averaging the temperature of weather stations, not the landscape between … all that swirling air of different temperatures is not measurable under current practices … and may never be."

The point is well made. Short of having a network of thermometers on the ground spaced a kilometre apart across the entire surface of the planet – land and sea – the "average" temperatures we're claiming to measure global warming against today would be subject to a margin of error many times higher than the 0.6°C temperature rise claimed over the past century.

"Temperature measurement of the entire climate is inherently difficult," writes Paul Rogers in response to a *New Scientist* claim of rising average temperatures.[260]

"It is claimed that average temperatures increased by 0.7°C in one hundred years. I have been in Texas where the temperature dropped 20°C in 5 minutes. There is so much natural variation in weather that it is an immensely difficult task to measure and compute the global average temperature. Indeed, it is immensely difficult even to define

260 http://www.newscientist.com/article/dn11649

what is meant by global average temperature. Air temperature, sea temperature, soil temperature. They all vary by season, and other known and unknown cyclical factors by location and above all by random fluctuation."

The claim that average temperatures have risen "globally" makes a neat TV or radio soundbite, but you can sympathise with those who call it "voodoo".

NASA's guesswork, however, isn't the only torpedo to have hit the global warming believers. Another high priest, Michael Mann, was mauled by an independent review of his scientific claims.

As previously noted, Mann was responsible for the infamous "hockey stick" graph that Al Gore used in his movie and which the IPCC relied on its major 2001 briefing on the science of global warming.

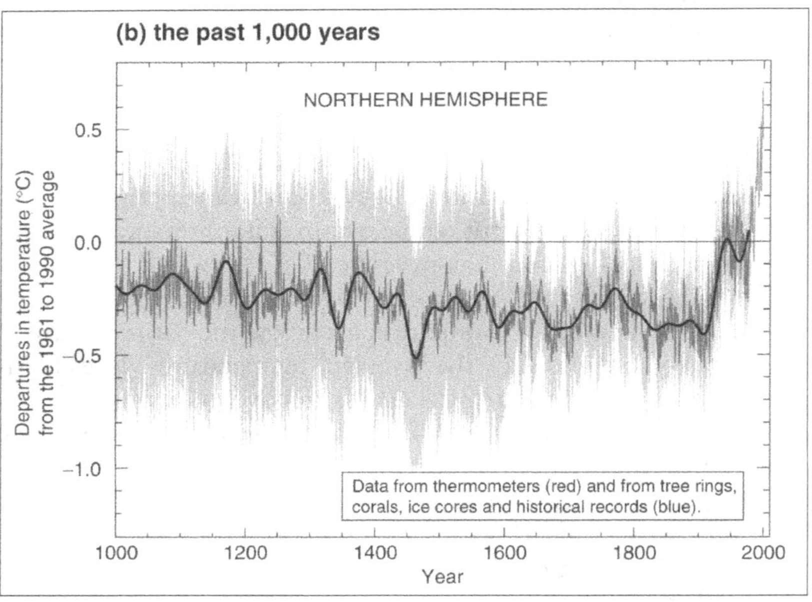

What I'll now explain is how Mann published such a mistake-ridden graph and got away with it. First, a bit of backstory. Mann's statement that 1998 was the hottest year in a millennium was based on research for a 1999 scientific paper.[261]

261 *Geophysical Research Letters*, 15 March 1999

In a news release,[262] Mann and his co author state that they "relied on three sets of 1,000-year-long tree-ring records from North America, plus tree rings from northern Scandinavia, northern Russia, Tasmania, Argentina, Morocco, and France."

Additionally, they studied ice cores from Greenland and the Andes mountains.

Three trees in the US, plus presumably a tree each in the other locations, and the two ice cores. On that small sample they built their claim.

You've already seen a raft of scientific studies proving that temperatures in the Medieval Warm Period far exceeded the global warming that's occurred in the 20th century. Yet Mann's news release, based on the imprecision of tree rings, waved the MWP aside as if it didn't exist.

"Even the warmer intervals in the reconstruction pale in comparison with mid-to-late 20th-century temperatures," said co-author Malcolm Hughes.

It was this data that became the basis of the hockey stick. But in mid 2006, Mann's hockey-stick got pulped and shredded by a committee of experts (the Wegman Panel) appointed to review whether he'd done his homework properly.

"The conclusions from MBH98 [Mann, Bradley, Hughes], MBH99 were featured in the Intergovernmental Panel on Climate Change report entitled *Climate Change 2001: The Scientific Basis*.[263] This report concerns the rise in global temperatures, specifically during the 1990s," noted the expert panel's report to the Senate.[264]

"The MBH 98 and MBH99 papers are focused on paleoclimate temperature reconstruction and conclusions therein focus on what appear to be a rapid rise in global temperature during the 1990s when compared with temperatures of the previous millennium. These conclusions generated a highly polarized debate over the policy implications of MBH98, MBH99 for the nature of global climate change, and whether or not anthropogenic actions are the source."

Looking at the methodologies used, the panel quickly focused on the tree ring data, explaining that tree rings were not necessarily reliable: "The width and density of tree rings are dependent on many confounding factors, making it difficult to isolate the climatic temperature signal."

---

262 http://www.agu.org/sci_soc/prrl/prrl9906.html

263 Graph taken from this report

264 http://www.climateaudit.org/pdf/others/07142006_Wegman_Report.pdf

Then the experts hammered Mann et al for skewed study results because of incorrect methodology:

"The proxy data [things that indicate past temperatures] are incorrectly centred, which inflates the variance of certain proxies and selectively chooses those decentred proxies as the temperature reconstruction.

"The controversy of Mann's methods lies in that the proxies are centred on the mean of the period 1902-1995, rather than on the whole time period. This mean is, thus, actually decentred low, which will cause it to exhibit a larger variance, giving it preference for being selected as the first principal component.

"The net effect of the decentring is to preferentially choose the so-called hockey stick shapes."

In plain English, the way Mann's study was constructed, it was always going to produce a dramatic hockey stick effect, no matter what data was entered. You could have entered last week's grocery list and still received "proof" of global warming!

"While this error would have been less critical had the paper been overlooked like many academic papers are, the fact that their paper fit some policy agendas has greatly enhanced their paper's visibility. Specifically, global warming and its potentially negative consequences have been central concerns of both governments and individuals.

"The 'hockey stick' reconstruction of temperature graphic dramatically illustrated the global warming issue and was adopted by the IPCC and many governments as the poster graphic. The graphic's prominence together with the fact that it is based on incorrect use of [data] puts Dr. Mann and his co-authors in a difficult face-saving position."

The expert panel gave the global warming guru the benefit of the doubt when it came to whether the mistakes were intentional, saying "It is not clear that Mann and associates realized the error in their methodology at the time of publication."

However, like critics of NASA's Goddard Institute, the expert panel was disappointed at the lack of verifiable scientific data:

"Because of the lack of full documentation of their data and computer code, we have not been able to reproduce their research."

Although Mann, Bradley and Hughes' work was "peer reviewed", the expert panel wasn't impressed at the quality of that either, suggesting his "peers" were far too cosy with him.

"In our further exploration of the social network of authorships in temperature reconstruction, we found that at least 43 authors have direct ties to Dr. Mann by virtue of co-authored papers with him. Our findings from this analysis suggest that authors in the area of paleoclimate studies are closely connected and thus 'independent studies' may not be as independent as they might appear on the surface."

This is an important point: journalists reading this, take note. Is the so-called "consensus" on global warming nothing more than a bunch of mates all backing each other up? It appears so.

Then the experts turned to the claim about the hottest years of the millennium.

"Overall, our committee believes that Mann's assessments that the decade of the 1990s was the hottest decade of the millennium and that 1998 was the hottest year of the millennium cannot be supported by his analysis."

Ouch.

"Both Esper et al. (2002) and Moberg et al. (2005) indicate that current global temperatures are *not* [my emphasis] warmer than the medieval warm period."

On the assumption that CO_2 is the driving force of climate change – another key plank to the IPCC/Mann/Hansen theories – the expert panel said, not necessarily:

"Mann et al. also infer that since there is a partial positive correlation between global mean temperatures in the 20th century and CO_2 concentration, greenhouse-gas forcing is the dominant external forcing of the climate system. Osborn and Briffa make a similar statement, where they casually note that evidence for warming also occurs at a period where CO_2 concentrations are high. A common phrase among statisticians is *correlation does not imply causation*. The variables affecting earth's climate and atmosphere are most likely to be numerous and confounding. Making conclusive statements without specific findings with regard to atmospheric forcings suggests a lack of scientific rigour and possibly an agenda."

Among the panel's recommendations – when so much is at stake, make sure the science really is settled, and accurate:

"Especially when massive amounts of public monies and human lives are at stake, academic work should have a more intense level of scrutiny and review. It is especially the case that authors of policy-related documents like the IPCC report, *Climate Change*

2001: The Scientific Basis, should not be the same people as those that constructed the academic papers."

Another area heavily criticised was the willingness of the cliquey, publicity-hungry paleoclimate community to swing in behind Mann even though he'd got it wrong, and continue defending his findings, which they are doing to this very day. According to the expert panel, the biggest weakness is that scientists like Mann, with no training in statistics, have been churning out studies claiming to be statistically-based, yet they never consulted experts in that field to cross-check.

"As statisticians," wrote the expert panel, "we were struck by the isolation of communities such as the paleoclimate community that rely heavily on statistical methods, yet do not seem to be interacting with the mainstream statistical community. The public policy implications of this debate are financially staggering and yet apparently no independent statistical expertise was sought or used."

It's an extreme note of warning to governments around the world as they consider carbon taxes and emission trading schemes based on IPCC reports that are scientifically incorrect, and academic studies that are badly flawed.

Although the expert panel report gave Mann and his colleagues the benefit of the doubt when it came to 'intent', there are those who have accused Mann and his supporters of bending the credibility of science beyond breaking point.

"Despite Mann's reticence to allow anyone to check his work, sceptics still began to emerge," notes *Coyoteblog's* Warren Meyer. "Take that big temperature bulge in the Middle Ages shown in the previous consensus view. This bulge was annoying to climate interventionists, because it showed that large variations in temperature on a global scale can be natural and not necessarily the fault of modern man.

"But Mann made this whole medieval bulge go away. How? Well, one of the early revelations about Mann's work is that all the data before 1450 or so comes from studying the tree rings of one single tree. Yes, that's one tree (1). Using the evidence of this one tree, Mann flattened the temperature over the 500 year period from 1000-1500 and made the Medieval warm period just go *poof*. Wow!," exclaims Meyer.

Not just one tree, but Mann and his team chose bristlecone pines, which they knew or should have known were unreliable. The rings of bristlecones not only grow more when it's warm, but also when there's more CO2 (which plants need for growth), so the tree ring results were prone to exaggeration when plugged into the infamous graph.

Which is why Al Gore was able to point to the Medieval Warm Period blips on Mann's hockey stick graph and reassure viewers it was no big deal.

Global warming critic, Britain's Viscount Christopher Monckton, is another to suggest a more cynical motive for the fake hockey stick graph:[265]

"The UN abolished the medieval warm period (the global warming at the end of the First Millennium AD). In 1995, David Deming, a geoscientist at the University of Oklahoma, had written an article reconstructing 150 years of North American temperatures from borehole data. He later wrote: 'With the publication of the article in *Science*, I gained significant credibility in the community of scientists working on climate change. They thought I was one of them, someone who would pervert science in the service of social and political causes. One of them let his guard down. A major person working in the area of climate change and global warming sent me an astonishing email that said: *'We have to get rid of the Medieval Warm Period'*.'

"So they did. The UN's second assessment report, in 1996, showed a 1,000-year graph demonstrating that temperature in the Middle Ages was warmer than today. But the 2001 report contained a new graph showing no medieval warm period. It wrongly concluded that the 20th century was the warmest for 1,000 years."

Monckton also ascribes a certain amount of Machiavellian preplanning to construct the false data used in the UN report (designed to scare political leaders and media) and Al Gore's movie (designed to scare the public).

"They said they had included 24 data sets going back to 1400. Without saying so, they left out the set showing the medieval warm period, tucking it into a folder marked 'Censored Data'."

Yeah, that'd do it.

Michael Mann, however, is not the only one trying to bury the evidence that 20th century global warming is not unique, or even the warmest of recent times. *New Scientist* magazine, which used

265 "The sun is warmer now than for the past 11,400 years", Christopher Monckton, *The Telegraph*, 5 November 2006, http://www.telegraph.co.uk/news/uknews/1533312/The-sun-is-warmer-now-than-for-the-past-11,400-years.html

to pride itself on its credibility, lists the attack on Mann's hockey stick as a "myth" used by sceptics.

"Most researchers would agree that while the original hockey stick can – and has – been improved in a number of ways, it was not far off the mark. Most later temperature reconstructions fall within the error bars of the original hockey stick. Some show far more variability leading up to the 20th century than the hockey stick, but none suggest that it has been warmer at any time in the past 1000 years than in the last part of the 20th century."[266]

That claim by *New Scientist*, as you should by now well and truly realise, is utterly untrue (see photo section for comparative graphs from the Wegman report). The Harvard-Smithsonian Centre for Astrophysics study proved the MWP was hotter, regardless of all the hockey stick graphs in the world.[267] Yet Goebbels would be proud of his global warming high priests and their useful idiots in the news media: Keep repeating the lie often enough...

Warren Meyer has a theory on why Mann's flawed studies still carry weight with earnest global warming believers.

"Since AGW supporters refuse to acknowledge flaws in Mann, it is almost certain that these flaws still exist in the other analyses (therefore making it unsurprising that new analyses show roughly the same results). Remember that Mann was replaced by Briffa as lead author of this section of the Fourth IPCC report, and it was Briffa who dropped 20-30 years of recent data from his historical reconstruction when it did not show the result that he wanted it to."

New Scientist magazine, like a loyal but ultimately dumb rat that's well and truly nailed its own tail to the mast of a sinking ship, continues to insist that scientists know Mann's work to be good and true, and that the MWP was definitely cooler than now.

"What is clear, both from the temperature reconstructions and from independent evidence – such as the extent of the recent melting of mountain glaciers – is that the planet has been warmer in the past few decades than at any time during the medieval period. In fact, the world may not have been so warm for 6000 or even 125,000 years."

Given that no one was photographing the melt of glaciers in the

---

266 http://www.newscientist.com/article/dn11646

267 "20th Century Climate Not So Hot", 31 March 2003, Harvard-Smithsonian Center for Astrophysics http://www.cfa.harvard.edu/press/archive/pr0310.html

past, *New Scientist* appears to be playing to public gullibility in running its claims.

Not that this, "deny everything" stunt hasn't been pulled by comedians on previous momentous occasions. Some of you may remember Saddam Hussein's former spokesman, nicknamed 'Comical Ali':

"It's as if a bad *Saturday Night Live* skit is playing in Baghdad," wrote *USA Today's* Maria Puente.[268]

"The Iraqi information minister stands in front of the cameras, a grim smile on his face, a military beret on his head, and declares forcefully, *"There are no American troops in Baghdad!"* Meanwhile, black smoke rises in the distance behind him, weapons fire can be heard all around, and American tanks rumble down streets only yards away.

"Pay no attention to those tanks", Mohammad Saeed al-Sahhaf may as well be saying. There are no invaders, no troops – only *"liars."*

"The infidels are committing suicide by the hundreds on the gates of Baghdad," he said at one news conference. *"We slaughtered them."*

"Who is this guy, and does he think he is fooling anybody?" asks Puente.

In a similar vein, one of *New Scientist's* "myth busters", Michael Le Page, found his essay allegedly debunking sceptics' "myths" shredded in the comments section by other scientists and observers, including this analysis from one of his critics.[269]

"Interesting article. I have been provided a couple of the research studies backing the IPCC study and it is curious how the scientists have covered their behinds:

"1. They state that the models they used were specified by the IPCC and were not of their own. The IPCC for these studies (through its Program for Climate Model Diagnosis and Intercomparison) specified exactly what models the scientists would use in their analysis of atmospheric convection and circulation.

"2. The scientists were told to keep radiative forcings (the radiation coming into and leaving our system) and other key factors constant throughout much of their analysis which is bogus in that atmospheric and oceanic dynamics will change with changing environs. Some astrophysicists have reported that the biggest cause of global warming has been due to increased solar radiation and the planet is

268 "Saddam's spokesman staying on message", 7 April 2003, *USA Today*, http://www.usatoday.com/news/world/iraq/2003-04-07-iraq-information-minister_x.htm

269 "Climate change: a guide for the perplexed" by Michael le Page, *New Scientist*, 16 May 2007, comment posted on website at http://www.newscientist.com/article/dn11462 by Michael Thomas

warming as a result – the requirement in these studies of constant radiative forcings means that this possible cause of global warming is disregarded by definition.

"3. Of the 22 climate models utilized by the IPCC in analyzing man-made global warming, *they all came to the same conclusions.* Given how little we know about atmospheric and ocean science, particularly in light of our infancy in modelling global systems and the lack of proven model specifications, the odds that all 22 would come to the same conclusions is truly astronomical.

"4. The operative analysis horizon was 100 years for most models and for some, they were taken to 300 years. Hey, the computer models these guys use can't even forecast the local weather for next week with any reasonable degree of accuracy!

"5. The scientists express a caveat regarding the models they were instructed to use by the IPCC, they state that "the robustness of this behaviour across all models cannot be confirmed due to lack of data".

"6. The classic study on rising sea levels by the University of Maine, Dept of Geophysics, neglected to take into account changing land mass levels, plate tectonics, in their seminal study. I approached them about this major oversight and they swallowed hard. A major oversight that brings into question the validity of their other findings.

"7. Your researcher's sources were weak, many of the points he made were refuted quite some time ago – sounded like his only source was the Al Gore movie. He also mentions the large number of scientists who support the man-made global warming theories. It is my understanding from the IPCC researchers that only 55 scientists actually signed the IPCC study – the remaining 1,945 didn't even get to see it before it was published and a large percentage have voiced their disagreement with its findings. This excludes the 19,000 scientists who signed the Oregon Petition saying this matter is bogus."

For the record, although *New Scientist* editorial staff including Michael Le Page responded to points raised by some *other* commenters, they made no efforts to respond to the detailed critique above.

But what do other scientists, no longer impressed by the global warming believers in their midst, really think? We're about to find out.

Chapter Thirteen

This Was The Moment When The Tides Turned

"Let all men know how empty and worthless is the power of kings. For there is none worthy of the name but God, whom heaven, earth and sea obey"

– King Canute,
upon proving even kings cannot turn back tides, c AD 1025

In 2008, the global warming gravy train came perilously close to jumping the rails and plunging into the abyss of irrelevance. President Obama set the scene with his hopelessly audacious boast that his election to the top job would freeze the melting icebergs in their tracks.

"I am absolutely certain that generations from now, we will be able to look back and tell our children that this was the moment when... the rise of the oceans began to slow and our planet began to heal."

As if on cue, however, in mid-December 2008 the University of Colorado published satellite data proving the rise of the oceans has indeed slowed, quite rapidly since 2005,[270] coinciding with the time George W. Bush was sworn in for his second term.

Additionally, after a decade of hype about global warming, new data from beneath the oceans is casting doubt on any long term effect at all:

"Some 3,000 scientific robots that are plying the ocean have sent home a puzzling message," reported National Public Radio in 2008. "These diving instruments suggest that the oceans have not warmed up at all over the past four or five years."[271]

---

270 http://wattsupwiththat.com/2008/12/05/satellite-derived-sea-level-updated-trend-has-been-shrinking-since-2005/

271 "The Mystery of Global Warming's Missing Heat", Richard Harris, NPR, 19 March 2008, http://www.npr.org/templates/story/story.php?storyId=88520025

Er, what was that? No warming?

"That could mean global warming has taken a breather. Or it could mean scientists aren't quite understanding what their robots are telling them."

The message from the robots is loud and clear, however. Despite the ongoing claims about the last decade being the warmest on record, it is not being reflected in the oceanic sensors, where it should be.

"Josh Willis at NASA's Jet Propulsion Laboratory says the oceans are what really matter when it comes to global warming," continued the NPR report.

"In fact, 80 percent to 90 percent of global warming involves heating up ocean waters. They hold much more heat than the atmosphere can. So Willis has been studying the ocean with a fleet of robotic instruments called the Argo system. The buoys can dive 3,000 feet down and measure ocean temperature. *Since the system was fully deployed in 2003, it has recorded no warming of the global oceans.* [My emphasis]

"There has been a very slight cooling, but not anything really significant," Willis told NPR.

It's yet another case of the real data not matching the computer projections that excite climate scientists. If genuine global warming is happening, it must leave real unambiguous evidence, otherwise we may as well be invoking "fairies" as the cause.

But what was it NASA's climate guru James Hansen has said about sea level rises?[272]

"As an example, let us say that ice sheet melting adds 1 centimetre to sea level for the decade 2005 to 2015, and that this doubles each decade until the West Antarctic ice sheet is largely depleted. This would yield a rise in sea level of more than 5 metres by 2095," intoned Hansen.

But although Hansen started with five, the excitement in his *New Scientist* article built as he threw in a fresh estimate of ten metres a century.

"There is growing evidence that the global warming already under way could bring a comparably rapid rise in sea level. The process

272 "Huge sea level rises are coming, unless we act now", James Hansen, *New Scientist*, 25 July 2007, http://www.newscientist.com/article/mg19526141.600-huge-sea-level-rises-are-coming--unless-we-act-now.html

begins with human-made greenhouse gases, which cause the atmosphere to be more opaque to infrared radiation, thus decreasing radiation of heat to space. As a result, the Earth is gaining more heat than it is losing: currently 0.5 to 1 watts per square metre. This planetary energy imbalance is sufficient to melt ice corresponding to 1 metre of sea level rise per decade, if the extra energy was used entirely for that purpose – and the energy imbalance could double if emissions keep growing.

"I find it almost inconceivable that 'business as usual' climate change will not result in a rise in sea level measured in metres within a century. Am I the only scientist who thinks so?"

Possibly, now, he just might be. According to a new study by the Scripps Institute of Oceanography and Colorado University's Institute of Arctic and Alpine Research,[273] Hansen has more chance of finding a snowball in global warming hell than he does of seeing the sea level rise by five or ten metres.

The study took on board the claims Hansen and other global warming believers have made about melting ice, but then did something that hasn't been done before – they checked to see how fast the glaciers would actually have to melt to achieve a multi-metre rise in sea levels, and whether this was actually possible.

"Despite projections by some scientists of global seas rising by 20 feet or more by the end of this century as a result of warming, a new University of Colorado at Boulder study concludes that global sea rise of much more than 6 feet is a near physical impossibility," begins a bulletin from the University.

"For Greenland alone to raise sea level by two metres by 2100, *all* of the outlet glaciers involved would need to move more than three times faster than the fastest outlet glaciers ever observed, or more than 70 times faster than they presently move," one of the Colorado team, Tad Pfeffer, explained.[274] "And they would have to start moving that fast today, not 10 years from now. It is a simple argument with no fancy physics."

The reason it's simple is because glaciers don't just melt overnight, no matter how hot it's been. The ice not only has to turn to water, but the water has to find its way out. The world's fastest outlet

273 *Science*, 5 September 2008

274 "Global Sea-Rise Levels By 2100 May Be Lower Than Some Predict, Says New CU-Boulder Study" University of Colorado news release, 4 September, 2008

glaciers, incidentally, are moving at a hefty 12 kilometres a year, so Greenland's glaciers, all of them, would have to start whizzing out of their valley beds at speeds approaching one kilometre a week, and continue moving that fast, day and night, for the rest of this century, just to achieve Hansen and Gore's fantasy of a two metre sea level increase.

Logistically then, the chances of a big sea level rise are almost non-existent, even with global warming.

"The gist of the study is that very simple, physical considerations show that some of the very large predictions of sea level rise are unlikely, because there is simply no way to move the ice or the water into the ocean that fast," said Pfeffer.

Their study, published in *Science*, indicates the NASA GISS chief's claim is possible, but only on paper, and arguably even then only after a long night of weed-smoking.

"We consider glaciological conditions required for large sea level rise to occur by 2100 and conclude increases of 2 metres are physically untenable. We find that a total sea level rise of about 2 metres by 2100 could occur under physically possible glaciological conditions but only if all variables are quickly accelerated to extremely high limits."

Like others, Pfeffer says the research is a warning to political leaders to make sure they're certain of the science before they start spending money like water.

"If we plan for 6 feet and only get 2 feet, for example, or visa versa, we could spend billions of dollars of resources solving the wrong problems."

The irony here is that global warming's high priests are being unintentionally skewered by colleagues working in the climate research field.

As if on cue, the journal *Science* has just published a major study suggesting nearly three quarters of the warming recorded in the Atlantic Ocean since 1980 has nothing to do with climate change, and everything to do with dust storms in Africa spreading out over the Atlantic like a cloud and keeping the sea cooler. An analysis of the past 26 years shows years with fewer dust storms and volcanic eruptions invariably resulted in higher sea temperatures.

"A lot of this upward trend in the long-term pattern can be explained just by dust storms and volcanoes," lead researcher Amato

Evan told *ScienceDaily*. "About 70 percent of it is just being forced by the combination of dust and volcanoes, and about a quarter of it is just from the dust storms themselves."

Only 30% of the ocean warming can be laid at the door of other factors, including climate change or solar forcing.

Dustier or more volcanic years mean fewer hurricanes because of the lower ocean temperatures (hurricanes need warm water to feed on). Conversely clearer years mean a warmer Atlantic and rising sea levels because of thermal expansion of the water, as well as warmer currents affecting Arctic sea ice.

"Volcanoes and dust storms are really important if you want to understand changes over long periods of time," Evan reported. "If they have a huge effect on ocean temperature, they're likely going to have a huge effect on hurricane variability as well."

So much for the hurricane images in Al Gore's movie.

There's another fairly major twist to this case of rising sea levels, however, and it spills over into the CO_2 emissions area as well, and it's this: undersea volcanoes. If one land-based volcano like Mt Pinatubo can have an impact on global climate in 1992, and one volcano can churn out more CO_2 in a single eruption than humans can in a year, imagine the impact of millions of undersea volcanic vents, quietly chuffing away far below the surface like a stove element.

Midway through 2008, taking advantage of reduced Arctic ice cover, scientists on a research ship found huge volcanic activity boiling the water near Greenland.

"The arctic seabed is as explosive geologically as it is politically, judged by the 'fountains' of gas and molten lava that have been blasting out of underwater volcanoes near the North Pole," reported *Canwest News*.[275]

" 'Explosive volatile discharge has clearly been a widespread, and ongoing, process,' according to an international team that sent unmanned probes to the strange, fiery world beneath the Arctic Ice.

"The team returned with images and data showing the red-hot magma has been rising from deep inside the Earth and has blown the tops of dozens of submarine volcanoes, four kilometres below the ice.

" 'Jets or fountains of material were probably blasted one, maybe

275 "Arctic seabed afire with lava spewing volcanoes", *Canwest News*, 25 June 2008

even two, kilometres up into the water,' says geologist Robert Sohn, of the Woods Hole Oceanographic Institute, who led the expedition.

"The 1,800-kilometre-long ridge, which cuts across the Arctic from Greenland to Siberia, is in international waters. It is one of the planet's 'spreading' ridges where molten rock rises up form inside the Earth, creating new crust. In the valley where two crustal plates are coming apart, which is about 12 kilometres across, they found dozens of distinctive, flat-topped volcanoes that appear to have erupted in 1999, producing the layer of dark, smoky, volcanic glass on the seabed."

I know I've said it before, but a seven year old could make the link between volcanoes four km underneath the sea ice erupting in Pompeii-sized explosions since 1999, and potential increased melting of the ice due to marginally-warmer seas in the area. Heat, after all, rises, and bubbles of hot gas under the ice would have an impact.[276]

More to the point, the subsurface warming of the ocean worldwide caused by volcanic eruptions would also account for at least some thermal expansion and therefore a rise in sea levels, albeit not major.

"We found more hydrothermal activity on this cruise than in 20 years of exploration on the mid-Atlantic Ridge," Charles Langmuir of Columbia University's Lamont-Doherty Earth Observatory was reported as telling a news conference.[277]

Another Woods Hole oceanographer, Henry J.B. Dick, says some of the volcanoes were massive structures like those on land, and chunks of the Earth's mantle, "mile-high walls of rock" had pushed through the sea bed, clawing in vain at the sky just two thousand metres above.[278]

The contribution of volcanic activity to melting polar ice sheets

---

276 Some climate scientists have argued that heat from undersea volcanoes has negligible impact on ice melt because the heat is 'contained' within stratified water layers and therefore doesn't reach the surface. There's considerable scepticism about this claim, because it then raises questions about whether other heat 'trapped' in the oceans can reach the surface, as claimed by climate computer models. Additionally, there's the amount of heat involved. Measurements of one volcanic vent in the ocean found it was pumping 100 megawatts of energy into the ocean per hour – the equivalent of a medium sized power station (Sellafield Nuclear Plant in the UK is rated at 200MW). See Little et al., 1987 *Journal of Geophysical Research* 92 (B3), 2587–2596.; Rona and Trivett, 1992; Ginster et al., 1994. The average warming effect of a volcanic vent has been measured at 4000 watts per square metre. See Pruis et al, "Turbulent heat flux in the deep ocean above diffuse hydrothermal vents", http://www.nwra.com/resumes/pruis/Pruisetal_deepsea_2004.pdf

277 "Undersea vents bubble under Arctic ocean", *USA Today*, 28 November 2008, http://www.usatoday.com/news/science/cold-science/2001-11-28-healy-science.htm

278 Ibid

is being taken seriously by some scientists at least. One study last year by the British Antarctic Survey centred on Pine Island Glacier on the West Antarctic ice sheet.

BBC News reported concerns that Pine Island is the canary down Antarctica's mineshaft.

"There is good reason to be concerned. Satellite measurements have shown that three huge glaciers here have been speeding up for more than a decade.[279]

"The biggest of the glaciers, the Pine Island Glacier, is causing the most concern.

"Julian Scott has just returned from there. He told the BBC: 'This is a very important glacier; it's putting more ice into the sea than any other glacier in Antarctica. It's a couple of kilometres thick, its 30km wide and it's moving at 3.5km per year, so it's putting a lot of ice into the ocean.'[280]

"It is a very remote and inhospitable region. It was visited briefly in 1961 by American scientists but no one had returned until this season when Julian Scott and Rob Bingham and colleagues from the British Antarctic survey spent 97 days camping on the flat, white ice."

But, as a study published in *Nature Geosciences* records,[281] global warming does not seem to be the culprit. Instead, the massive glacier is sitting close to an active volcano *buried under the ice*, providing plenty of incentive for the ice to scoot towards the coast.

That particular volcano is on land, albeit under an ice sheet, but the area is active and it's quite possible other vents exist under the sea ice shelves, just like they do in the Arctic.

An ABC Australia TV crew got a chance to see an active underwater volcano that's poked through the ocean surface, at White Island off the New Zealand coast.

---

279 "Antarctic glaciers surge to ocean", BBC, 24 February 2008, http://news.bbc.co.uk/2/hi/science/nature/7261171.stm

280 Whilst true that Pine Island Glacier, or 'PIG' as it is affectionally known, is putting a lot of ice into the ocean, and that measurements of the glacier itself show it is not being replenished to the same extent by fresh snow, this is too narrow a construction to support the idea that all of its ice will be a net gain to the sea, and hence sea levels. Snow generated by evaporation caused by warm currents lapping the glaciers will not necessarily re-dump in the precise spot it is being lost from. It can dump anywhere across the Antarctic. The only way to judge sea level risk from Antarctica is via a 'total ice mass' approach, which as we've shown recorded some sizeable gains. In other words, doomsday stories centred on how much ice PIG is losing are far too simplistic,

281 "A recent volcanic eruption beneath the West Antarctic ice sheet" by Hugh Corr and David Vaughan, *Nature Geosciences*, February 2008, also referenced at http://www.sciencedaily.com/releases/2008/01/080120160720.htm

Reporter Dr Jonica Newby describes the scene:[282]

"Narration: Two kilometres below our feet is a hot boiling Hades of magma – and all around us are what are known as vents – gaping holes through which superheated steam and deadly hydrogen sulfide gas shoot free of the earth's crust.

Dr de Ronde: Keep coming, can you feel the heat?

Jonica: Oh yeah. Boy, that's hot.

Dr de Ronde: That's superheated steam.

Jonica: That's hot.

Dr de Ronde: That's great! Come stand there, you'll be right.

Now you can feel what it's really like. Otherwise it's just too touristy.

Jonica: I think he's trying to kill me.

Narration: Dr De Ronde's extreme tour isn't over though until we reach the lip of the main crater.

Dr de Ronde: Check this out.

Jonica: Wow! My God.

Dr de Ronde: It's amazing, eh;

Jonica: It's incredible.

Dr de Ronde: Isn't that the most amazing thing you've ever seen?

Jonica: It's like a giant cauldron, I've never seen anything like it.

Dr de Ronde: It's mind boggling isn't it?

Narration: While we've known about these volcanic systems for a long time, it's only very recently we realised how far they extend underwater.

In the late 80's, New Zealand scientists discovered the volcanoes of the Kermadec arc ... a 1300 km long stretch of continental collision ... where the vast Pacific plate grinds beneath the Australasian plate ... releasing all that pent up force as heat ... in 33 spectacular submarine arc volcanoes.

And the jewel in the crown is the massive Brothers Volcano.

It's the reason I've been brought here. White island is the closest thing you can see to Brothers on land.

Dr de Ronde: Imagine this volcano with all its vents and all its glory, make it 3 times bigger, stick it on the bottom of the sea, stick 2 k's of water on top and that's Brothers.[283]

There are thousands of volcanoes, and millions of vents, under the

282 "Undersea vents in the Pacific Ring of Fire", *ABC Australia*, 14 February 2008 http://www.abc.net.au/catalyst/stories/2008/02/14/2160790.htm?site=catalyst

283 For more detail of the investigation of the Brothers volcano, NOAA has assembled video and photo footage here: http://oceanexplorer.noaa.gov/explorations/05fire/logs/may09/may09.html

ocean. All of them piping emissions at temperatures of 300°C (576°F) or higher into the oceans around them. It's possibly no coincidence that satellite maps of areas where sea levels have increased due to thermal expansion show the greatest sea level rises over known volcanic zones like the Pacific's Ring of Fire.

There's also the not unrelated issue of underwater volcanoes pushing out CO_2. Off the coast of Hawaii in 1997, unbeknown to most of us, the underwater volcano Loihi blew its stack in what scientists called "A Mount St. Helen-size event":[284]

"Pele's Dome, an area on the southern rim of the volcano that previously had been considered very stable, has simply vanished into a giant pit, which we have named the Pele's Pit Crater," Hawaiian volcanologist Alexander Malahoff explained. "What we learn from this event will have profound implications for virtually everything we now know about undersea volcanism–including the effects of volcanic carbon dioxide emissions on climate, the possible generation of tsunamis that could strike coastal areas, and the impacts on the microscopic organisms that live in and near sea floor vents.

"Loihi is also a big carbon dioxide polluter," Malahoff added.

It would be fair to say then, that geoscience is still in its infancy in terms of working out all the factors that influence climate and life. And this is one of the fault lines within science itself, between those who argue global warming should be tackled at all costs, and the growing numbers of scientists defecting to a view that acknowledges we've been warming, but increasingly sees it as naturally driven and therefore impossible to stop.

We're seeing this dispute manifest in a growing scientific backlash against the UN's climate change theories, and that's where we now turn.

---

284 "Collapsed undersea volcano gives view of island's birth", NOAA feature article, http://www.oar.noaa.gov/spotlite/archive/spot_nurp.html

Chapter Fourteen

Unsettled Scientists

"The scientists are virtually screaming from the rooftops now. The debate is over! There's no longer any debate in the scientific community about this."

– Al Gore, former "next President of the United States"

In early 2009, a fascinating poll was published claiming to be one of the most detailed surveys of climate researchers:[285]

"A group of 3,146 earth scientists surveyed around the world overwhelmingly agree that in the past 200-plus years, mean global temperatures have been rising, and that human activity is a significant contributing factor in changing mean global temperatures.

"Peter Doran, University of Illinois at Chicago associate professor of earth and environmental sciences, along with former graduate student Maggie Kendall Zimmerman, conducted the survey late last year.

"The findings appear today in the publication *Eos*, Transactions, American Geophysical Union.

"In trying to overcome criticism of earlier attempts to gauge the view of earth scientists on global warming and the human impact factor, Doran and Kendall Zimmerman sought the opinion of the most complete list of earth scientists they could find, contacting more than 10,200 experts around the world listed in the 2007 edition of the American Geological Institute's Directory of Geoscience Departments.

"Two questions were key: have mean global temperatures risen compared to pre-1800s levels, and has human activity been a significant factor in changing mean global temperatures.

285 "Survey: Scientists agree human induced global warming is real", http://www.eurekalert.org/pub_releases/2009-01/uoia-ssa011609.php

"About 90 percent of the scientists agreed with the first question and 82 percent the second."

On the face of it, you'd think this was good news for Hansen and Mann, but look again.

Of more than 10,000 scientists contacted, only 3,146 replied. Of those 3,146, 18% (or 566 scientists directly working in the field) don't believe that "human activity has been a significant factor in changing mean global temperatures".

These, according to the survey organisers (global warming believers), are "experts" in the field. If nearly 20% of those at the coal face who responded don't believe in human-caused global warming, it's hard to see how Al Gore and Bill McKibben can claim "the science is settled".

But it gets worse for global warming believers, because that last survey was the "good news". Now here's the bad. As I mentioned at the start of this book, the list of dissenting scientists who no longer believe in anthropogenic global warming has ballooned out to more than 31,000 – nearly 10,000 of those working at Ph.D. level.[286]

In a speech to the US Senate two years ago, Senator James Inhofe questioned news media claims of a "global warming consensus" among scientists.[287]

"Key components of the manufactured 'consensus' fade under scrutiny," he said.

"We often hear how the National Academy of Sciences (NAS) and the American Meteorological Society (AMS) issued statements endorsing the so-called 'consensus' view that man is driving global warming. But what you don't hear is that both the NAS and AMS never allowed member scientists to directly vote on these climate statements.

"Essentially, only two dozen or so members on the governing boards of these institutions produced the 'consensus' statements. It appears that the governing boards of these organizations caved in to pressure from those promoting the politically correct view of UN and Gore-inspired science. The Canadian Academy of Sciences reportedly endorsed a 'consensus' global warming statement that was never even approved by its governing board.

286 http://www.petitionproject.org/index.html

287 http://epw.senate.gov/public/index.cfm?FuseAction=Minority.Blogs&ContentRecord_id=595F6F41-802A-23AD-4BC4-B364B623ADA3

"Rank-and-file scientists are now openly rebelling. James Spann, a certified meteorologist with the AMS, openly defied the organization when he said in January that he does 'not know of a single TV meteorologist who buys into the man-made global warming hype.' In February a panel of meteorologists expressed unanimous climate scepticism, and one panellist estimated that 95% of his profession rejects global warming fears.

"In August 2007, a comprehensive survey of peer-reviewed scientific literature from 2004-2007 revealed 'Less Than Half of all Published Scientists Endorse Global Warming Theory'.

" 'Of 539 total papers on climate change, only 38 (7%) gave an explicit endorsement of the consensus. If one considers 'implicit' endorsement (accepting the consensus without explicit statement), the figure rises to 45%. However, while only 32 papers (6%) reject the consensus outright, the largest category (48%) are neutral papers, refusing to either accept or reject the hypothesis. This is no consensus,' according to an August 29, 2007 article in *Daily Tech*," said Senator Inhofe.

It's a standard propaganda trick – capture the mouthpieces of an organisation in order to boost the apparent legitimacy of the project. By getting major scientific organisations to endorse global warming as real, the global warming believers built a 'consensus' to convince the public, the media and political leaders with. But the consensus has no more substance than a Hollywood film set; all façade and nothing else. What real legitimacy do scientific associations have to endorse global warming theory, if they have not actually permitted their members the right to vote on the claim?

But the tide hasn't just turned on the sea level issue. Since Inhofe's 2007 speech, growing numbers of scientists have found their voices again. A US Senate minority report released at the end of 2008 contains the views of 650 scientists rubbishing the global warming believers, and those in the news media who continue to take the global warming hysteria seriously. What do they have to say? Let's find out:

"I am a sceptic...Global warming has become a new religion." – Nobel Prize Winner for Physics, Ivar Giaever.

"Since I am no longer affiliated with any organization nor receiving any funding, I can speak quite frankly....As a scientist I remain sceptical...The main basis of the claim that man's release of greenhouse gases is the cause of the warming is based almost entirely upon climate models.

We all know the frailty of models concerning the air-surface system." – Atmospheric Scientist Dr. Joanne Simpson, the first woman in the world to receive a PhD in meteorology, and formerly of NASA, who has authored more than 190 studies and has been called "among the most preeminent scientists of the last 100 years."

[Warming fears are the] "worst scientific scandal in the history...When people come to know what the truth is, they will feel deceived by science and scientists." – UN IPCC Japanese scientist Dr. Kiminori Itoh, an award-winning PhD environmental physical chemist.

"The IPCC has actually become a closed circuit; it doesn't listen to others. It doesn't have open minds... I am really amazed that the Nobel Peace Prize has been given on scientifically incorrect conclusions by people who are not geologists." – Indian geologist Dr. Arun D. Ahluwalia at Punjab University and a board member of the UN-supported International Year of the Planet.

"So far, real measurements give no ground for concern about a catastrophic future warming." – Scientist Dr. Jarl R. Ahlbeck, a chemical engineer at Abo Akademi University in Finland, author of 200 scientific publications and former Greenpeace member.

"Anyone who claims that the debate is over and the conclusions are firm has a fundamentally unscientific approach to one of the most momentous issues of our time." – Solar physicist Dr. Pal Brekke, senior advisor to the Norwegian Space Centre in Oslo. Brekke has published more than 40 peer-reviewed scientific articles on the sun and solar interaction with the Earth.

"The models and forecasts of the UN IPCC "are incorrect because they only are based on mathematical models and presented results at scenarios that do not include, for example, solar activity." – Victor Manuel Velasco Herrera, a researcher at the Institute of Geophysics of the National Autonomous University of Mexico

"It is a blatant lie put forth in the media that makes it seem there is only a fringe of scientists who don't buy into anthropogenic global warming." – U.S Government Atmospheric Scientist Stanley B. Goldenberg of the Hurricane Research Division of NOAA.

"Even doubling or tripling the amount of carbon dioxide will virtually have little impact, as water vapour and water condensed on particles as clouds dominate the worldwide scene and always will." – . Geoffrey G. Duffy, a professor in the Department of Chemical and Materials Engineering of the University of Auckland, NZ.

"After reading [UN IPCC chairman] Pachauri's asinine comment [comparing sceptics to] Flat Earthers, it's hard to remain quiet." – Climate statistician Dr. William M. Briggs, who specializes in the statistics of forecast evaluation, serves on the American Meteorological Society's Probability and Statistics Committee and is an Associate Editor of Monthly Weather Review.

"The Kyoto theorists have put the cart before the horse. It is global warming that triggers higher levels of carbon dioxide in the atmosphere, not the other way round…A large number of critical documents submitted at the 1995 U.N. conference in Madrid vanished without a trace. As a result, the discussion was one-sided and heavily biased, and the U.N. declared global warming to be a scientific fact," Andrei Kapitsa, a Russian geographer and Antarctic ice core researcher.

"I am convinced that the current alarm over carbon dioxide is mistaken…Fears about man-made global warming are unwarranted and are not based on good science." – Award Winning Physicist Dr. Will Happer, Professor at the Department of Physics at Princeton University and Former Director of Energy Research at the Department of Energy, who has published over 200 scientific papers, and is a fellow of the American Physical Society, The American Association for the Advancement of Science, and the National Academy of Sciences.

"Nature's regulatory instrument is water vapour: more carbon dioxide leads to less moisture in the air, keeping the overall GHG content in accord with the necessary balance conditions." – Prominent Hungarian Physicist and environmental researcher Dr. Miklós Zágoni reversed his view of man-made warming and is now a sceptic. Zágoni was once Hungary's most outspoken supporter of the Kyoto Protocol.

"For how many years must the planet cool before we begin to understand that the planet is not warming? For how many years must cooling go on?" – Geologist Dr. David Gee, the chairman of the science committee of the 2008 International Geological Congress who has authored 130 plus peer reviewed papers, and is currently at Uppsala University in Sweden.

"Gore prompted me to start delving into the science again and I quickly found myself solidly in the sceptic camp…Climate models can at best be useful for explaining climate changes after the fact." – Meteorologist Hajo Smit of Holland, who reversed his belief in man-made warming to become a sceptic, is a former member of the Dutch UN IPCC committee.

"The quantity of CO_2 we produce is insignificant in terms of the natural circulation between air, water and soil... I am doing a detailed assessment of the UN IPCC reports and the Summaries for Policy Makers, identifying the way in which the Summaries have distorted the science." – South African Nuclear Physicist and Chemical Engineer Dr. Philip Lloyd, a UN IPCC co-coordinating lead author who has authored over 150 refereed publications.

"Many [scientists] are now searching for a way to back out quietly (from promoting warming fears), without having their professional careers ruined." – Atmospheric physicist James A. Peden, formerly of the Space Research and Coordination Center in Pittsburgh.

"All those urging action to curb global warming need to take off the blinkers and give some thought to what we should do if we are facing global cooling instead." – Geophysicist Dr. Phil Chapman, an astronautical engineer and former NASA astronaut, served as staff physicist at MIT (Massachusetts Institute of Technology)

"Creating an ideology pegged to carbon dioxide is a dangerous nonsense...The present alarm on climate change is an instrument of social control, a pretext for major businesses and political battle. It became an ideology, which is concerning." – Environmental Scientist Professor Delgado Domingos of Portugal, the founder of the Numerical Weather Forecast group, has more than 150 published articles.

"CO_2 emissions make absolutely no difference one way or another.... Every scientist knows this, but it doesn't pay to say so...Global warming, as a political vehicle, keeps Europeans in the driver's seat and developing nations walking barefoot." – Dr. Takeda Kunihiko, vice-chancellor of the Institute of Science and Technology Research at Chubu University in Japan.

"The [global warming] scaremongering has its justification in the fact that it is something that generates funds." – Award-winning Paleontologist Dr. Eduardo Tonni, of the Committee for Scientific Research in Buenos Aires and head of the Paleontology Department at the University of La Plata.

"Whatever the weather, it's not being caused by global warming. If anything, the climate may be starting into a cooling period." Atmospheric scientist Dr. Art V. Douglas, former Chair of the Atmospheric Sciences Department at Creighton University in Omaha, Nebraska, and is the author of numerous papers for peer-reviewed publications.

"But there is no falsifiable scientific basis whatever to assert this

warming is caused by human-produced greenhouse gasses because current physical theory is too grossly inadequate to establish any cause at all." – Chemist Dr. Patrick Frank, who has authored more than 50 peer-reviewed articles.

"The 'global warming scare' is being used as a political tool to increase government control over American lives, incomes and decision making. It has no place in the Society's activities." – Award-Winning NASA Astronaut/Geologist and Moonwalker Jack Schmitt who flew on the Apollo 17 mission and formerly of the Norwegian Geological Survey and for the U.S. Geological Survey.

"Earth has cooled since 1998 in defiance of the predictions by the UN-IPCC....The global temperature for 2007 was the coldest in a decade and the coldest of the millennium...which is why 'global warming' is now called 'climate change.'" – Climatologist Dr. Richard Keen of the Department of Atmospheric and Oceanic Sciences at the University of Colorado.

"I have yet to see credible proof of carbon dioxide driving climate change, yet alone man-made CO2 driving it. The atmospheric hot-spot is missing and the ice core data refute this. When will we collectively awake from this deceptive delusion?" – Dr. G LeBlanc Smith, a retired Principal Research Scientist with Australia's CSIRO.

So, is the science really "settled"? More to the point, is it settled enough that it justifies doubling petrol and food prices, and hitting you and your family with tax bills of thousands of dollars each year above what you already pay, yet with no guarantee that your extra taxes will make any dent in global temperatures at all?

In a now-famous December 2007 letter to UN Secretary-General Ban Ki-moon, more than a hundred dissenting scientists, many from within the UN's IPCC, made exactly this point:[288]

"It is not possible to stop climate change, a natural phenomenon that has affected humanity through the ages. Geological, archaeological, oral and written histories all attest to the dramatic challenges posed to past societies from unanticipated changes in temperature, precipitation, winds and other climatic variables. We therefore need to equip nations to become resilient to the full range of these natural phenomena by promoting economic growth and wealth generation.

288 http://www.nationalpost.com/news/story.html?id=164002

"The United Nations Intergovernmental Panel on Climate Change (IPCC) has issued increasingly alarming conclusions about the climatic influences of human-produced carbon dioxide (CO_2), a non-polluting gas that is essential to plant photosynthesis. While we understand the evidence that has led them to view CO_2 emissions as harmful, the IPCC's conclusions are quite inadequate as justification for implementing policies that will markedly diminish future prosperity. In particular, it is not established that it is possible to significantly alter global climate through cuts in human greenhouse gas emissions. On top of which, because attempts to cut emissions will slow development, the current UN approach of CO_2 reduction is likely to increase human suffering from future climate change rather than to decrease it.

"The IPCC Summaries for Policy Makers are the most widely read IPCC reports amongst politicians and non-scientists and are the basis for most climate change policy formulation.

"Yet these Summaries are prepared by a relatively small core writing team with the final drafts approved line-by-line by government representatives. The great majority of IPCC contributors and reviewers, and the tens of thousands of other scientists who are qualified to comment on these matters, are not involved in the preparation of these documents. The summaries therefore cannot properly be represented as a consensus view among experts.

"Contrary to the impression left by the IPCC Summary reports:

*Recent observations of phenomena such as glacial retreats, sea-level rise and the migration of temperature-sensitive species are not evidence for abnormal climate change, for none of these changes has been shown to lie outside the bounds of known natural variability.

*The average rate of warming of 0.1 to 0.2 degrees Celsius per decade recorded by satellites during the late 20th century falls within known natural rates of warming and cooling over the last 10,000 years.

*Leading scientists, including some senior IPCC representatives, acknowledge that today's computer models cannot predict climate. Consistent with this, and despite computer projections of temperature rises, there has been no net global warming since 1998. That the current temperature plateau follows a late 20th-century period of warming is consistent with the continuation today of natural multi-decadal or millennial climate cycling.

"In stark contrast to the often repeated assertion that the science of climate change is "settled," significant new peer-reviewed research has cast even more doubt on the hypothesis of dangerous human-caused global warming. But because IPCC working groups were generally instructed[289] to consider work published only through May, 2005, these important findings are not included in their reports; i.e., the IPCC assessment reports are already materially outdated.

"The UN climate conference in Bali has been planned to take the world along a path of severe CO2 restrictions, ignoring the lessons apparent from the failure of the Kyoto Protocol, the chaotic nature of the European CO2 trading market, and the ineffectiveness of other costly initiatives to curb greenhouse gas emissions. Balanced cost/benefit analyses provide no support for the introduction of global measures to cap and reduce energy consumption for the purpose of restricting CO2 emissions. Furthermore, it is irrational to apply the "precautionary principle" because many scientists recognize that both climatic coolings and warmings are realistic possibilities over the medium-term future.

"The current UN focus on "fighting climate change," as illustrated in the Nov. 27 UN Development Programme's Human Development Report, is distracting governments from adapting to the threat of inevitable natural climate changes, whatever forms they may take.

"National and international planning for such changes is needed, with a focus on helping our most vulnerable citizens adapt to conditions that lie ahead. Attempts to prevent global climate change from occurring are ultimately futile, and constitute a tragic misallocation of resources that would be better spent on humanity's real and pressing problems," concluded the scientists in their letter to the Canadian prime minister.

Realising they were losing the PR battle as sceptics turned up the heat (metaphorically), the global warming believers hit back in March 2009 with what they claimed was the latest, state-of-the-art analysis of global warming science.

It came at a gathering of climate change believers in Copenhagen, dubbed an "emergency summit" ostensibly for the purpose of focusing political attention on what needs to be achieved at the main Copenhagen Treaty meeting this coming December.

289 http://ipcc-wg1.ucar.edu/wg1/docs/wg1_timetable_2006-08-14.pdf

The *Observer's* Robin McKie, who waxed so lyrically but so inaccurately about the opening of the Northwest Passage, was back in full flight with his preview of the mini-Copenhagen conference:[290]

"Scientists will warn this week that rising sea levels, triggered by global warming, pose a far greater danger to the planet than previously estimated. There is now a major risk that many coastal areas around the world will be inundated by the end of the century because Antarctic and Greenland ice sheets are melting faster than previously estimated.

"Low-lying areas including Bangladesh, Florida, the Maldives and the Netherlands face catastrophic flooding, while, in Britain, large areas of the Norfolk Broads and the Thames estuary are likely to disappear by 2100. In addition, cities including London, Hull and Portsmouth will need new flood defences.

"It is now clear that there are going to be massive flooding disasters around the globe," said Dr David Vaughan, of the British Antarctic Survey. "Populations are shifting to the coast, which means that more and more people are going to be threatened by sea-level rises."

Sounds horrific! There must be some devastating new evidence. Indeed, McKie claimed there was. He explained that the 2007 IPCC report had "underestimated" the extent of sea level rise because not enough was known about the rate of Greenland and Antarctic icecap melt.

"However, we are now getting a much better idea of what is going on in Greenland and Antarctica and can make much more accurate forecasts about ice-sheet melting and its contribution to sea-level rises," British glaciologist David Vaughan told McKie.

The *Observer's* 'science' editor explained that scientists have been watching Greenland and Antarctic sea ice "dwindle and disappear", thanks to satellite imagery. Well that's a dubious claim from the start, but it's not the main point. McKie warns that if sea ice melts, then land-based ice sheets are no longer held in place.

"Without sea ice to prop them up, the land sheets tip into the water and disintegrate at increasing rates, a phenomenon that is now being studied in detail by researchers," McKie wrote.

"It is becoming increasingly apparent from our studies of Greenland and Antarctica that changes to sea ice are being transmitted

290 "Scientists to issue stark warning over dramatic new sea level figures", Robin McKie, *The Observer*, 8 March 2009, http://www.guardian.co.uk/science/2009/mar/08/climate-change-flooding

into the hearts of the land-ice sheets in a remarkably short time," added Vaughan.

As a result, continued McKie, those land sheets are breaking up faster and far more melt water is being added to the oceans than was previously expected.

"These revisions suggest sea-level rises could easily top a metre by 2100 – a figure that is backed by the US Geological Survey, which this year warned that they could reach as much as 1.5 metres.

"In addition, in September, a team led by Tad Pfeffer at the University of Colorado at Boulder published calculations using conservative, medium and extreme glaciological assumptions for sea-level rise expected from Greenland, Antarctica and the world's smaller glaciers and ice caps. They concluded that the most plausible scenario, when factoring in thermal expansion due to warming waters, will lead to a total sea level rise of one to two metres by 2100."

Again, another very dubious claim. We've quoted Tad Pfeffer's study in this book, where he said at most the sea level will rise two metres, but only if *all* the outlet glaciers in Greenland suddenly *started moving 70 times faster than they already are* into the ocean, and only if such speed began last year and lasted for the rest of this century. Which hasn't happened.

You can see then, that journalists who believe in global warming are happily spouting whatever a believer tells them.

So let's see, regardless of the hype and panic being pitched at mini-Copenhagen, what the latest scientific studies actually say about Greenland's ice melt.

According to a University of Sheffield study published a year ago, a string of warm summers several years ago provoked "the most extreme Greenland ice melting in 50 years."[291]

OK, so we know it's melting faster than ever (or it was two years ago when it was warmer), but how fast is that exactly?

Well, according to late 2008 data from the GRACE satellite tracking system, regarded as "ideal" for keeping tabs on the Greenland icecap, the ice melted so fast between 2003 and 2008 that "it is now estimated that Greenland is accountable for a half millimetre rise in the global sea level per year," reported scientists from the Center

291 "Record Warm Summers Cause Extreme Ice Melt In Greenland", *ScienceDaily* 16 Jan 2008, http://www.sciencedaily.com/releases/2008/01/080115102706.htm

for Space Research, in the journal *Geophysical Research Letters*.[292]

Earth to Copenhagen: take a note, half a millimetre a year equates to 50 millimetres a century, or two inches of sea level increase, which could pose a threat to garden gnomes on the English coast, but that's about all. What's this one to two metres of sea level increase you are rabbitting on about?

"However," warned the scientists, "it is not yet clear whether the ice will continue to melt at this rate during the next few years, as ice loss varies greatly from summer to summer."

Varies greatly? I didn't read that in the Copenhagen coverage.

"Long term observations are needed to compile a reliable estimate of Greenland's contribution to the rising sea level during the next century," reported the scientists. "Satellite data of this kind are ideal for measuring areas such as Greenland, where the extreme conditions make local measurements very difficult."

If that's not enough to convince you that the mini-Copenhagen conference blatantly lied to journalists when organisers decreed Greenland ice melt had almost become "irreversible", try this just-published paper in *Science*: "Ice loss in Greenland has had some climatologists speculating that global warming might have brought on a scary new regime of wildly heightened ice loss and an ever-faster rise in sea level. But glaciologists reported at the American Geophysical Union meeting that Greenland ice's Armageddon has come to an end."[293]

Come to an end? You heard it right. It's the stunning admission in a major scientific journal published six weeks *before* the alarmist one-metre sea-level rise claims of mini-Copenhagen. You'd think a report like that, signalling an end to massive Greenland ice melt, would have made world headlines. The journalists must have been asleep that day.

"So much for Greenland ice's Armageddon," reported Richard Kerr in the *Science* paper, describing the results of the Geophysical Union conference. " 'It has come to an end,' glaciologist Tavi Murray of Swansea University in the United Kingdom said during a session at the meeting. 'There seems to have been a synchronous switch-off' of the speed-up, she said. Nearly everywhere around

292 "GRACE observes small-scale mass loss in Greenland," Wouters B et al, *Geophys. Res. Lett.*, October 2008 DOI: 10.1029/2008/GL034816

293 "Galloping Glaciers of Greenland Have Reined Themselves In", Richard A. Kerr, *Science* 23 January 2009: Vol. 323. no. 5913, p. 458 DOI: 10.1126/science.323.5913.458a

southeast Greenland, outlet glacier flows have returned to the levels of 2000. An increasingly warmer climate will no doubt eat away at the Greenland ice sheet for centuries, glaciologists say, but no one should be extrapolating the ice's recent wild behaviour into the future."

In other words, all that Greenland melt alarm we've been subjected to turns out to have been nothing more than a passing warm spell, making claims of a one metre sea level hike this century even more impossible to meet than they already were.

No danger then of Greenland's icecap inundating "Bangladesh, Florida, the Maldives and the Netherlands", with the Thames estuary thrown in for good measure?

"The speed-up [of Greenland ice melt] has stopped across the region," confirmed the UK Met Office Hadley Centre's Vicky Pope in February this year.[294]

Fundamental question: why did the organisers of mini-Copenhagen lie, and why didn't scientists issue news releases denouncing Copenhagen's false claims? Although there are numerous references in Google News archives to Tavi Murray warning in earlier years about rapid Greenland melting, not one newspaper anywhere in the world appears to have picked up her announcement that Greenland's meltzkrieg has "come to an end".[295]

The answer is probably because the purpose of mini-Copenhagen was political: to scare world leaders into believing global warming was steaming ahead at record pace requiring drastic and decisive action at full-Copenhagen this coming December. Introducing those most recent inconvenient scientific studies that blow global warming out of the water would not serve the purpose of mini-Copenhagen organisers.

And let's face it, there were plenty of inconvenient facts waiting to get in their way. The Hadley Centre's Vicky Pope, in her *Guardian* opinion article, also denounced bogus claims about Arctic ice melt in general:

294 "Scientists must rein in misleading climate change claims," Vicky Pope, *The Guardian*, 11 February 2009 http://www.guardian.co.uk/environment/2009/feb/11/climate-change-science-pope

295 Some global warming believers cling desperately to a claim that ice loss in Greenland hit record levels in 2007 and 2008, based on a study of photos taken by satellites. Apart from being imprecise, it doesn't take account of how much is being added back by snowfall. It's also worth remembering that there's a certain time lag built in between the warming that causes the melt, and the actual melt happening. It's likely some of this melt lags the warming by years or possibly decades, depending on which glaciers or parts of the larger ice sheet are involved.

"Recent headlines have proclaimed that Arctic summer sea ice has decreased so much in the past few years that it has reached a tipping point and will disappear very quickly. The truth is that there is little evidence to support this. Indeed, the record-breaking losses in the past couple of years *could easily be due to natural fluctuations in the weather,* [my emphasis] with summer sea ice increasing again over the next few years."

Again, you'd think comments like that from the chief climate advisor at one of the temples of global warming belief, the UK Met Office Hadley Centre, would be carried worldwide. But not a single news organisation other than the *Guardian* appears to have run it, according to a search of Google News archives.

Proof, if ever it was needed, that media coverage is overwhelmingly biased towards supporting claims of catastrophic global warming.

One of the other scare theories proposed at mini-Copenhagen is that surface meltwater in Greenland is falling through holes in the ice and creating slippery rivers between the bottom of the glaciers and the bedrock, essentially scooping ice out to sea like a child on a water slide.

So, if that's the spin at Copenhagen for the media's benefit, what are the latest scientific studies actually saying about meltwater runoff?

A University of Washington/Woods Hole Oceanographic Institution joint study last year established that while spectacular, meltwater may be largely insignificant.[296]

"New findings indicate that while surface melt plays a substantial role in ice sheet dynamics, it may not produce large instabilities leading to sea level rise," reported glaciologist Ian Joughin, who added that a year's worth of data showed surface meltwater was only responsible for a tiny percentage of the movement of six outlet glaciers under surveillance.

Their study, which used the RADARSAT satellite and ground-based GPS sensors, described meltwater as "inconsequential" when it comes to making outlet glaciers move, although it does help lubricate the large ice sheets. However, even then their study has found that Greenland continues to be dynamic – even though ice is falling into the sea and melting, most of this "is balanced by

296 "Greenland ice may not be headed down too slippery a slope", *ScienceDaily*, 20 April 2008, http://www.sciencedaily.com/releases/2008/04/080417142507.htm

snowfall each winter". Again, the overall sea-level contribution of Greenland, which has been melting at its fastest rate in 50 years, is pinpointed at between "one to two inches per century".

The meltwater can be quite stunning as it drains off the ice however. It forms on top of the ice sheet, gathering in lakes and puddles, and generally soaks through the ice to the rock below. One lake being surveyed was nearly four kilometres wide and 12 metres (40 feet) deep. After spending time installing monitoring equipment, the researchers were shocked, in a moment reminiscent of the fate befalling the cartoon 'squirrel rat' in the movie "Ice Age", to discover a massive crack opening up through which the entire lake "drained in 90 minutes with a fury comparable to Niagara Falls".

According to *ScienceDaily*, the research team have resolved not to flounder around on ice sheet lakes in their 3 metre runabout boat with a 5hp engine, ever again, lest they too be sucked down the plughole.

You'd think with such vast volumes of meltwater going through that it would indeed create those slippery rivers they were hyping at mini-Copenhagen. But apparently not.

The lake drainage just described briefly lifted the ice sheet in that area by six metres (20 feet), and doubled the speed of the ice sheet's progress across the bedrock for a short period of time.[297] But not for long.

Instead, the water carves out its own escape route rather than lifting the entire sheet for too long, and in most cases the water actually refreezes underneath, helping the ice sheet to grip the bedrock again.

"Scientists from around the world are closely monitoring the Greenland ice sheet, as accelerated glacial melting is believed to cause rising sea level," reported *ScienceDaily* last year.[298]

"The theory is that increased volumes of meltwater accelerate the movement of ice to warmer low-lying areas and, consequently, even more intensified glacial melting. Utrecht University researchers, however, insist that this is not how the process actually works long term.[299]

297 "Ice sheet plumbing system found", *ScienceDaily*, 18 April 2008, http://www.sciencedaily.com/releases/2008/04/080417142503.htm

298 "Intensified Ice Sheet Movements Do Not Affect Rising Sea Levels", *ScienceDaily*, 11 July 2008, http://www.sciencedaily.com/releases/2008/07/080708093615.htm

299 "Large and Rapid Melt-induced Velocity Changes in the Ablation Zone of the Greenland Ice Sheet", R S W van de Wal et al, *Science*, 2008, 321 (5885): 111 DOI: 10.1126/science.1158540

"Since the early 1990s, Utrecht University scientists have tracked the movement of the West Greenland ice sheet using GPS measurements. During warmer weather, the ice appears to move – over the course of a few days – as much as four times faster, because the meltwater acts as a lubricant."

However, as the water refreezes or carves out an escape channel through the gravel, the ice sheet stops moving.

"Ultimately," reports the study, "this is not a cause of accelerated sea level rise."

It's getting very hard to see just where this one to 1.5 metre sea level rise before 2100 is going to come from.

Chapter Fifteen

Exposing The 'Icons' Of Global Warming

"Brothel owners in Bulgaria are blaming global warming for staff shortages"

– News, 26 March 2007

And so here we are, two thirds of the way through the book and with most of the science now knocked on the head; it's worth recapping some of the main points on this fascinating journey and reviewing what it means for ordinary people, rather than scientists and environmentalists. I began the book as a climate change sceptic, knowing a little of what appears within these pages but unaware of most of it. My scepticism on global warming wasn't strongly held; I suspect like many people I could have described it as a 'nagging doubt' or a growing disbelief that something didn't quite add up. A decade ago, I was a global warming believer, but that's changed over time. As a magazine editor I have entertained articles by others on the subject expressing their own doubts, and contributed the occasional column or editorial scoffing at some of the wilder claims or highlighting interesting alternative explanations, but this is my first comprehensive effort to fully research global warming.

Listening to commentators here and there, stumbling across the occasional article, you get a feel for the lie of the land but not necessarily an appreciation of the hard evidence. In researching and writing *Air Con*, I've gained that appreciation for the hard evidence.

Global warming believers might choose to doubt me on this, but too bad: if the evidence had convinced me that catastrophic anthropogenic global warming is an odds-on chance, I'd have written this book with a statement saying so. I don't have a particular dog in the ring, and I could probably sell more books preaching to the choir than running against the tide.

But as an investigative journalist, I'm bound to say I no longer believe there's a hope of catastrophic global warming caused by humans. I say this not just because of the data, but because of the dishonest misrepresentations of the scientific data to deliberately scare the public and politicians.

One of the things we're trained to look for in my branch of the profession is suspicious behaviour. If data is faked or twisted, and can be proven so, then the credibility of those responsible is in doubt. In something like global warming, which should be based entirely on objective scientific methods and empirical evidence, it is even more disturbing that the global warming industry has effectively lied to the public, from Al Gore down.

To any trained investigator, police, FBI or otherwise, that's a red flag with bells on. If the science is right, and the claims of global warming are true, then the data doesn't need twisting.

It's as simple as that. It actually doesn't matter at this point whether the global warming sceptics have been lying through their teeth (I've found no evidence of that in the data highlighted in this book), because it's the IPCC and their hangers on who are making the positive claim: anthropogenic global warming is happening and the public must pay for it.

If the side making the positive claim has to fake data to shore up its position, it doesn't say much for its position or the integrity of its claim. In a court case, if the prosecution was found lying during the presentation of its evidence, the judge would be entirely within their right to throw out the prosecution case without even hearing the defence argument. And occasionally, that happens.

Even the mealy-mouthed decision to change the phrase "global warming" to "climate change" by the United Nations and its client scientists is dishonest. The central premise of their theory and alarmism is that Planet Earth's temperature is rising, and humans are causing it. "Climate change" is meaningless. Climate changes all the time. Naturally, seasonally, decadally, centennially – even in multi-thousand year cycles. By calling it "climate change" the UN is trying to turn any weather event to its propaganda advantage.

Has the global warming industry made a compelling case that human-induced warming is occurring to any significant level? No, not even close.

I'm staggered that some journalists, particularly in the world's major TV networks, have allowed this farce to continue unchallenged as long as it has.

If I've appeared overly harsh so far on politicians and journalists, forgive me. I've worked as a media advisor to a senior Labour cabinet minister[300] so I've seen politics firsthand, and of course my bread and butter is journalism. I feel I've earned the right to administer the occasional rebuke. Having said that, both the politics and media professions entail members being jacks of all trades and masters of none. In journalism's case I'm reminded of those hucksters of cheap fabrics: 'never mind the quality, feel the width!'.

Daily deadlines and shrinking staff levels and resources in newsrooms worldwide have turned many of my colleagues into rats on treadmills, always fighting to get a story through by deadline and not having enough time to genuinely research it ourselves. Instead, we rely on 'experts' to provide context to the facts we're presenting to readers. Lobby groups recognised long ago that by getting themselves on the newsroom's 'expert' contact list, they could very quickly manipulate the news agenda to suit their purposes.

Then it comes down to a battle of the 'experts'. If a lobby can present a frightening, easily digestible message in a few seconds – human pollution is causing global warming and the planet is sick – and they have the 'experts' with appropriate letters after their name, then an industry based on "bad news sells" is almost guaranteed to bite.

When you throw in the social liberal leanings of most journalists, whose motivation as young students was similar to beauty pageant contestants – to 'help make the world a better place' – then an environmental issue like global warming will catapult to the top of the bulletins because it assuages our inner liberal guilt and makes us feel as editors that we're 'helping to make the world a better place'.

Having compassion for the planet, however, is not an excuse for abject stupidity. At the end of the day, the claims from silver-tongued lobbyists still have to stack up, otherwise our news bulletins descend entirely into propaganda.

Investigative journalists, like fraud investigators, are taught to 'follow the money'. Human behaviour is for the most part driven by personal gain either in kind or in cash. And where large amounts of

300 New Zealand's Minister of Overseas Trade, Mike Moore, who later headed the World Trade Organisation.

taxpayer cash, and tens of millions of jobs, are at stake, it becomes even more important to constantly check on the motivations of lobbyists and the truth behind what they're saying.

Unfortunately, many newsrooms only do half the job. They think that if they discover a financial interest that this is the end of the story. In this way, some of the critics of global warming belief have been sidelined by media and left-wing blogs because of a perception that their "links to industry" automatically invalidate anything they might say.

It is good to point out potential conflicts of interest, but it is foolish to assume that's all you need to prove to discredit an argument. If it was as simple as that, then I could have opened this book with the line: "Prominent global warming believer James Hansen of NASA was paid $250,000 by political interests to help him promote public belief in global warming".

But would that have been the end of it? No. I still needed to show where Hansen and his supporters are wrong. I actually have to deal with arguments, not just make *ad hominem* attacks and presume that proves it.

There are, for example, numerous global warming believers who demonise Professor Richard Lindzen, often with a dismissive wave of their hands:

"He is closely associated with the fossil fuel industry and is hardly a credible source."[301]

I could equally say of James Hansen, "he is closely associated to the Democratic Party and is hardly a credible source". Or I could say of virtually all climatologists: "He/she is being state-funded to research this, and the government has a direct financial interest in the outcome, so he/she is hardly a credible source".

The news media fall all over themselves to quote Greenpeace, yet Greenpeace sells guilt. It makes money from hyping environmental Armageddon.

Al Gore has extensive business interests in the carbon trading area, and got pinged by *WorldNetDaily*[302] for paying carbon credits to cover the huge energy usage of his home, to a carbon credit company that he himself was a major shareholder in and benefi-

301 http://scienceblogs.com/islandofdoubt/2009/01/dear_john.php?utm_source=sbhomepage&utm_medium=link&utm_content=channellink

302 "Gore's 'carbon offsets' paid to firm he owns", *WorldNetDaily*, 2 March 2007 http://www.worldnetdaily.com/news/article.asp?ARTICLE_ID=54528

ciary of! In other words, whilst he could say he was paying carbon credits, he was getting a chunk of it back behind the scenes. In the southern hemisphere the leader of the New Zealand Green Party is a significant shareholder in a wind turbine company that stands to benefit from government policy changes advocating alternative energy development. Various Green Party members were shareholders or employees of a company manufacturing those hideous mercury-filled compact fluorescent lights – a company that just coincidentally won a multi-million dollar supply contract from the Green-leaning government at the time.

Do we discount the entire merits of the UN's global warming argument on the basis that these people have these conflicts, or do we simply acknowledge where they're coming from, and continue to examine the detail of what they're claiming?

Who we associate with may colour our perspective on the problem, but ultimately the argument stands or falls on its merits: is that person's claim true/false?

Be extremely sceptical of news articles or websites that dismiss someone's argument on global warming (pro or con) purely through an *ad hominem* attack. If that's the best they can say, then they've said nothing at all. Whilst I've taken a swipe at some people in this book, and am about to take some more in the chapters ahead, I'd like to think I presented the evidence that justified the swipe.

In the case of "follow the money", it's important to recognise that both sides in this debate have a financial motivation, but arguably that of global warming believers is bigger. In the scientific community there's a standing joke, "No problem = no funding". In other words, to keep getting the research grants scientists have to keep coming up with problems requiring research or solution. In most cases, public tax money funds this and those projects that are seen as the most important get the most funding. Global warming is sex on a stick in the science community right now.

"They give an endless amount of money to the side which agrees with the IPCC," complained one Swedish geologist in an interview recently.[303] "The European Community, which has gone far in this thing: If you want a grant for a research project in climatology, it is written into the document that there *must* be a focus on global warming.

303 "Claim that sea level rising is a total fraud", Interview with Nils-Axel Mörner, *EIR*, 22 June 2007

"All the rest of us, we can never get a coin there, because we are not fulfilling the basic obligations. That is really bad, because then you start asking for the answer you want to get. That's what dictatorships did, autocracies. They demanded that scientists produce what they wanted."

Politicians, and particularly the United Nations, can see that if the public buy into global warming belief they'll get away with charging higher taxes on families and corporates, because it's seen as "a good cause".

Whether the public's willingness to keep believing in global warming survives the current world recession and expense of daily life, remains to be seen.

Recognising that global warming belief is under threat not just from science but from economic reality, believers are redoubling their efforts to put it back on top of the agenda for political leaders, thanks to President Obama's promises.

As Karla Bell at the GHG blog writes:[304]

"Remember the target for Kyoto was only -5%, a first step, now the U.S is talking about an 80% reduction in GHG emissions, so all sectors must be included. The Pew centre estimates of global emissions by sector are as follows Agriculture 14%, Transport 13%, Energy Efficiency is the solution to energy supply and commercial and residential building combined to be 34%. The grand total is 61%."

In other words, virtually no sector of the economy is going to be left unscathed by the coming tax burden to fix a problem that actually may not exist.

If the cost to the US of meeting Kyoto targets was estimated at $300 billion a year in today's money, consider what it's going to cost every western taxpayer to fund the production cuts, new technology and carbon and emissions taxes of the new regime: untold trillions of dollars over the next three decades would be a conservative estimate. If life on earth is genuinely under threat from human-induced warming, then making those changes is a small price to pay. But if promoters are lying about the size of the problem, or the problem is entirely natural in origin and can't be altered by anything we do, then we are destroying civilisation as we know it and there will be no coming back from the precipice.

---

304 http://ghgblog.com

So far I've taken you through some of the key areas of global warming debate. It's worth revisiting the highlights – what I call 'the icons of global warming' – in this chapter, before we go on and start following the money to find out who's pushing this agenda and why.

ICON ONE: RECORD ARCTIC SEA ICE LOSS

It is true that the Arctic lost a lot of summer ice in the mid 2000s. It is also true that the ice grew back.

The scientific studies show a trend to equilibrium between the two poles, so that while one pole is losing ice the other gains it, and there is no significant net change. Proof of this was borne out by figures released from the University of Illinois' Arctic Centre at the end of 2008, revealing that overall, global sea ice levels were nearly the same in 2008 as they were when satellite monitoring began in 1979.

In other words, during three decades of supposedly rampant global warming, overall sea ice levels haven't changed much. The university, fielding calls as a result of *Daily Tech's* revelations of this, has tried to explain away the problem by saying *global* ice coverage is not relevant to the *global* warming issue.[305]

"One important detail about the article in the *Daily Tech* is that the author is comparing the GLOBAL sea ice area from December 31, 2008 to same variable for December 31, 1979. In the context of climate change, GLOBAL sea ice area may not be the most relevant indicator.

"Almost all global climate models project a decrease in the Northern Hemisphere sea ice area over the next several decades under increasing greenhouse gas scenarios. But, the same model responses of the Southern Hemisphere sea ice are less certain. In fact, there have been some recent studies suggesting the amount of sea ice in the Southern Hemisphere may initially increase as a response to atmospheric warming through increased evaporation and subsequent snowfall onto the sea ice."

All very well and good, but that seems to illustrate how the planet balances itself out. Given that the southern hemisphere is the engine room for planetary climate and the oceans absorb more heat than

305 http://arctic.atmos.uiuc.edu/cryosphere/global.sea.ice.area.pdf

land does, the increasing southern sea ice extent[306] increases albedo (reflectivity of heat and light back out to space) and helps compensate for heat elsewhere.

Kilimanjaro in Africa is losing its ice and snow because there's no longer enough moisture in the air – not because of global warming, so clearly a healthy planet needs a certain amount of warmth to generate evaporation off the oceans and forests and re-deposit it as snow on mountains and poles. That's how sea ice grows, and how we avoid runaway warming.

Historical studies (looking at periods pre-dating anthropogenic global warming) have shown the poles balance each other out, and the latest satellite observations confirm that. The University of Illinois has been forced to appeal to computer model predictions, which have so far been extremely unreliable indicators of what will happen.[307]

The opening of the Northwest Passage was hailed by Britain's *Observer* and many other media as "unprecedented in human history". As you've seen, it's only "unprecedented" if you ignore boats sailing through it in 1903, 1942 and 2000. Scientific studies have shown that the Arctic Circle has been warmer in the past 1000 years than it is now, so it's no great stretch to suggest the Northwest Passage has probably opened on numerous occasions.

Explorers' reports show the Arctic was melting a hundred years ago, long before CO2 emissions became a problem, further indicating a natural cause.

Finally, you've just read in the last chapter an admission from the Hadley Centre in the UK Met Office that claims of Arctic sea ice loss have been dramatically overplayed, and that the summer sea ice melt in the last few years may well have been "natural" after all. In 2009 the Hadley Centre issued a strong warning to fellow global warming believers: stop making exaggerated and false claims to support public belief in AGW.

SUMMARY: So far, the Arctic is failing to live up to its poster child status as proof of anthropogenic global warming. The studies have not proven that the warming is human caused, and unrelated to

306 Sea ice 'extent' – area of coverage – is relevant to reflecting heat and light back into space (albedo), and thus the extent of sea ice is important. The mass of sea ice – its thickness and its extent overall – tells us whether we are losing or gaining ice in total, but it's not as relevant to the cooling effect of albedo.

307 http://www.warwickhughes.com/hoyt/scorecard.htm

natural solar and climate cycles. Further, global warming advocates have admitted that Arctic ice loss may also be happening because of pollutant dust and soot given off by industry and land use changes, which cloaks the ice with a more heat absorbent darker coating and provokes increased melt. Whilst human-caused, this aspect has nothing to do with CO_2 or the greenhouse effect.

AGW STATUS: No proof. Likely natural, with some input from human-caused dust and soot, rather than industrial greenhouse gas emissions.

ICON TWO: HOW MUCH CAN A POLAR BEAR BEAR?

As we've seen, last year's decision to list polar bears as an "endangered" species because of global warming was politically-motivated, not scientific. In the early 1980s, there were around 5,000 bears remaining in the wild because of unrestricted hunting, and the annual slaughter of Harp seal pups, one of their main food supplies, by furriers.

With the sensible decision to restrict killing of polar bears, and stop bashing baby seals to death with clubs, the baby seals were instead available a la carte for hungry bears.

Polar bear numbers rose from an estimated 5,000 to around 25,000 today, thanks to these policy changes.

Studies have shown that when cooler periods created more sea ice in the seventies and eighties (sunspot minima), the seal populations dropped, meaning the bears had less food. During warmer temperatures as the bears have been enjoying, most of their populations have grown.

The two groups of bears described as "thriving" are in areas that have become warmer, whilst the two populations known to be declining are in areas that became colder.

Projections of future ice loss in the Arctic as a result of global warming, and based on those ubiquitous computer models (garbage in/garbage out), convinced US federal authorities that despite population growth from 5,000 to 25,000 the bears are now endangered.

A scientific audit of the methodology used to reach that decision described the studies they relied on as "erroneous…unscientific and of no consequence to decision makers".

Concerned members of the public, upon learning of the polar bears' impending doom, nonetheless promised to "turn off the lights" and stop eating hamburgers, in a bid to save the bears – don't ask me how this will save a bear, I'm just the messenger![308]

Sceptical environmentalist Bjorn Lomborg made the point that adopting the Kyoto Protocol in toto would save 0.06 polar bears from extinction and, estimating that around 15 bears from the worst hit population are being lost to climate change each year, recommended perhaps a better answer than listing them as "endangered" and turning off lights would be to revoke hunting permits on that particular bear population, which allow 49 bears to be killed every year currently.

SUMMARY: The bear population is not in danger. Some small groups of bears are facing hardship for a number of reasons, but most of the world's 25,000 bears are doing nicely, thank you. Predictions of massive Arctic ice loss may not come to pass based on latest data, but even if they do, the polar bear community has lived through hotter warm periods in the past.

AGW STATUS: No proof that bears are suffering from anthropogenic global warming, and no evidence that bears are suffering significantly from general warming cycles, beyond their capacity to adjust.

ICON THREE: ANTARCTIC WARMING

While the North Pole has been shedding some ice, satellite measurements suggest the great southern continent has been gaining it, except for one portion – the Antarctic Peninsula and adjacent West Antarctic ice sheets.

There can be no denying that temperature readings on the peninsula show a gradual warming since the 1940s, but the difficulty lies in ascertaining whether the warming is part of a regional natural climate cycle, or human-induced global warming.

A string of scientific studies show the West Antarctic is naturally vulnerable to warm ocean currents caused mostly by the tropical El Niño Southern Oscillation to the north, which dictates much of the planet's weather. Those warm currents have helped weaken sea ice and raise temperatures, but they are localised effects.

The far larger East Antarctic region remains substantially colder

308 Global warming believers are actively pushing the UN to adopt a vegetarian position and run ad campaigns discouraging people from eating meat in order to "do their bit" for the planet. See www.letsactnow.org for an example of this spin.

and, according to the satellite measurements, may be adding more ice volume than West Antarctica is losing. Certainly Antarctica overall has added sufficient ice to compensate for losses at the North Pole.

East Antarctica is not vulnerable to the warm ocean currents affecting the west, and in fact extra moisture caused by oceanic evaporation is re-dumping itself as snow on Antarctica and adding to ice volume and extent.[309]

Global warming believers say their computer models now add this as an expected side effect of global warming. Cynics point out it's nature's way of regulating temperatures and providing balance.

Steig and Mann's *Nature* paper earlier this year, setting the agenda for Copenhagen by suggesting in widely reported comments that Antarctica was warming, appears to be in serious danger after closer inspection: according to critics Steig et al were too selective in choosing their data (in essence they made calculated 'guesses' about temperatures from areas where no weather data was available, and chose a basket of indicators most likely to support global warming), and when all relevant weather data for Antarctica was taken into account it actually shows the continent has been cooling overall since 1980, at a rate of up to a full degree Celsius per century. A detailed rebuttal of the *Nature* paper has been carried out[310], prompting one scientist to comment, "Unless Steig provides the requested data, *Nature* should withdraw this work or even that Issue/Volume. I certainly will not be submitting any work to this journal in the future if this is the way they referee submissions."

Other commenters compared it to the "hockey stick" controversy, and blamed a weak peer-review process: "If all the peers reviewing a paper accept the premise and conclusion already, such as global warming, they're most likely to just scan quickly over a paper before giving it their blessing, as appears to have happened with Steig et al. This is the sort of problem I recall that was addressed in the Wegman report to Congress regarding the 'hockey stick'. How embarrassing it must be to have a peer-reviewed paper published and then to have significant flaws pointed out, flaws which should have been found in a robust peer review process."

309 An added twist to this is the recent discovery of an Antarctic Circumpolar Wave, which drives warm and cold patches of water (up to several thousand kilometers long) around the ice continent every 8 or 9 years. The ACW is thought to play a role in rainfall patterns in southern New Zealand and Australia. See: http://oceancurrents.rsmas.miami.edu/southern/antarctic-cp.html

310 http://wattsupwiththat.com/2009/02/28/steigs-antarctic-heartburn/

Recently released data from a joint study by Victoria University's Antarctic Research Centre in New Zealand and the Northern Illinois University has shown the West Antarctic ice sheet has always been vulnerable to melt – vanishing and re-growing dozens of times with long warm intervals in between (all of them natural in origin).

"Antarctica's ice sheets have grown and collapsed at least 40 times over the past five million years," VUARC director Tim Naish told journalists in March this year. "The biggest finding from our project is that it was not possible to sustain a West Antarctic Ice Sheet for long periods."[311]

Those long periods – up to 200,000 years at a stretch – were warmer than we currently enjoy, and are thought to have been caused by solar forcings and orbital variations, acting on ocean currents and climate oscillations.

Even so, it appears the much larger East Antarctic sheet has remained stable and will continue to do so.

"It may be nibbled away at the edges," said Naish, "but a number of other lines of evidence make it unlikely that we will lose half the East Antarctic ice sheet."

Key point? Massive melt in the West Antarctic is not new, has always been natural, and the Antarctic ice sheets began receding not last century but around 7,000 years ago, so human-caused CO_2 levels are unlikely to be a major factor.

SUMMARY: There is no study yet available that shows the massive East Antarctic region is in any serious danger of warming enough to melt. Predictions about existing meltwater are based on the melt continuing for a century unabated which, given the cyclical cooling and warming associated with various climate oscillations and the current Antarctic cooling, appears to be a big ask.

With average temperatures of more than 50 degrees Celsius below zero, interior ice-sheet melting is not an option, and sublimation (changing ice to water vapour via evaporation without melt) won't happen as long as warmer oceans keep evaporating seawater which then freezes and drops back as snow, then turns into ice again.

AGW STATUS: Michael Mann's hopeful claims notwithstanding, there is scarcely evidence of warming, let alone human-induced warming, in Antarctica's main ice fields. Not proven.

311 "Antarctic data set to spur sea level concerns", preliminary data from The Andrill Project, *Dominion Post*, 5 March 2009, http://www.stuff.co.nz/environment/1397735

ICON FOUR: MODERN WARMING IS 'UNPRECEDENTED'

Since the UN IPCC's report of 2001, this has been the ongoing message sold to the news media, political leaders and the public. If global warming believers can fool people into accepting that the modern era is the warmest ever, the public will be more likely to go along with the decisions made by the UN and political leaders.

Unfortunately, the claim is utterly untrue.

More than six thousand boreholes drilled around the world tell the same story: the planet was several degrees hotter in medieval times than it is now. Virtually all scientists agree that humans could not have caused that level of climate change back then, so the process had to be natural.

Knowing this is the only answer, climate scientists at the forefront of the global warming belief essentially faked data for the UN IPCC's 2001 report with a 'hockey stick' shaped graph purporting to show gradually declining temperatures over the past thousand years until the industrial revolution kicked in. At that point, the temperature graph, and carbon emissions graph, both shot upwards.

A scientific audit of this graph revealed the data it was based on was so heavily flawed that the graph was effectively useless and, in the opinion of some, a fraud, even though Al Gore had made it the centrepiece of his *An Inconvenient Truth* movie.

Data correctly displaying the true size of the medieval warm period was manipulated to hide its strength, thus allowing climate scientists and tame media organisations to claim that the 20th century is the warmest period in human history.

A review of 240 studies by Harvard University and the Smithsonian's Centre for Astrophysics sank this claim by confirming the Medieval Warm Period was far warmer, less than a thousand years ago.

Even Michael Mann himself, whilst still trying desperately to argue his way out of the hole, now concedes that "in the context of the long-term reconstructions, the early 20th century appears to have been a relatively cold period while the mid 20th century was comparable in warmth, by most estimates, to peak Medieval warmth (i.e., the so-called "Medieval Warm Period")."[312]

So we have the guru now admitting modern warmth is at the same level as the MWP. Where's that on the hockey stick graph he drew?

---

312 http://www.realclimate.org/index.php/archives/2004/12/myths-vs-fact-regarding-the-hockey-stick/

Even in Greenland, one of the 'icons' of Arctic melt, the latest studies have revealed a similar rapid melt occurred there back in the 1920s as a result of temperatures the same as they are today. In other words, Greenland's modern warm spell is about the same and had the same effects as one 90 years ago, at a time when greenhouse gas emissions were not an issue.

"They recently recognised from using weather station records from the past century that temperatures in Greenland had warmed in the 1920s at rates equivalent to the recent past," reported *ScienceDaily*.[313]

One of Greenland's largest glaciers, the Kangerdlugssuaq, "lost a piece of floating ice that was nearly the size of New York's Manhattan Island" in that 1920s warm spell.[314]

SUMMARY: This is the icon that actually sinks the global warming belief system. Climate scientists and the UN's IPCC actually misled the public by providing false information. If human-induced global warming is true, its believers should not have had to fake scientific data.

AGW STATUS: Does not provide evidence of AGW.

ICON FIVE: BUT THE PLANET IS STILL HEATING UP

Again, based on current data, not true. Overall, temperatures have been on a downward slope since 1998. Between January 2007 and February 2008, global average temperatures plummeted – and I use that word advisedly – by between 0.59°C and 0.75°C, wiping out on paper virtually an entire century's gains due to "global warming".[315]

And this wasn't sceptics running the numbers, this is the UK Met Office's Hadley Centre, NASA and the University of Alabama – three major temperature monitoring organisations.

As proof, however, of the kind of obfuscation and downright fudging thrown into the mix by global warming believers, look at *RealClimate's* attack on the Hudson Institute's Dennis Avery for quoting those figures in a CNN debate:[316]

---

313 "Current melting of Greenland's ice mimics 1920s-1940s event", *ScienceDaily*, 13 December 2007, http://www.sciencedaily.com/releases/2007/12/071210094332.htm

314 This discovery also raises questions about Michael Mann's assertion that the "early 20th century appears to have been a relatively cold period".

315 Of course, these average temperatures are close to meaningless in real terms, given the amount of climate variation, and a one year drop doesn't mean it won't rise again. Nonetheless, it shows what goes up can come down.

316 http://www.realclimate.org/index.php/archives/2009/01/cnn-is-spun-right-round-baby-right-round/

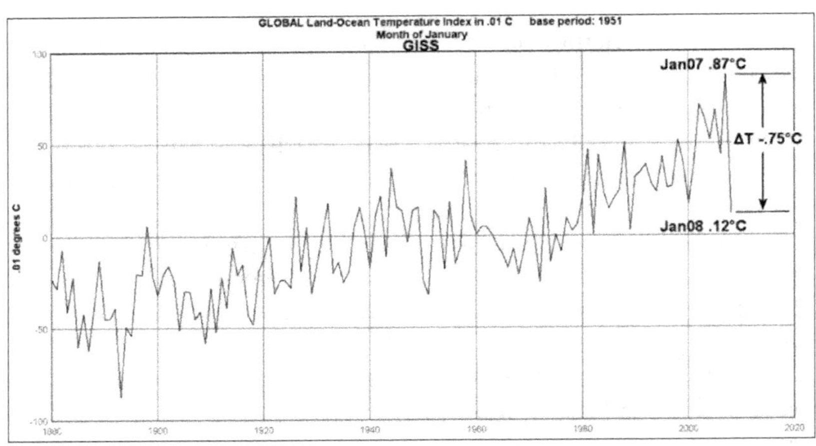
GLOBAL Land-Ocean Temperature Index in .01 C base period: 1951
Month of January
GISS
.01 degrees C
Jan07 .87°C
ΔT -.75°C
Jan08 .12°C

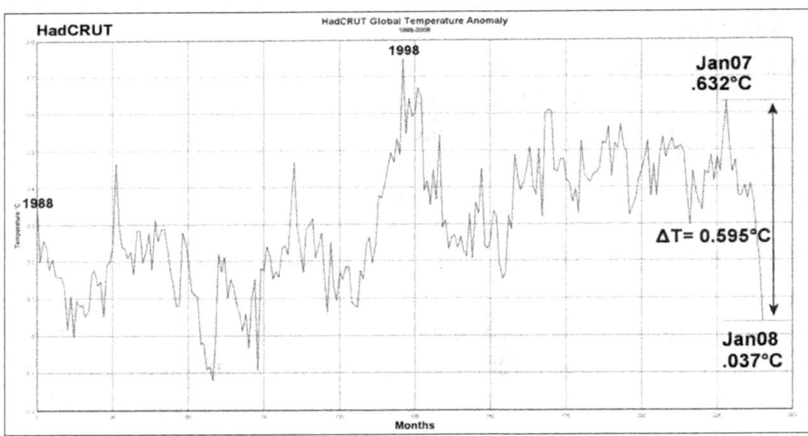
HadCRUT
HadCRUT Global Temperature Anomaly
1998
1988
Jan07
.632°C
ΔT= 0.595°C
Jan08
.037°C
Months

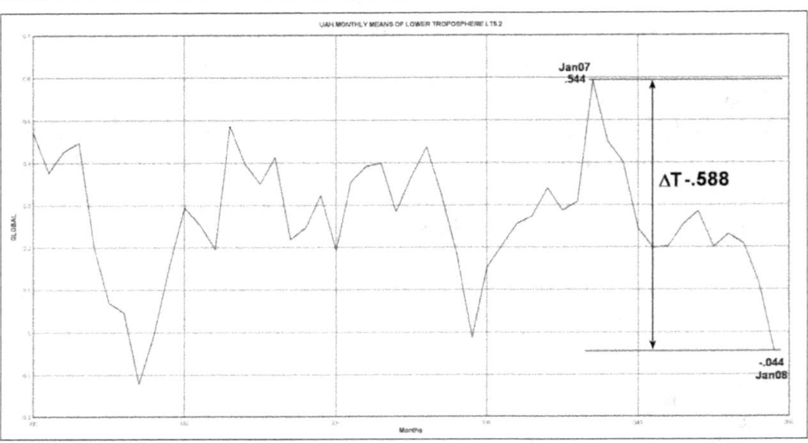
UAH MONTHLY MEANS OF LOWER TROPOSPHERE LT5.2
Jan07
.544
ΔT-.588
-.044
Jan08
GLOBAL
Months

"As for his great cherry pick (0.6° C in two years – we're doomed!), this appears to simply be made up," exclaimed *RealClimate*. "Even putting aside the nonsense of concluding anything from a two year trend, if you take monthly values and start at the peak value at the height of the last El Niño event of January 2007 and do no actual trend analysis, I can find no data set that gives a drop of 0.6°C.

"The issue being not that it hasn't been cooler this year than last, but why make up numbers? This is purely rhetorical of course, they make up numbers because they don't care about whether what they say is true or not."

Pot, meet kettle. *RealClimate* is the blogsite of global warming's high priesthood, and includes contributions from Michael Mann, Eric Steig and NASA's Gavin Schmidt. Those paragraphs just quoted were contributed by Schmidt.

Now, as proof that the NASA GISS scientist got it seriously wrong, I've reprinted the dataset graphs Schmidt claims don't exist on the adjacent page.

Embarrassingly for Schmidt and his "I can find no data set that gives a drop of 0.6°C", one of the graphs was from his own unit, and *that* graph shows the highest rate of cooling, -0.75°C. Yet these are the people the media trust on global warming, the ones politicians have been told are competent.

Schmidt's writings are frequently wheeled out on global warming believers' sites to 'debunk' "climate change deniers", which sounds great in theory but as you've seen turns to custard when they try and put it in practice.

These, just to make this point as strongly as I can, are the guys in white coats telling the entire world, "Trust us, we know what we're doing, we're from the Government", and whose work is the basis of a multi-trillion dollar planned burden on the already recession-hit economies of the world.

Do we really trust them, *that* much?

And here's the rub: according to NASA, NOAA and the Hadley Centre, 2008 was cooler than 2007, making 2008 the coldest year of the century and showing that climate theory based on CO_2 emissions is wrong – carbon levels have continued to rise, but temperatures are going down.

"We now have data showing that, from 2000 to 2007, greenhouse gas emissions increased far more rapidly than we expected, primarily because developing countries like China and India saw a huge upsurge in electric power generation, almost all of it based on coal," said Stanford's Chris Field, one of the authors of the 2007 IPCC Fourth Assessment Report Summary for Policymakers and its attendant studies.[317]

So just to make the point clear, greenhouse gas emissions have continued to go through the roof, increasing "far more rapidly than we expected", yet the planet is cooling down. The global computer models didn't predict that, because it goes against the whole grain of a greenhouse theory which argues the gases will trap heat no matter what, and that our only protection against rising heat is to lower emissions rapidly. Clearly that isn't true. Temperatures have dropped from their 1998 highs, and the latest climate data from Antarctica reveals an overall cooling trend since 1980, while CO_2 and methane emissions have surged upwards.

The *Boston Globe's* Jeff Jacoby nicely summed up the dilemma facing global warming believers:[318]

"The United States has shivered through an unusually severe winter, with snow falling in such unlikely destinations as New Orleans, Las Vegas, Alabama, and Georgia. On Dec. 25, every Canadian province woke up to a white Christmas, something that hadn't happened in 37 years. Earlier this year, Europe was gripped by such a killing cold wave that trains were shut down in the French Riviera and chimpanzees in the Rome Zoo had to be plied with hot tea. Last week, satellite data showed three of the Great Lakes – Erie, Superior, and Huron – almost completely frozen over. In Washington, D.C., what was supposed to be a massive rally against global warming was upstaged by the heaviest snowfall of the season, which paralyzed the capital.

"Meanwhile, the National Snow and Ice Data Center has acknowledged that due to a satellite sensor malfunction, it had been underestimating the extent of Arctic sea ice by 193,000 square miles – an area the size of Spain. In a new study, University of Wisconsin researchers Kyle Swanson and Anastasios Tsonis conclude that global warming could be going into a decades-long remission. The current global cooling "is nothing like anything we've seen since 1950," Swanson told

317 "Amid rising anxiety level, scientists warm to Denmark", *Canberra Times*, 10 March 2009, http://www.canberratimes.com.au/news/opinion/editorial/general/amid-rising-anxiety-level-scientists-warm-to-denmark/1454492.aspx?storypage=0

318 "Where's the Global Warming?", Jeff Jacoby, *Boston Globe*, 8 March 2009

Discovery News. Yes, global cooling: 2008 was the coolest year of the past decade – global temperatures have not exceeded the record high measured in 1998, notwithstanding the carbon-dioxide that human beings continue to pump into the atmosphere."

Adding fuel to the fire, or should that be ice to the fridge, a new scientific study suggests global warming may be officially over, at least for a while.[319]

The suggestion comes after a study of the major ocean oscillations over the past century, including the Pacific Decadal Oscillation, El Nino, the North Atlantic Oscillation and others. The researchers found that when those oscillations were in sync – striking at or near the same time – it actually provoked a global "climate shift".

As alluded to above, however, they found the global cooling now gripping the planet strongly suggests Earth has a way of regulating its climate that scientists had not anticipated.

"This cooling, which appears unprecedented over the instrumental period, is suggestive of an internal shift of climate dynamical processes that as yet remain poorly understood," explain Swanson and Tsonis in their groundbreaking paper.

Discover magazine, one of the homes of global warming belief, summarised the latest study like this:[320]

"In their paper, Swanson and Tsonis then look at the past few years. They see a peak in synchronization in 2001 and 2002, and they also observe that in the years since, the temperature change has been on average flat (although much warmer than at the beginning of the century). They estimate that all the warming due to carbon dioxide should have driven the temperature up 0.25 degrees C since then. The fact that it hasn't leads them to propose that the oceans and atmosphere have changed the way they handle heat. The oceans may have absorbed more heat due to a change in circulation, or the atmosphere may radiate more heat away by clouds. If this hypothesis is true, then it's possible that the climate will remain in this new stage for some years to come before it shifts again."

In fact, Swanson and Tsonis are picking a cooling trend through to 2025, followed by a brief upward blip then cooling for the rest of the century.

Isn't that what I've been arguing in *Air Con* already? – we're seeing

---

319 Swanson, K. L., and A. A. Tsonis (2009), "Has the climate recently shifted?", *Geophys. Res. Lett.*, doi:10.1029/2008GL037022, Draft avail from http://www.uwm.edu/~kswanson/publications/2008GL037022_all.pdf

320 http://blogs.discovermagazine.com/loom/2009/03/04/checking-george-will-the-perils-of-time-travel/

temperature regulating mechanisms kick in (higher snowfall, rainfall etc) as a response to the warming that's taken place, and those mechanisms are helping balance it out, stripping the atmosphere in particular (and hence the 'greenhouse') of some of its heat.

Finally on this point, the "greenhouse effect" implies that the Earth takes on some of the attributes of a hot house. As the theory goes, the troposphere (lower atmosphere), should be heating drastically, because it's trapping the surface heat via the CO_2, methane and water vapour. That's the theory. In practice however satellite measurements and weather balloon temperature readings have failed to show much temperature increase in the atmosphere at all. All the computer models used by the UN IPCC have strong atmospheric warming built in, so the non-appearance of global warming in the atmosphere, where it is supposed to have been for the past ten years, has been quite a behind the scenes embarrassment for global warming believers.

The problem came to a head just over a year ago when the University of Rochester, University of Alabama (Huntsville, UAH) and University of Virginia checked the climate change 'forecasts' from 22 of the leading computer models against the actual hard temperature data collected by various sensors on balloons, satellites and surface stations in the tropics.

"The models forecast that the troposphere should be warming more than the surface and that this trend should be especially pronounced in the tropics," explained Dr John Christy, director of UAH's Earth System Science Centre to *ScienceDaily*.[321]

"When we look at the actual climate data, however, we do not see accelerated warming in the tropical troposphere...for those layers of the atmosphere, the warming trend we see in the tropics is typically less than half what the models forecast."

Not to be beaten, a team at Yale University hit back in support of the computer models, by suggesting that if the hard data and the computer models disagreed, it was probably the hard data that was faulty – *all* the weather balloons, and *all* the satellite readings, were wrong. To get around this, they chose to ignore temperature readings from weather balloons and satellites altogether, and estimate temperatures based on tropical wind speeds.

321 "New study increases concerns about climate model reliability", study by David Douglass et al, published in the *International Journal of Climatology*, reported via *ScienceDaily*, 12 December 2007, http://www.sciencedaily.com/releases/2007/12/071211101623.htm

"By measuring changes in winds, rather than relying on problematic temperature measurements, [they] estimated the atmospheric temperatures near 10km [altitude] in the tropics rose about 0.65 degrees Celsius per decade since 1970," reported *ScienceDaily* last year,[322] "probably the fastest warming rate anywhere in Earth's atmosphere. The temperature increase is in line with predictions of global warming models."

Yes, with that kind of scientific alchemy I'm sure it would be. Ignore the actual temperature readings in favour of sticking your finger in the air and testing the wind, then report to the world you "estimated" temperatures had risen 2.5°C!

Crisis at the UN averted, global warming panic permitted to continue based on yet more dodgy data, whilst the "problematic temperature measurements" are quietly tucked away. Remember, it is the UN's preferred computer climate models that continue to be the basis of this year's big increase in global warming scare stories in the lead-up to Copenhagen.

And don't forget also how the computer models fail to properly account for the cooling effect of clouds, and may be overestimating global warming this century by 75%.[323]

Echoing that finding, a major new study by the US National Oceanic and Atmospheric Administration (NOAA) sheepishly reveals that "nearly half" the recorded warming previously attributed to global warming in North America and Canada may in fact be natural in origin.[324]

A *Canwest News* report on the study notes a couple of highlights:

– "Variations within North America 'are very likely influenced by variations in global sea surface temperatures through the effects of the latter on atmospheric circulation, especially during winter.' The term 'very likely' is defined as a chance of 90% or more." [This would fit with the recent study that dust storms and volcanoes have a much larger impact on the Atlantic ocean temperature than climate change does]

– "It's 'unlikely' that patterns of drought have changed due to global

322 "Apparent problem with global warming climate models resolved", *Nature Geoscience*, 18 May 2008, reported via *ScienceDaily*, 30 May 2008, http://www.sciencedaily.com/releases/2008/05/080530144943.htm

323 "Cloud and radiation budget changes associated with tropical intraseasonal oscillations", Spencer, R. W., W. D. Braswell, J. R. Christy, and J. Hnilo (2007), *Geophys. Res. Lett.*, 34, L15707, doi:10.1029/2007GL029698., http://www.drroyspencer.com/Spencer_07GRL.pdf

324 "Recent NOAA Study: Climate change not all man-made", reported http://wattsupwiththat.com/2009/04/02/recent-noaa-study-climate-change-not-all-man-made/#more-6736

warming caused by human pollution. Rather, natural shifts in ocean currents are probably to blame. For instance, the current drought in Texas and the southwest are due to La Nina, a Pacific Ocean current that starts and stops periodically (such as El Nino), and cuts off the movement of moist air inland. Warmer temperatures from greenhouse gases, however, would worsen the basic drought."

In other words, yet again the science is trending towards a back-down on the extent of warming caused by humans.

SUMMARY: The blizzards that hit the US, Europe and Asia at the start of 2009 were reportedly the worst in decades, with a further late dumping of snow in March. Not withstanding a hot regional blip in Australia, the odds are not good that 2009 will be a record-breaking warm year. On current trends it may be cooler again than 2008. Global warming believers write this off as an effect of the La Niña weather pattern. Global warming sceptics agree, but add that La Niña/El Niño are natural weather cycles that help maintain planetary climate balances. The full extent of the sun's drop in activity isn't expected to hit until late 2009 and into 2010, because of the thermal lag in the oceans. In other words, the cooling is likely to increase.

AGW STATUS: Whilst admittedly they're a 'weather event' rather than a climate trend, it's hard to see how average global temperatures that plunge more in a year than the planet has warmed in a century, can be listed as proof of anthropogenic global warming. According to the environmental monitors, greenhouse gas emissions have continued to increase through 2007/08 (albeit that the rate of increase has slowed), so if CO2 is the great forcer, creating climate feedbacks, why are we cooling down? The only obvious explanation is that *natural* climate forcings and feedbacks are far stronger, and provoke far greater swings either way, than human contributions. The lack of atmospheric warming, in defiance of the 'greenhouse' effect, is another dead giveaway. The planet is not currently heating up, despite what the computer models are screaming.

ICON SIX: GLACIER MELT

Glaciers in many parts of the world are melting. This is frequently cited by the news media and alarmist researchers as evidence of "global warming". Unfortunately, the bit of the story that gets lost in translation is that many of the big glaciers are melting now because of warming which began two hundred years ago or even up to a thousand or ten thousand years ago.[325]

Hard as it may be to comprehend, most glaciers don't turn on a dime, retreating or advancing depending on this year's weather. The melting we are seeing today is not related to current CO_2 emissions, or human-induced global warming. In New Zealand, where some 50 alpine glaciers in a temperate climate are under scrutiny, many have grown because of increased snowfalls caused by El Niño systems.

However, the largest NZ glaciers are shrinking overall, as a direct result of the warming that pulled the planet out of the Little Ice Age. Those glaciers have however not yet retreated to their pre-LIA levels.

In Africa, Tanzania's Mt Kilimanjaro has been used as an example of the effects of human-induced global warming. The glacier on the mountain's summit is shrinking, and the slopes no longer carry as much snow as they once did. But scientific studies have now proven that Kilimanjaro's ice loss has nothing to do with CO_2, and everything to do with deforestation.

Tropical rainforests give off vast amounts of moisture. That moisture normally rises into the atmosphere, and where it rises above the freezing level it becomes snow or hail. When this happens, the snow or hail drops back out of the clouds again but, as it comes down into warmer temperatures, it melts to become rain, mist or even just evaporates before most of it hits the ground, in this ongoing climate cycle. Where Kilimanjaro fits in that cycle is that its 5,892 metre (19,330 feet) summit was above freezing level so the snow could fall and actually stay on the ground as snow. Hence the appearance of a tropical glacier.

But, with a lack of moisture caused by deforestation, there's not enough snowfall on the mountain to compensate for the loss there has always been through sublimation, which is where the wind and dry air suck moisture content (and thus volume) out of the ice.

The sublimation gets worse the drier the air is, so the lack of moisture in the air has contributed to an acceleration of the disappearance of Kilimanjaro's white suit.

Whilst Kilimanjaro has certainly suffered as a result of humans cutting down the rainforests over the centuries, it has nothing to do with CO_2 emissions, greenhouse gases or global warming.

---

325 The proviso to this is that glaciers which terminate in the ocean are more rapidly affected by warmer ocean currents.

If the forests were allowed to grow back, so would Kilimanjaro's white cloak.

SUMMARY: Many of the big glaciers currently melting are doing so in response to 19th century warming or earlier – not human related. This is part of natural planetary climate cycles that see glaciers advance and recede on a regular basis over a period of centuries. Some glaciers in New Zealand, Norway and Alaska are now actually growing again.

AGW STATUS: No evidence of significant glacier melt caused by human-induced global warming.

ICON SEVEN: SEA LEVELS ARE RISING

Well, yes, but not by anywhere near enough to cause mayhem in 90 years' time.

There's also the very embarrassing fact that global warming believers have not factored tectonic plate movements into their models as a possible cause of sea level changes. If you can imagine a waterproof tarpaulin held at four corners, with water poured into the middle of it, the water would cause the tarp to dip in the middle where the water gathered. Assuming everyone stood statue-still, the water level would stabilise. But if someone crawled underneath the tarp and pushed the centre upward, the water level would rise because of the displacement.

That's what happens on the ocean floor, every hour of every day. Most of the earth is a molten ball of liquid rock and iron that burns at intense heat, contained underneath a thin crust of rock which makes up the sea floor and continents. As we all know, heat causes things to expand, and pressure is constantly being released undersea. It's estimated some 95% of the world's volcanic activity is taking place under the ocean. Magma will cause a bulge underneath the sea floor, and sea levels will rise accordingly under the laws of physics and fluid dynamics.

To paint a mental image for you, the sea at its deepest point is only seven miles (11 kilometres) deep. That's less than the distance from your home to the supermarket for many readers. If you stand on a hill on a fine day, you can often see a hundred kilometres (60 miles) into the distance. Eleven kilometres is a short hop in comparison. The *average* depth of the oceans, however, is only three kilometres (two miles). The driveway to my property is 500 metres long, thus the average depth of the world's oceans is six times the length of my driveway.

Now, imagine yourself in the space shuttle, looking down at earth. The relative depth of the oceans, from that height, is similar to the film of water left on the floor when you spill a glass of water. In comparison to the size of the planet, the oceans are extremely shallow. The Earth's diameter is just over 12,700 km. The oceans make up about six kilometres (an average of three km's deep all around) of that 12,700 km. The sea bed, the rock the oceans sit on, is only an average of four kilometres deep. Beneath that, the putty like layer of Earth's outer mantle.

So, continuing my driveway analogy, if I hopped in my car I would only have to travel 14 lengths of my driveway to have covered the distance from the ocean surface right down through the sea bed into Earth's mantle.[326]

In the great scheme of things, we live on a very thin crust.

We have little data on how much the sea floor is moving up or down because we don't have continuous real-time monitoring of the ocean floor worldwide. However, one study off the southeast coast of India reveals the sea floor over the 140 km long survey area rose more than half a metre in just 25 years, between 1975 and 1999,[327] coincidentally the same period of time we've had satellites monitoring the oceans.

The ocean floor clearly moves up and down, but without any accurate planet-wide measuring system, it's a data void that makes all other predictions about sea level rise meaningless – how do we know half the measured sea level rise to date is not geological, rather than climate, related? We don't.

It makes the entire computer modelling programme for climate change irrelevant in terms of sea levels.[328]

NASA's James Hansen, Al Gore's adviser as you've already seen,

---

326 Again, depths are averages. There's actually an area in the mid Atlantic of several thousand square kilometres where the Earth's mantle *IS* the seabed, having pushed through the crust to lie exposed on the sea floor. See "Scientists to study gash on Atlantic seafloor", *LiveScience*, 1 March 2007, http://www.msnbc.msn.com/id/17407745/

327 "Monitoring Changes in Seafloor Morphology Using Multi-date Bathymetry data: A Case Study of the Gulf of Mannar, Southeast Coast of India," http://www.gisdevelopment.net/application/nrm/ocean/ma03009pf.htm

328 The UN IPCC, at page 408 of its Fourth Assessment Report, claims to 'correct' for possible seismic ocean floor rises by selecting tide gauges in areas unaffected by seismic activity. Here's the fundamental problem with that claim – all the oceans of the world are linked. A sea floor rise mid-Pacific might not register as a significant seismic incident 6,000 kilometres away in Peru, but it will push up sea levels regardless for all the reasons outlined above.

has put his credibility on the line by suggesting we are heading for a five metre rise in sea level "by 2095". But what was it the scientists found when they examined this wild claim?

"For Greenland alone to raise sea level by two metres by 2100, *all* of the outlet glaciers involved would need to move more than three times faster than the fastest outlet glaciers ever observed, or more than 70 times faster than they presently move," one of the Colorado team, Tad Pfeffer, explained.[329] "And they would have to start moving that fast today, not 10 years from now. It is a simple argument with no fancy physics."

Pfeffer made that statement late last year, based on data taken at the height of a major melt period in Greenland. The glaciers there will have to move 70 times faster *than they already are*, which they have not done.

Satellite samplings suggest an average sea level rise of 3mm a year,[330] or 3cm a decade, which works out at 30cm a century. This is where the IPCC gets its estimates of a 20cm to 43cm increase in sea levels by 2100. The UN's own IPCC Fourth Assessment Report concedes that net sea level rise observed so far may only be as high as 1.3mm a year, equivalent to 13cm a century.[331]

Leaving aside the possibility that sea floor rise may be partially responsible, much of the observed sea level increase is not the result of actual ice melt, but what's known as thermal expansion (water expands in volume when it is heated).

With the sun at its highest level of activity in 11,400 years over the past century, it's a fair bet that the sea level has risen because of thermal expansion.

In fact, a report in *Science* magazine says exactly that:[332]

"The 3.2 ± 0.2 millimetre per year global mean sea level rise observed by the Topex/Poseidon satellite over 1993-98 is fully explained by thermal expansion of the oceans," reported the study authors, who added after further research, "the 20th century sea level rise estimated from tide gauge records may have been overestimated."

Another study reveals that during what were allegedly the hot-

329 http://www.colorado.edu/news/r/c3cb8187d1bf611c77bbf951ffc3e96a.html

330 http://sealevel.colorado.edu/

331 UN IPCC AR4, page 419, figure 5.21

332 "Sea Level Rise During Past 40 Years Determined from Satellite and in Situ Observations", C Cabanes et al, *Science* 26 October 2001: Vol. 294. no. 5543, pp. 840 – 842 DOI: 10.1126/science.1063556,

test years on Earth in living memory, 1992 through 2003, the East Antarctic ice shelf gained an incredible 450 billion tonnes in ice – according to satellite measurements.[333] This, say scientists, is the result of higher snowfalls and helps offset sea level rises – part of the planet's way of balancing climate issues.

One man highly sceptical of the sea-level issue is Sweden's Nils-Axel Mörner, head of the Paleogeophysics and Geodynamics Department at the University of Stockholm, who accuses the IPCC of selective data collection.[334]

"Tide gauging is very complicated, because it gives different answers for wherever you are in the world. But we have to rely on geology when we interpret it. So, for example, those people in the IPCC [Intergovernmental Panel on Climate Change], choose Hong Kong, which has six tide gauges, and they choose the record of one, which gives 2.3 mm per year rise of sea level. Every geologist knows that that is a subsiding area. It's the compaction of sediment; it is the only record which you shouldn't use.

"And that is just ridiculous. Not even ignorance could be responsible for a thing like that. So tide gauges, you have to treat very, very carefully."

The thing about tide gauges[335], and even satellites for that matter, is working out how to correctly factor in all sorts of localised climate variations. For example, when the weather presenter on TV is talking about "Highs" and "Lows", these are atmospheric pressure systems. High pressure zones, the ones that bring the good weather, push down on the sea surface and, in effect, lower it. That's why they're called "high pressure".

Low pressure systems, on the other hand, allow the sea level to rise and – in the right conditions – cause storm surges. The rule of thumb is that for every one millibar drop in atmospheric pressure on the weather map, the sea level in the affected area rises one centimetre. Thus, a "Low" of 960 millibars, which is a drop of 50 mb off normal, will generally raise sea levels by half a metre. The deeper the low, the bigger the incoming tide.

333 "Snowfall driven Growth in East Antarctic Ice Sheet Mitigates Recent Sea Level Rise", Davis et al, *Science*, 24 June 2005, Vol 308, no 5730, pp. 1898-1901

334 "Claim that sea level rising is a total fraud", Interview with Nils-Axel Mörner, *EIR*, 22 June 2007

335 Many of the world's tide gauges are attached to wharves whose piles sink, gradually, into the mud. Even if the wharf is only sinking a millimeter a year, that's a millimeter that will be interpreted by scientists as a rise in sea level. Expressed simply, tide gauges were never designed to be accurate in millimeters, but in centimeters.

To correct for things like that, climate scientists must consciously enter data on atmospheric pressures for the areas where they're monitoring sea levels, otherwise the figures are meaningless.

When new satellite data released in 2009 showed a 30% drop off in the rate that sea levels were rising, from 3mm a year down to 2mm a year since 2005 (equating to a 20 centimetre rise over this coming century on current tracking), that too provoked a series of 'why?' questions, given that accelerating CO_2 levels should have been enough to punch sea levels higher regardless, under current computer modelling theories.

While some wags have dubbed it 'the Obama effect' and proof that the president has filled his first election promise, other scientists reckon the original satellite estimations of sea level rise were too high.

"The satellite system has undoubtedly shown a rise since 1992, but it has levelled off," *Daily Tech*[336] quoted NZ-based climatologist and IPCC expert reviewer Vincent Gray. "They had some bad calibration errors at the beginning."

Daily Tech added that a Flanders University study backs up Gray, with findings of "no perceptible increase in sea level over its entire 15 year period".

The mini-Copenhagen conference in March 2009 received worldwide headlines with its claim we should brace for a sea level rise higher than a metre by 2100 because of increased melt from Greenland and Antarctica, but we've already seen the studies refuting that.

What wasn't so widely reported was a paper presented at mini-Copenhagen itself, which reveals climate scientists hugely "overestimated" the likelihood of irreversible ice melt on the Greenland icecap.

"The giant Greenland ice sheet may be more resistant to temperature rise than experts realised," reported the *Guardian* based on a new study by the University of Bristol.[337]

"The finding gives hope that the worst impacts of global warming, such as the devastating floods depicted in Al Gore's film *An Inconvenient Truth*, could yet be avoided."

It turns out the tipping point for Greenland is double what James Hansen and the rest of the alarmist team had been preaching – "it

336 "Defying predictions, sea level rise begins to slow", Michael Asher, *Daily Tech*, 15 December 2008
337 "Greenland ice tipping point 'further off than thought'," *The Guardian*, 10 March 2009, http://www.guardian.co.uk/environment/2009/mar/10/greenland-ice-sheet-climate-change

would take an average global temperature rise of 6ºC to push Greenland into irreversible melting," the new study points out.

Up to now, global warming believers and the UN IPCC have relied on a tipping point temperature increase of three degrees.

The scientists on the study say previous computer models used by the UN IPCC were too "simple", and did not factor in enough ancient data – such as a period 125,000 years ago when Greenland's temperature was "5ºC higher than today", and the ice sheet remained.

That piece of news might come as a shock, because if you recall Al Gore's movie he tried to tell the world it's never been as hot as it is today in 650,000 years.

Whilst warmer temperatures, if they persisted all century, would cause Greenland to shed some water, the rate of melt will not be as large as currently claimed, because current claims are based on a tipping point that's far too low, it turns out.

There goes that one metre sea level increase. Again.

SUMMARY: Sea levels are increasing, but there's no evidence of runaway rises or in fact any impact at all from human-induced global warming. On current tracking based on the highest figures of what we're actually measuring, sea levels are likely to rise about 30cm (12 inches) by 2100. There is no credible scientific support for scenarios of a two metre increase, let alone the extreme claims of five to ten metre rises that have been suggested by NASA's GISS unit.

AGW STATUS: Most of the ocean level rise can be accounted for through thermal expansion of the water, and that's consistent with increased solar activity over the past century heating up the oceans. Some of the sea level rise may be attributable to tectonic shifting of the sea floor and volcanic heating, but that data was overlooked when climate scientists were constructing their computer models for the IPCC. There is no unambiguous scientific evidence that CO_2 gas emissions from human enterprise are having any significant effect on the sea level.

These, then, are summaries of the data we've reviewed so far. Having established how weak the scientific case for human-caused global warming actually is, let's now turn attention to those promoting it and why.

Chapter Sixteen

The Real Agenda

"A world government [can] only be created out of war or crisis – an emergency that provide[s] an appropriate combination of the motivations of fear and opportunity"

– Herman Kahn, Council on Foreign Relations, 1993[338]

So if the climate science is wrong, what's the real motivation here? Why are so many people prepared to spend so much money convincing the public there's a secular Armageddon awaiting them?

You don't have to look past the usual suspects, Greed and Power. The battle to convince you of the reality of "climate change" is intricately entwined with the collapse of the world financial markets and the growing push for a de-facto world government.

Nobel Peace Prize winner and former Secretary of State Henry Kissinger told CNBC that the election of President Obama, combined with a unique set of crises, have created a political 'perfect storm'.

"The president-elect is coming into office at a moment when there is upheaval in many parts of the world simultaneously...His task will be to develop an overall strategy for America in this period when, really, a new world order can be created. It's a great opportunity, it isn't just a crisis."[339]

Yeah, it's the old 'New World Order' theory again, and I'd be tempted to write it off as well, except for the inconvenient fact that globalists keep on repeatedly talking about it. Clearly, it's on their agenda whether we choose to believe it or not.

---

338 "World Federal Government", essay by Herman Kahn and Anthony J Wiener, in *Uniting the Peoples and Nations*, published by World Federalist Association 1993

339 "Kissinger: Obama primed to create 'New World Order', *WorldNetDaily*, 6 January 2009

"Global issues demand global solutions"[340], is the buzzphrase being applied both to climate change and to the financial crisis. You'll spot it in a number of places, as I have footnoted, or perhaps in its new manifestation as "a global New Deal", but perhaps the most disturbing reference is this one:

"Global problems demand global solutions. And the United Nations is, truly, the world's only global institution," UN Secretary-General Ban Ki-moon told his audience at a meeting organized by the World Affairs Council.[341]

"Polls show that even larger majorities – 74 per cent to be exact – believe the United Nations should play a larger role in the world."

Now that former New Zealand Prime Minister, climate change priestess and globalisation powerhouse Helen Clark has taken up her new post heading the UN Development Programme this year, you can expect that organisation's dream of reining in the USA under UN hegemony and creating Kissinger's 'new world order' to come a whole lot closer.[342]

Speaking in San Francisco, the birthplace of the United Nations, Ban told his audience that climate change is the opportunity for massive new growth in the UN's power and influence.

"San Francisco is the birthplace of the United Nations, which was created to save this world from the scourge of war. I'm here to discuss the future of our planet Earth, and this can become the birthplace of a new movement to save it for future generations," he said.

It sounds so noble, but the origin of the business end of the global warming industry – the real agenda – is a tale of sleaze, power-hungry bureaucrats and an ungodly alliance of big business and non-governmental organisations (NGOs) who saw a chance to coat-tail on climate change panic and control world markets and environmental policies.

You'll recall that NASA's James Hansen was one of the first to publicly hype the global warming threat at a 1988 congressional hearing. You'll also recall Hansen served as a science consultant for Al Gore, and how Hansen received a $250,000 grant from the Heinz Foundation, run by 2004 presidential hopeful John Kerry's wife, Teresa Heinz-Kerry.

340 http://www.britishcouncil.org/bern_manifesto_on_climate_change-4.pdf

341 "UN best-placed to tackle global problems in today's world – Ban Ki-moon", UN news release, 26 July 2007, http://www.un.org/apps/news/story.asp?NewsID=23345&Cr=San&Cr1=Francisco

342 For a biography of Helen Clark, see *Absolute Power* by Ian Wishart, Howling At The Moon Publishing, 2008, ISBN 9780958240130, http://www.ianwishart.com, or Amazon.com

Well, that same James Hansen was also a global warming consultant to a company called Enron,[343] which – as you'll recall – went belly-up in 2002 owing more than $6 billion.

The deeper reasons for Enron's collapse are irrelevant here, but that company's central role in the business model for climate change is something everyone needs to understand.

Enron was a giant American energy company that was diversifying out of coal and into natural gas, solar and wind power. It owned the largest gas pipeline in the world outside of Russia, and it knew soon after Hansen's original testimony in 1988 that the writing was on the wall for excessive carbon-burning fuels.

Natural gas has a significantly lower carbon footprint than coal, so Enron also knew that if it could corner the gas market and make coal uneconomic as a competitor, Enron's shareholders stood to make huge amounts of money through the company's gas, wind and solar operations worldwide.

But Enron needed help in Washington to make it happen. The Clinton administration, with Al Gore in the veep-seat, took office early 1993, and Enron wasted no time saying hello. Gore had played a key role in the 1988 global warming hearings and was keen to use his new-found power to push the alarm-bells on climate change.

The Democrats had been working feverishly since 1988 to develop a 'green' business model. Senators John Heinz and Tim Wirth co-sponsored "Project 88", which was intended to provide a pathway for businesses to make money out of environmental issues, by buying and selling excess emissions credits for sulphur dioxide and nitrogen oxide – a couple of components in acid rain. Carbon dioxide was not, at that stage, on anyone's horizon. Project 88 became the 1990 Clean Air Act.

The Heinz name has already featured in this tale, but remember Tim Wirth's as well.

In 2005, *Investigate* magazine published a popular backgrounder on the Enron/Kyoto Protocol link, by weather forecaster Ken Ring:[344]

"Without Enron there would have been no Kyoto Protocol. About 20 years ago Enron was owner and operator of an interstate network of natural gas pipelines, and had transformed itself into a billion-

343 "Why Enron wants Global Warming", Patrick J. Michaels, Cato Institute, http://www.cato.org/pub_display.php?pub_id=3388

344 "The Kyoto Conspiracy: how Enron hyped global warming for profit", Ken Ring, *Investigate*, October 2005

dollar-a-day commodity trader, buying and selling contracts and their derivatives to deliver natural gas, electricity, internet bandwidth, whatever.

"The 1990 Clean Air Act amendments authorized the Environmental Protection Agency to put a cap on how much pollutant the operator of a fossil-fuelled plant was allowed to emit. In the early 1990s Enron had helped establish the market for, and became the major trader in, EPA's $20 billion-per-year sulphur dioxide cap-and-trade program, the forerunner of today's proposed carbon credit trade. This commodity exchange of emission allowances caused Enron's stock to rapidly rise.

"Then came the inevitable question, what next? How about a carbon dioxide cap-and-trade program? The problem was that CO_2 is not a pollutant, and therefore the EPA had no authority to cap its emission. Al Gore took office in 1993 and almost immediately became infatuated with the idea of an international environmental regulatory regime. He led a U.S. initiative to review new projects around the world and issue 'credits' of so many tons of annual CO_2 emission reduction."

Among the new appointees to the Clinton administration was former Colorado Democrat senator Tim Wirth, the new Undersecretary of State for Global Affairs. Wirth worked closely with Gore on climate and environmental issues, and Enron boss Ken Lay cultivated a relationship with Wirth to push the idea of some kind of market for carbon credits.

"Under law a tradable system was required, which was exactly what Enron also wanted because they were already trading pollutant credits," says Ken Ring. "Thence Enron vigorously lobbied Clinton and Congress, seeking EPA regulatory authority over CO_2. From 1994 to 1996, the Enron Foundation contributed nearly $1 million dollars – $990,000 – to the Nature Conservancy, whose Climate Change Project promotes global warming theories. Enron philanthropists lavished almost $1.5 million on environmental groups that support international energy controls to 'reduce' global warming.

"Executives at Enron worked closely with the Clinton administration to help create a scaremongering climate science environment because the company believed the treaty could provide it with a monstrous financial windfall. The plan was that once the problem was in place the solution would be trotted out," writes Ring.

Among the bedfellows Enron roped into the cause were the Heinz Foundation and the Pew Center, whose climate change division heavily hyped the global warming paranoia. Enron pumped serious money into groups like Pew, and stooped to some serious dirty tricks in the process.

A 1998 letter signed by Enron's then-CEO, Ken Lay, begged "President Clinton, in essence, to harm the reputations and credibility of scientists who argued that global warming was an overblown issue. Apparently they were standing in Enron's way," writes the Cato Institute's Patrick J. Michaels.

"The letter, dated Sept 1, asked the president to shut off the public scientific debate on global warming, which continues to this date. In particular, it requested Clinton to 'moderate the political aspects' of this discussion by appointing a bipartisan 'Blue Ribbon Commission'," continues Michaels.

"The purpose of this commission was clear: high-level trashing of dissident scientists. Setting up a panel to do this is simple – just look at the latest issue of *Scientific American*, where four attack dogs were called out to chew up poor Bjorn Lomborg. He had the audacity to publish a book demonstrating global warming is overblown.

"Because of the arcane nature of science, it's easy to trash scientists. Imagine a 1940 congressional hearing to discredit Einstein. 'This man actually believes the faster you drive, the slower your watch runs. Mr. Einstein, then why weren't you here yesterday?' The public, listening on radio, immediately concludes this Princeton weirdo is just another academic egghead. End of reputation.

"The proposed commission was billed as an 'educational effort' that would lead to 'subsequent policy actions', which the letter itself recommended. These included a directive to 'establish the rules for crediting early, voluntary emissions reductions [of carbon dioxide].' And who was going to sell these credits? Enron, of course."

Enron poured huge amounts of money into greasing the right political palms. Internal Enron memoranda disclose some of these activities, such as an August 4, 1997 meeting at the Whitehouse with Clinton and Gore to drum out an approach to the upcoming Kyoto conference that December.[345]

"In an August 1997 memo by Mr. Lay to all Enron employees,

345 "Controlling hypocritical authority", Christopher Horner, *National Review*, 23 April 2002, http://www.nationalreview.com/comment/comment-horner042302.asp

the chairman said Mr. Clinton and Mr. Gore had 'solicited' his view on how to address the issue of global warning 'in advance of a climate treaty to be negotiated at an international conference.' That memo said Mr. Clinton agreed a market-based solution, such as emissions trading, was the answer to reducing carbon dioxide in the atmosphere."[346]

Ken Lay would later give key staff a briefing on his Washington sojourn, noting in an internal Enron memo that "the Kyoto agreement, if implemented, would 'do more to promote Enron's business than almost any other regulatory initiative outside of restructuring the energy and natural gas industries in Europe and the United States'."[347]

Tim Wirth was doing his bit, having been delegated the task as lead negotiator for the US in Kyoto, and when Vice President Al Gore signed off on Kyoto in December 1997, Enron thought all its Christmases had come at once.

"This agreement will be good for Enron stock!!" exclaimed one of Enron's men in Kyoto, who in the same memo to head office peppered the main points of the newly-agreed Kyoto Protocol with phrases like, "we won", "another victory for us", and "exactly what I have been lobbying for".[348]

But alas, it wasn't to be. Despite pouring almost $5 million into political campaign donations for both Democrats and Republicans, Enron's dreams of the US ratifying Kyoto were shattered when the Clinton administration realised it couldn't get the numbers to pass the law.

The reason was simple: adopting Kyoto would raise fuel prices by more than 50%, and electricity prices would almost double. The cost to the US economy annually was estimated at as much as $400 billion, yet the gains according to Clinton's science advisers would be negligible, with almost no reduction in world temperatures by 2050.

Kyoto wasn't actually about winding back warming, it was about commercialising fear of warming. The multinationals involved were financially supportive of high-profile environmental groups like Greenpeace, so it was a nice cosy relationship between big business and big Green.

---

346 "Enron Gave Cash To Democrats, Sought Pact Help," Jerry Seper, *The Washington Times*, 16 January 2002

347 Dan Morgan, "Enron Also Courted Democrats," Dan Morgan, *The Washington Post*, 13 January 2002

348 "Outside View: Caught En Flagrente Kyoto," Christopher Horner, *United Press International*, 31 January 2002

Money – the $400 billion – doesn't just "disappear" however. If Kyoto was ratified, energy companies like Enron and other large multinationals would suddenly become very rich. There are foreign exchange markets, but we don't all have to buy foreign exchange. There are gold markets, but we don't all have to buy gold. Imagine if there was a *compulsory world-wide market* in carbon credits, however, that everyone effectively had to be represented in by international decree – imagine the money that dealers in a carbon credits 'exchange', clipping the ticket on both sales and purchases, could make?

No one could buy or sell without going through this new market, and those who controlled the market controlled the world.

Kyoto, the first step towards this market, was little more than a direct raid on the pockets of consumers under the guise of "a good cause", and savvy politicians could see through it.

Gore was disappointed at Kyoto's defeat in the US, but Tim Wirth was even more so. He quit the Clinton administration to take up a new globalisation job at the invitation of CNN founder Ted Turner, as president of Turner's billion-dollar United Nations Foundation.

Another Clinton official, Eileen Claussen, quit to set up an environmental lobby group part funded by Enron. Having failed to get bang from its Democrat political donation bucks, Enron's documents show the energy giant turned its lobbying guns on Republicans.

Ken Lay used Tim Wirth as his middleman to lobby the CEO of multinational Alcoa, Paul O'Neill, who became George W. Bush's Treasury Secretary in 2000. O'Neill tried to resurrect Kyoto but was shot down by fellow Republicans, and that was the end of Enron's dream.[349]

Except the dream isn't dead. The identities of the players may have changed, in that Ken Lay and some of his Enron colleagues are now behind bars, but like rust the United Nations, its supporters and big energy don't sleep. The election of President Obama on a ticket of turning back the tides and implementing a cap on carbon production, and an emissions trading scheme, shows that the idea is once again firmly on the agenda and more likely than ever to become global policy.

349 "Enron's Ken Lay saw Kyoto Treaty 'Green' Only As The Color Of Money", Robert D. Novak, *New Hampshire Union Leader*, 21 January 2002

If it does, it will be the largest legalised bank robbery in history.

The Paris-based International Energy Agency last year issued its World Energy Outlook.[350] The independent Citizen Power Alliance costed it.[351]

"Assuming an average 3.3% global economic growth over the 2010-2050 period, governments and the private sector would have to make additional investments of US$45 trillion in energy, or 1.1 percent of the world's gross domestic product," reports the CPA.

Hard on the heels of the world financial collapse, and growing calls for state control of economies in a semi-socialist partnership with global businesses (the so-called 'Third Way'), emissions trading schemes may yet replace banking derivatives as the new financial weapons of mass destruction for multinational corporates.

It is multinationals, global players, who will benefit most, because like Al Gore's cunning way of paying himself carbon credits, businesses with plants in both the developed and developing world can shift emissions credits around in a way that small to medium-sized businesses cannot.

Global players are already setting up the infrastructure they need once the Copenhagen Treaty later this year sets up a compulsory global cap and trade emissions scheme (cap carbon output, sell credits).

The multinationals will be able to use the vast sums of cash sucked out of the leading Western economies to invest in developing economies where the rules will not be so stringent. In essence, not only is it a massive transfer of wealth from the developed world to the third world, but it serves a political purpose as well.

Most members of the United Nations come from the developing world. Most members of the United Nations stand to see massive investment in their countries if emissions schemes are introduced, because manufacturing there will be less restricted by carbon limits and conversely multinationals will make money by building new, greener power plants there. Politically, the UN not only sees this as levelling the playing field and creating a truly global economy, but it also sees it as a power shift away from the US.

The idea of finally bringing all the nations of the world together is not a new one. Various empires throughout history have attempted

350 IEA, World Energy Outlook Executive Summary, http://www.iea.org/Textbase/npsum/WEO2008SUM.pdf

351 "45 trillion needed to combat warming", CPA, http://cpagroup.pbwiki.com/%2445-trillion-needed-to-combat-warming-

to do just that. But every empire needs an Emperor, someone or something regarded as the ultimate authority and point of arbitration. In the 21st century, that figurehead is the United Nations general assembly, the nearest thing we have to a world parliament.

Since the UN was formed in 1945, thousands of treaties have been ratified by various of its 192 member countries. And each one of those treaties involves the nation signing it ceding a portion of its sovereignty to the UN.

As you saw at the start of this chapter, current UN Secretary-General Ban Ki-moon is making no secret of his ambitions for the UN to become a world government entity, but is that a pipe dream, a fantasy?

Not if you listen to some of the world's leading politicians.

If Enron saw Kyoto as a chance to make squillions, French President Jacques Chirac correctly identified the political aim:

"For the first time, humanity is instituting a genuine instrument of global governance...by acting together, by building this unprecedented instrument, the first component of an authentic global governance, we are working for dialogue and peace...to organise our collective sovereignty over this planet, our common heritage," Chirac told world leaders.[352]

"Today, at The Hague, the international community, represented by the world's environmental ministers, has a moral and political duty to move forward in the right direction."

And the direction, just as you saw a few paragraphs ago, is about transference of wealth.

"Are we going to allow the gap between rich and poor to grow ever wider," asked Chirac, "with the former adapting their activities at the expense of colossal defensive investments while the latter have no choice but to submit, for want of the means to modify their practice and policies? Once again, the question is, do we want to control and regulate the process of globalisation in order to make it fairer?"

The answer to Chirac's rhetorical question wasn't far behind.

"Solidarity between North and South means that the North cuts its emissions so that the South can develop while maintaining control over the growth in its own emissions."

352 Speech by Jacques Chirac, President of France, to the VIth Conference of the Parties to the UN Framework Convention on Climate Change", The Hague, 20 November 2000. http://www.sovereignty.net/center/chirac.html

Chirac revealed the aim was for a Protocol that provides compulsory targets and enforceable penalties, under UN authority, against countries that don't comply.

Chirac's speech caused ripples of concern that the cat had been let out of the bag, with columnist Henry Lamb noting that the final agreement thrashed out at The Hague would not be so audacious in public.

"It will not likely repeat Jacques Chirac's reference to global governance. But like a stone once thrown, his words cannot be recalled, and his words will have an impact wherever they land...Those sceptics who have been unwilling to believe that the UN is, indeed, contriving a world government need only to listen to the President of France."

European Environment Commissioner Margot Wallstrom only stoked the fires further when she too admitted Kyoto's purpose was much bigger than merely environmental issues.

"[Kyoto] is not a simple environmental issue where you can say it is an issue where the scientists are not unanimous," she said. "This is about international relations, this is about economy, about trying to create a level playing field for big businesses throughout the world."[353]

In 2008, a UN organised workshop entitled, "Environmental democracy, transparency and global governance: The Road from Rio to Copenhagen", examined the need for political globalization to fulfil the goals of combating climate change.[354]

The phrase "global governance" did not emerge from nowhere, and nor is it used accidentally. In international geopolitics, "accidents" are few and far between. Although the concept of global governance has been around for decades, it has gained momentum since the early 1990s. Revolutionaries, like rust, never sleep.

In 1994, the United Nations Development Program published its "Human Development Report", containing barely hidden references to a future world government run by the UN.[355]

"WE STRONGLY BELIEVE that the United Nations must become the principal custodian of our global human security.

353 "WTO To Face U.S.-E.U. Kyoto Dispute", Christopher C. Horner , *The Washington Times*, 21 September 2002

354 Michael Stanley-Jones Environmental Information Management Officer UNECE Aarhus Convention Secretariat 13th International Anti-Corruption Conference Workshop 1.4: Governing the Climate Change Agenda – Making the Case for Transparency Athens, 30 October 2008

355 http://hdr.undp.org/en/media/hdr_1994_en_overview.pdf

Towards this end, we are determined to strengthen the development role of the United Nations and to give it wide-ranging decision-making powers in the social-economic field by establishing an Economic Security Council.

"We must seek a new role for the United Nations so that it can begin to meet humanity's agenda not only for peace but also for development.

"There must be a 'New World Social Charter' where the world will redistribute wealth as it cannot survive one-quarter rich and three-quarters poor, and where the UN must become the principal custodian of global human security and help with basic education, healthcare, immunization, and family planning," states the report.

"A major restructuring of the world's income distribution, production and consumption patterns may therefore be a necessary precondition for any viable strategy for sustainable human development."

To fund this grand idea of world government requires money, but the UNDP (which as of April 21 2009 is being run for the first time by a political leader, in the form of NZ's globalisation powerhouse Helen Clark) had that sussed as well:

"Global taxation may become necessary in any case to achieve the goals of global human security. *Some of the promising new sources include tradable permits for global pollution*, [my emphasis] a global tax on non-renewable energy, demilitarization funds and a small transaction tax on speculative international movements of foreign exchange funds."

I know I'm being as subtle as a sledgehammer in a china shop, but follow me here for a moment. The year was 1994, and some kind of global tax on pollution emissions was already being mooted – not to save the planet from pollution but to fund the United Nations dream of governing the world.

Elaborating further on the funding options, the report notes:

"A second logical source of funds for a global response to global threats is a set of fees on globally important transactions or polluting emissions. This is probably some way off, but even at this stage it is worth considering some of the more promising options, two of which are discussed in chapter 4.

"One is a tax on the international movements of speculative capital suggested by James Tobin, winner of the Nobel Prize for Economics (special contribution, p. 70). Tobin suggests a tax rate of 0.5% on such transactions, but even a tax of 0.05% during 1995-2000 could raise $150 billion a year. Such a tax would be largely invisible and totally non-discriminatory."

It is worth noting that at 0.5%, the transaction tax would raise US$1.5 trillion a year in 1994 terms for the United Nations, or 150 times the amount of its 1994 budget of $10 billion.

In 2000, *Boston Globe* editor and global warming believer Ross Gelbspan delivered a speech also calling on introduction of the Tobin Tax:[356]

"Finally, I think we need to use a tax on international currency transactions to finance the transfer of climate-friendly technologies to the developing world. Virtually all developing countries would love to solar; virtually none can afford it. The currency transaction tax was conceived by Dr. James Tobin, a Nobel Prize-winning economist at Yale, as a way to stabilize volatile capital flows."

Spookily, and entirely 'coincidentally' no doubt, there is talk in 2009 of the Obama administration of introducing the Tobin Tax as a result of the financial market meltdown. The President's key economics advisers on the stimulus package are disciples of Tobin's ideas.

"After a three-decade run," reported *Bloomberg*[357] in February, "the free-market philosophies of Friedman that shaped U.S. policy are being eclipsed by the pro- government ideas of Tobin, the late Yale economist and Nobel laureate who brought John Maynard Keynes into the modern era."

Not bad foreshadowing from the UN Development Programme 15 years ago. Another prediction on the money was carbon taxes:

"Another [option] is a global tax on energy: a tax of $1 on each barrel of oil (and its equivalent on coal) during 1995-2000 would yield around US$66 billion a year," said the UN report.

Call me a hardbitten cynical investigative journalist if you will (because, frankly, I am), but as I mentioned a moment ago experience has trained me not to believe in 'coincidence theory' when it comes to geopolitics. In my view, anthropogenic global warming theory is nothing more than a propaganda stunt: manufacture the appearance of a crisis, then present the public with a solution. Cli-

356 "Ross Gelbspan on Global Warming", Foreign Policy In Focus Presentation, 12 July, 2000, http://www.fpif.org/commentary/2000/gelbspan_presentation_body.html

357 "Yale's Tobin Guides Obama From Grave as Friedman Is Eclipsed", *Bloomberg* Wire Service, 27 February 2009, http://www.bloomberg.com/apps/news?pid=20601087&sid=ajz1hV_afuSQ&refer=home

mate change theory, driven by the UN IPCC, is the UN's Trojan Horse gift to the citizens of the world.

The specifics of this December's proposed Copenhagen Treaty are already being hammered out behind closed doors in Europe and at the UN. However, a London School of Economics report by Nicholas Stern (of the infamous 'Stern Review' fame), gives a hint of what's in store.[358]

Firstly, the Treaty will make it clear that humans are causing global warming and, more specifically, that it is wealthy western humans who must take the blame and bear the financial cost of fixing it.

"Equity" says Stern, "must take account of the fact that it is poor countries that are often hit earliest and hardest, while rich countries have a particular responsibility for past emissions."

His reasoning is only correct if, in fact, global warming is caused by humans. If it isn't, then the proposed plan to make Western taxpayers build clean factories in the Third World is nothing more than bank robbery, where your retirement savings, your children and grandchildren will be the victims.

One analysis suggests even modest greenhouse gas emission cuts in the US would cost households more than US$3,300 a year, and kill 2.4 million jobs in the US economy.[359]

If you harbour any doubts that consumers will ultimately pay, Stern says the Copenhagen Treaty is likely to begin to address where the final bill lies, "detailing how to allocate responsibilities on emissions between producers and consumers, as this is an important element of the equity story.

"Current targets are expressed in terms of production, but it is not clear whether producers or consumers should be responsible for emissions associated with products consumed."

Regardless of what Copenhagen declares, it is an economic point of fact that the cost of producing a product is passed on to consumers.

Stern recommends a 'cap and trade' worldwide emissions trading system, supplemented by regulation and carbon taxes in individual countries, although he warns against relying too much on carbon taxes because they don't achieve the UN goal of wealth transference to the poorer nations:

"Importantly, from the standpoint of both equity and global

358 "Key Elements of a Global Deal on Climate Change", Nicholas Stern et al, LSE, 2008
359 http://www.heartland.org/events/NewYork09/background.html

efficiency, a carbon tax would not automatically channel finance for low-carbon development towards developing countries."

Even one of the world's ultimate greens, Professor James Lovelock of 'Gaia Hypothesis' fame, admits the UN push for a worldwide carbon emissions trading scheme is just a "scam" on the gullible public:

"Most of the 'green' stuff is verging on a gigantic scam. Carbon trading, with its huge government subsidies, is just what finance and industry wanted. It's not going to do a damn thing about climate change, but it'll make a lot of money for a lot of people and postpone the moment of reckoning," Lovelock told *New Scientist* magazine.[360]

Lovelock also ridicules the move to biofuels and windfarms:

"I am not against renewable energy, but to spoil all the decent countryside in the UK with wind farms is driving me mad. It's absolutely unnecessary, and it takes 2500 square kilometres to produce a gigawatt – that's an awful lot of countryside."

At least Lovelock can see the green spin for what it is. Empty rhetoric.

"What then energizes all these false prophets of doom and their demands for immediate drastic action?," asks Paul Driessen in his essay *Prophets, False Prophets and Profiteers*.[361] "Simply put, profits and power.

"Just the 12 largest environmental lobby groups in the U.S. had a combined budget of $2 billion in 2003. Collectively, the global environmental movement has a war chest of up to $8-billion a year. That buys a lot of influence, but apparently it's never enough. As National Audubon Society chief operating officer Dan Beard has admitted, 'What you get in your mailbox is a never-ending stream of crisis-related shrill material designed to evoke emotions, so that you will sit down and write a check.'

"Global warming is big business. The U.S. government ladled out $15 billion on global warming research and 'education' between 1992 and 2000. The United Nations spent billions more, as did the European Union, and big foundations provided hundreds of millions more.

"Unfortunately, most government money goes to researchers who support the position that human-caused climate change is a serious

360 "One last chance to save mankind", *New Scientist*, 23 January 2009, http://www.newscientist.com/article/mg20126921.500-one-last-chance-to-save-mankind.html?full=true

361 http://www.cgfi.org/2004/12/15/prophets-false-prophets-and-profiteers/

problem. Foundation money does likewise, for operations like the Pew Charitable Trusts' Global Climate Change Center, and an International Institute for Sustainable Development $700,000 study of 'how farmers in India may be vulnerable' to problems supposedly caused by 'economic globalization and climate change'."

So the money, lots of it, is already being spent on programmes and studies favourable to the idea of human-caused global warming. How does that affect ordinary householders like you? By enabling climate alarmists to bamboozle you every night with bad news climate stories, and blind you with their science. More importantly, the more people 'buy the lie', the easier it is to convince politicians to increase your taxes in the name of saving the planet.

In academic papers, "scientists" are discussing openly the need for propaganda and manipulation of the public to achieve climate change goals:[362]

"There is a largely invisible and dormant, but exceedingly powerful resource in every community that can be harnessed to promote climate change mitigation and adaptation goals: opinion leaders. If activated, popular opinion leaders – not necessarily famous media figures such as Oprah Winfrey but rather the person down the block, the person many of us in the neighbourhood turn to for advice when making the right decision really counts – are a potentially important asset in accelerating individual behaviour change and fostering citizen demand for carbon regulation."

In other words, they want to scare you so much you'll be begging for them to save you from environmental apocalypse. But maybe they are more sophisticated than that now. As the paper admits, the public are starting to wise up to the failure of the scare stories, and the authors suggest a strategic change:

"Going forward, they recommend that climate campaigners use fear appeals in limited and selective ways."

Those new ways include information overload, or what they call "big messy programs", so that everywhere you turn for information, the internet, radio, TV, pop-culture, the information has been provided by climate change believers:

"Communication initiatives with many communication activities, by many sources, delivered through many channels – are most likely

362 "No More Business as Usual", *Science Communication*, March 2009, pp299-304 http://scx.sagepub.com/cgi/reprint/30/3/299?rss=1

to be successful. The Kahlor and Rosenthal study suggests that, to the extent possible, climate change communication efforts should explicitly engage a plurality of information channels."

They don't want you to hear both sides of the debate, however:

"In any topic as complex as climate change, not all knowledge has equal value in informing the important decisions that people face. Because of the inherent limits of communication in improving people's knowledge of any complex issue, communication planners must make every possible effort to identify the information most worth knowing and focus their communication outreach accordingly.

"Having less knowledge, if what is known is more worth knowing, can have greater individual and societal value than having more knowledge that is less worth knowing."

Surely it would have been easier simply to say: "trust us, we know best what's good for you, don't you worry your pretty little heads about the details." Or perhaps that was just too blunt.

Instead, they want you to think that concern about climate change is all your idea, and that when you have become sufficiently brainwashed to believe in human-caused global warming, you'll know who to call:

"Although it may be trite, it is not hyperbolic to say that never before have our skills and insights been more desperately needed... [and there are] important opportunities to use our skills in ways relevant to one of the most serious challenges ever to face human civilization."

Cue majestic pipe organ.

Little wonder that President Obama promised to put "science" back on its rightful pedestal; that's code for "the new religion", provided they're Obama's 'tame' scientists.

But it wasn't only Enron, scientists and European leaders yanking the Democratic Party's chain. Another massive political donor to the Democrats, and President Obama himself, is shady billionaire George Soros, who makes no secret of wanting a single world government under UN auspices.

His fingers, as you'll see, are in every little pie.

Chapter Seventeen

The Audacity Of Dope

"The horde of 'journalists' in attendance had come not as news reporters but as advocates and propagandists. They were there to regurgitate and retail as gospel whatever globaloney the UN and its proponents dished out"

– *William Jasper, 2001*[363]

Punch the name "George Soros" into Google News, and chances are you'll find a reference to "billionaire philanthropist" somewhere.

"When George Soros speaks, the global financial community listens. Yesterday, when the 76-year-old financier and philanthropist gave warning that the world was facing its most serious financial crisis since the Second World War, the statement struck home," wrote Susan Thompson in the *Times of London*. [364] Admittedly the *Times* also quoted those who called Soros an "economic war criminal" and one who "sucks the blood from the people".

The media often talk of Soros as, "the man who broke the Bank of England", a reference to his 1992 attack on the British pound in the financial markets that pushed Britain's economy to the brink and caused double digit interest rates.

Less frequently, the media talk of Soros the convicted criminal, an insider trader, and even more infrequently do they mention Soros' involvement with drugs.

Neil Clark at *The New Statesman*, bastion media organ of the Left, nonetheless had the wit to see George Soros as he really is, not as Soros tells us it is:[365]

363 *The UN Exposed* by William F Jasper, The John Birch Society 2001, page 100

364 "Business big-shot: George Soros", *The Times*, 22 January 2008

365 "Profile – George Soros", Neil Clark, *The New Statesman*, 2 June 2003, http://www.newstatesman.com/200306020019

"Soros likes to portray himself as an outsider, an independent-minded Hungarian émigré and philosopher-pundit who stands detached from the US military-industrial complex. But take a look at the board members of the NGOs he organises and finances.

"At Human Rights Watch, for example, there is Morton Abramowitz, US assistant secretary of state for intelligence and research from 1985-89, and now a fellow at the interventionist Council on Foreign Relations; ex-ambassador Warren Zimmerman (whose spell in Yugoslavia coincided with the break-up of that country); and Paul Goble, director of communications at the CIA-created Radio Free Europe/Radio Liberty (which Soros also funds).

"Soros's International Crisis Group boasts such 'independent' luminaries as the former national security advisers Zbigniew Brzezinski and Richard Allen, as well as General Wesley Clark, once NATO supreme allied commander for Europe. The group's vice-chairman is the former congressman Stephen Solarz, once described as 'the Israel lobby's chief legislative tactician on Capitol Hill' and a signatory, along with the likes of Richard Perle and Paul Wolfowitz, to a notorious letter to President Clinton in 1998 calling for a 'comprehensive political and military strategy for bringing down Saddam and his regime'.

"Take a look also at Soros's business partners. At the Carlyle Group, where he has invested more than $100m, they include the former secretary of state James Baker and the erstwhile defence secretary Frank Carlucci, George Bush Sr and, until recently, the estranged relatives of Osama Bin Laden. Carlyle, one of the world's largest private equity funds, makes most of its money from its work as a defence contractor.

"Soros may not, as some have suggested, be a fully paid-up CIA agent. But that his companies and NGOs are closely wrapped up in US expansionism cannot seriously be doubted," writes Neil Clark.

Soros, a hard agnostic bordering on atheist, claims his motivation is to create an "Open Society" based on the ideas of philosopher Karl Popper, and indeed his main fundraising arm is the Open Society Institute. During the Cold War, he and his "Open Society" ideas were useful to the West in helping infiltrate communist countries and funnel money to dissident groups. By a process of capturing hearts and minds, often through bribery of key people, he helped bring about grass roots revolutions.

What's being lost sight of is that he's a modern Pied Piper, whose alluring tunes are now being used with devastating effect to mesmerise key power blocs in the West for use in his schemes. His supporters call him a "philanthropist". Those not so easily taken in are less effusive in their praise.

"Generally the sad conclusion is that for all his liberal quoting of Popper," remarks Clark in *The New Statesman* article, "Soros deems a society 'open' not if it respects human rights and basic freedoms, but if it is 'open' for him and his associates to make money. And, indeed, Soros has made money in every country he has helped to prise 'open'. In Kosovo, for example, he has invested $50m in an attempt to gain control of the Trepca mine complex, where there are vast reserves of gold, silver, lead and other minerals estimated to be worth in the region of $5bn. He thus copied a pattern he has deployed to great effect over the whole of eastern Europe: of advocating 'shock therapy' and 'economic reform', then swooping in with his associates to buy valuable state assets at knock-down prices."

You would think alert investigative journalists in the Western media would be all over Soros like a rash. But they're not, and there's a reason for that, which *New Statesman's* Neil Clark alludes to:

"More than a decade after the fall of the Berlin Wall, Soros is the uncrowned king of eastern Europe... With his financial stranglehold over political parties, business, educational institutions and the arts, criticism of Soros in mainstream eastern European media is hard to find."

Here is a man who has essentially purchased the intelligentsia and the powerful of Europe, and the story of how he did it, and how he's now captured the US as well, is central to unravelling the real agenda behind the global warming scare.

In a recent *Financial Times* profile was this wonderful gem:[366]

"Strobe Talbott, now the president of the Brookings Institute and a former deputy secretary of state, said: "[Soros] likes to think of himself as an outsider who can come in from time to time, including to the Oval Office, where I took him on a couple of occasions. But simply hobnobbing with the powerful isn't important."

No, hobnobbing wouldn't do it. *Owning them outright* appears to be the Soros style.

366 "The credit crunch according to Soros", Chrystia Freeland, *Financial Times*, 30 January 2009 http://www.ft.com/cms/s/2/9553cce2-eb65-11dd-8838-0000779fd2ac.html

According to CNS News, George Soros used a loophole in political donation laws to provide a young Barack Obama with $60,000 in funds to win his Illinois senate seat.[367]

"Not only did Soros donate to Obama's campaign, but four other [Soros] family members – Jennifer, sons Jonathan and Robert and wife Susan – did as well. Because of a special provision in campaign finance laws, the Soroses were able to give a collective $60,000 to Obama during his primary challenge."

Describing that situation as "unique", CNS News explained some other unusual aspects of the Obama/Soros relationship.

"Obama is one of only a handful of candidates to get a personal contribution from George Soros. The others include Senate Minority Leader Tom Daschle [Obama appointee, withdrew], D-S.D.; Sens. Barbara Boxer, [Chair, Senate Environment Ctte] D-Calif., Hillary Clinton, [Obama appointee] D-N.Y.; Bob Graham, D-Fla.; John Kerry, [Chair, Senate Foreign Relations Ctte and Obama envoy] D-Mass.; Patrick Leahy, [Chair, Senate Judiciary Ctte] D-Vt.; U.S. Rep. Tom Lantos [deceased]; and former Vermont governor Howard Dean."

Of those politicians, Clinton, Kerry, Graham and of course Obama became presidential candidates. Even now, Soros has purchased influence with a sizeable chunk of the Obama administration.[368]

"Why did George support Obama?" his spokesman, Michael Vachon, asked rhetorically of CNS News in 2004. "Because when they met in Chicago a couple of months ago, it was apparent that Barack Obama was an emerging national leader, and he would be an important addition to the Senate."

"Vachon said Obama was the only candidate this election cycle Soros had met personally, with the first powwow in March [2004]. Asked why Soros hasn't sought out a meeting with Kerry, the man he is pulling for to defeat President Bush on Nov. 2, Vachon said it was just a matter of Soros keeping his distance.

"George is a major funder of an independent 527 [political lobby fundraising] group, and it probably makes more sense for him and Kerry to keep each other at arm's length," Vachon said.

Those meetings with Obama caught the attention of the Illinois Republican Party, said spokesman Jason Gerwig.

367 "Unlike Kerry, Obama Covets Soros' Support", Robert B. Bluey, *CNSNews.com*, 28 July 2004, http://archive.newsmax.com/archives/articles/2004/7/28/164147.shtml

368 In case you're wondering, Soros also funded John McCain, http://www.worldnetdaily.com/index.php?fa=PAGE.view&pageId=56177

"Barack Obama and his liberal voting record have gotten a free ride," Gerwig said. "His aspirations seemed to be focused more nationally now than they do on Illinois, especially if you look at some of the money he's taken from Soros and from left-coast liberals."

Not that Soros isn't liberal enough for all of them. With his businesses using tax haven banks tied to Colombian drug cartels, his purchase of a major stake in a Colombian bank, and Soros never really explaining where he made his fortune (although he categorically denies it is drug money), there are already concerns about just what the billionaire really represents. Those concerns have been heightened in many circles by his wish to see hard drugs legalised in a process he calls "harm minimisation" – another buzzphrase doing the rounds at the moment.

"Away from the scrutiny or even the notice of the establishment press, Soros has emerged as a counter-culture hero," reports Cliff Kincaid at AIM.[369]

"The drug culture magazine, *Heads*, calls him "Daddy Weedbucks," ran an excerpt from his book, *Soros on Soros*, and declared that 'he drops the bucks exactly where they're needed'. The September-October issue of the drug culture magazine *High Times* recognizes the stakes, noting that there are 'ten reasons to get rid of Bush' and that one is that there will be 'No legalization of pot' under Bush. The implication of the article was that the situation would change under Kerry.

"None of this is being reported, however, by the major media. His partner, Peter Lewis, whitewashed by the *Post* as 'one of the country's 10 most generous philanthropists', was actually arrested in New Zealand for 'importing' drugs, including hashish and marijuana."

As the newspaper *TGIF Edition* revealed, Soros' networks not only helped fund a NZ Government sponsored conference on drugs in February 2009, but they managed to stack the speakers list with representatives of other groups he was funding, all pushing the same "it's time to legalise drugs" line.

Soros' paid lobby groups and officials have even gone so far as to try and ensure the Taliban are permitted to keep harvesting opium poppies so as to ensure heroin remains available for supply.[370]

369 "The Hidden Soros Agenda: Drugs, Money, the Media, and Political Power, Special Report," Cliff Kincaid, *Accuracy in Media*, 27 October 2004, http://www.aim.org/special-report/the-hidden-soros-agenda-drugs-money-the-media-and-political-power/

370 "Afghan Opium Pleases Taliban and Soros", by Ramtanu Maitra, Executive Intelligence Review, 22

Before you go scampering to "independent" media watchdog sites like Media Matters for rebuttal, however, be aware that Media Matters and most of the left-wing 'media watchdogs' are or have been funded by George Soros.

Examples listed by Kincaid suggest Soros has bought off many US investigative journalists:

- The Center for Investigative Reporting, $75,000
- The Center for Public Integrity, headed by former CBS news producer Charles Lewis, $246,000
- The Fund for Investigative Journalism, $200,000
- Indymedia, $70,000
- The Committee to Protect Journalists (presumably not from the likes of Soros), $220,000
- MediaChannel.org, a "media issues supersite, featuring criticism, breaking news and investigative reporting from hundreds of organisations worldwide", headed by former ABC and CNN journalist Danny Schecter, $160,000

And that's just scratching the surface. All those agencies have been compromised by Soros money. Key issues for Soros include the legalisation of drugs worldwide, protesting the Iraq war (he's been a big funder of antiwar protests), and the environment.

"He has handed $3.1 million to the left-wing Tides Foundation, which funds organizations, such as the Sea Shepherds, Earth First! and the Ruckus Society, that have condoned or engaged in eco-terrorism," reports *Investors Business Daily*.[371]

When Sinclair Broadcasting, a US media company that owns 58 TV stations and reaches up to 22% of American households, planned to run a documentary in the leadup to the 2004 election on [Soros protégé] John Kerry, Soros' bought-and-paid-for journalists rode in to attack.

"The power of the Soros-supported media network," writes Cliff Kincaid, "was demonstrated in mid-October [2004] when a controversy emerged over Sinclair Broadcasting airing parts of *Stolen Honor*, a film raising questions about the detrimental impact of John Kerry's 1971 anti-war testimony on U.S. Vietnam POWs being held by the communists. Kerry had branded U.S. soldiers as war

---

August 2008, http://larouchepub.com/other/2008/3533opium_pleases_soros.html

371 "George Soros: The Man, The Mind And The Money Behind MoveOn", 20 September 2007 http://www.ibdeditorials.com/IBDArticles.aspx?id=275181103776079

criminals, and POWs interviewed in *Stolen Honor* said this resulted in more torture to them.

"The Democratic Party, the Kerry campaign, and various groups denounced Sinclair for planning to air *Stolen Honor*. MediaChannel.org, Common Cause, the Alliance for Better Campaigns, Media Access Project, Media for Democracy, and the Office of Communication of the United Church of Christ held an anti-Sinclair news conference. They denounced Sinclair for allegedly abusing the public airwaves by planning to air 'propaganda'. All of these organizations – except for the possible exception of the Office of Communication of the United Church of Christ – are funded by Soros.

"Media Matters, a left-wing media watchdog group that was also pressuring Sinclair to abandon plans to air the testimony of the former POWs, was 'developed' with help from the Center for American Progress, funded by Soros."

All of which makes Soros a powerbroker, and a driving force for changing the world the way he wants it.

Soros has advocated a tax on global financial transactions as a means of financing world government. In one of the UN's own TV shows, that's a theme they too were pushing according to this account from the watchdog group Accuracy in Media:[372]

"The *World Chronicle* U.N. television program, which features interviews with U.N. officials, 'is available free of charge to authorized broadcasters who agree to give credit to the United Nations each time a program is aired,' its website says. 'Guests are interviewed by a panel of journalists from international news organizations accredited to the United Nations,' it adds. Guests have included Kofi Annan, Ted Turner and Dan Rather. One *World Chronicle* program, recorded on May 3, 2004, included the topic of global taxes on U.S. citizens. Tony Jenkins was the moderator and James Wurst of *U.N. Wire* and Louis Hamann of the CBC were on the show to question a U.N. official. Hamann asked if a global tax on financial transactions could become a new source of foreign aid. The official replied that we have to pay 'much attention' to 'innovative financing sources'. Hamann pressed, 'But do you think the idea of a global tax could ever come to be?' The official said there was no consensus now but that 'times may change'."

372 "Journalists exposed on the UN payroll," Cliff Kincaid, *Accuracy in Media*, 15 February 2005, http://www.aim.org/special-report/journalists-exposed-on-the-un-payroll-george-soros-ted-turner-pay-for-journ/

You've seen George Soros' tentacles in liberalisation of drugs, even heroin, under markets which he presumably or other multinationals could invest in. You've seen him buying influence with politicians of every hue. You've seen him pay off the news media and fund journalism prizes, so that his press continues to be positive. You've seen he has an absolute global, one world government agenda. So it's no surprise to find him heavily involved in promoting the global warming scare because, as we all know, "global problems demand global solutions".

"Last Friday," reported the *New York Times*,[373] "in an interview with Bill Moyers[374] on PBS, George Soros, who has made billions of dollars based on his ability to read the ebb and flow of markets, suggested that investing in alternative energy technologies, refurbishing aging electricity grids and pursuing household energy efficiency, among other green strategies, could yet save the global economy.

"Mr. Soros, whose prescient[375] book *The New Paradigm for Financial Markets: The Credit Crisis of 2008 and What It Means*, was published in May [2008], told Moyers that the business of green could serve as the new 'motor of the world economy' – echoing a refrain he has used before."

Echoing, too, the refrain of the European Commission, Enron and others. Forget the climate (and this is ultimately why they're not that interested in the science), this whole global warming debate is about political power and a new world order.

And if you want that phrase straight from the Soros' mouth, try this for size from his interview, where Bill Moyers was asking the solution to the current financial crisis:

"Recapitalize the banks. And then work on a better world order where we work together to resolve problems that confront humanity like global warming. And I think that dealing with global warming will require a lot of investment."

Get a sense that Soros and his five children might be planning to make a sizeable buck out of global warming? He sees so many opportunities for business:

373 http://greeninc.blogs.nytimes.com/2008/10/14/george-soros-on-the-green-energy-economy/

374 Moyers has benefited from Soros cash, see http://www.aim.org/media-monitor/bill-moyers-and-big-money/

375 Or was Soros one of the many who seemed to know in advance? See: "Who predicted world economic collapse?", TBR.cc, 11 October 2008, http://briefingroom.typepad.com/the_briefing_room/2008/10/who-predicted-w.html

"Instead of consuming, building an electricity grid, saving on energy, rewiring the houses, adjusting your lifestyle where energy has got to cost more until ... you introduce those new things. So it will be painful. But at least we will survive and not cook."

"You're talking," probed Moyers, "about this being the end of an era and needing to create a whole new paradigm for the economic model of the country, of the world, right?"

"Yes," confirmed Soros. For anyone still doubting his agenda, you can always read his 1993 book, *Toward a New World Order*.

But anyone outside the loony Left who thinks Soros is some kind of green saint should think again. His financial opportunism has extended to slashing Brazilian rainforest in favour of growing sugarcane plantations for biofuel, as the *Washington Post* has noted.[376]

"U.S. companies and investors – including George Soros and agribusiness giants Archer Daniels Midland and Cargill – are staking out territory in Brazil, expecting even greater growth in biofuels."

The amount of Soros' Brazil investment was US$900 million.[377] The size of his sugarcane plantation was 150,000 hectares. Given Soros' massive financial patronage of politicians and NGOs, it is possible, but purely speculation on my part, that he may have played a role in the decision by the US, New Zealand and others to pass laws mandating the use of biofuels, even though biofuels have been found to cause more carbon damage than they prevent.

For a smoking gun direct link to one of the key global warming characters in this book, however, you don't have to go past the Soros Foundation's 2006 annual report. At page 143 it lists a grant of $720,000 spent on the category of "politicisation of science". On page 123, meanwhile, it says the Foundation came to the rescue of one James Hansen:[378]

"**Scientist Protests NASA's Censorship Attempts**

"James E. Hansen, the director of the Goddard Institute for Space Studies at NASA, protested attempts to silence him after officials at NASA ordered him to refer press inquiries to the public affairs office and required the presence of a public affairs representative at any interview.

"The Government Accountability Project, a whistleblower protec-

376 "Losing Forests to Fuel Cars: Ethanol Sugarcane Threatens Brazil's Wooded Savanna", *The Washington Post*, 31 July 2007

377 http://news.mongabay.com/bioenergy/2007/06/soros-invests-us900-million-in.html

378 http://www.soros.org/resources/articles_publications/publications/annual_20070731/a_complete.pdf

tion organization and OSI grantee, came to Hansen's defense by providing legal and media advice.

"The campaign on Hansen's behalf resulted in a decision by NASA to revisit its media policy."

The story of Soros' link to Hansen, global warming's grand poobah, was broken by *Investor's Business Daily*,[379] but not one major media outlet picked it up, such is Soros' grip on the media, and Hansen's iconic status.

For his part, Hansen admitted receiving free legal and media assistance from GAP, but had no idea whether the GAP had been funded to the tune of $720,000 for the task.

Soon after Obama won the election last November, news agency *Bloomberg* ran a profile piece on the George Soros-funded think-tank driving Obama's policies, the "Center for American Progress".[380] *Bloomberg* reporter Edwin Chen explained that CAP was the originator of Obama's plan to turn back the tides.

"Obama has endorsed much of a CAP plan to create "green jobs" linked to alleviating global climate change. CAP also is advocating the creation of a 'National Energy Council' headed by an official with the stature of the national security adviser and who would be charged with 'transforming the energy base" of the U.S. In addition, CAP urges the creation of a White House 'office of social entrepreneurship" to spur new ideas for addressing social problems.

"To help promote its ideas," writes Chen, "CAP employs 11 full-time bloggers who contribute to two Web sites, *ThinkProgress* and the *Wonk Room*; others prepare daily feeds for radio stations. The center's policy briefings are standing-room only, packed with lobbyists, advocacy-group representatives and reporters looking for insights on where the Obama administration is headed."

Don't be misled by the blogging division. CAP's *total* staff is 180 and its budget $27 million. As with all good Soros propaganda producers, CAP "devotes as much as half of its resources to promoting its ideas through blogs, events, publications and media outreach" according to Chen.

CAP also funds the blogsite *ClimateProgress.org*, which weighed in to loudly blame human-caused global warming for the Australian

379 "The Soros Threat To Democracy", *Investor's Business Daily*, 24 September 2007 http://www.ibdeditorial.com/IBDArticles.aspx?id=275526219598836

380 "Soros-Funded Democratic Idea Factory Becomes Obama Policy Font", Edwin Chen, *Bloomberg*, 18 November 2008, http://www.bloomberg.com/apps/news?pid=washingtonstory&sid=aF7fB1PF0NPg

bushfires[381] which, as we've already seen, had nothing to do with global warming.

As *Accuracy in Media* documents, Soros associate Eric P. Schwartz was one of Obama's go-betweens for the Obama-Biden transition team in its negotiations with left-wing NGOs "to develop a range of pro-UN policies."[382]

An exposé of George Soros is not the aim of this book, but what I have tried to do is illustrate that behind all the scare stories on a range of fronts – from the need to give up the war on drug trafficking to the need to tax you thousands of dollars more per year because of your 'carbon footprint' – lies a left-wing billionaire (one of several in his group) with an agenda and the means to pull it off. For every left wing media commentator who protests about global warming sceptics being linked to big energy, simply laugh at them and ask about the global warming industry's links to the planet's biggest financial shark, George Soros, a man who buys political, scientific and media influence like no one else on Earth.

A man whose family businesses stand to make more money than Enron ever dreamed of, if he can successfully scare you into believing in global warming theory; and a collection of politicians who stand poised to gradually introduce a new world government, just as Soros has planned it.

That's what this is all about. There's still time to say 'no' to Copenhagen later this year. There's still time for your friends, family and neighbours to read this evidence for themselves.

It's a story the global warming industry doesn't want you to read. It's a story Soros and the United Nations certainly don't want you to read, and little wonder when you see in the next chapter some of the sillier ideas they have.

381 "What's climate got to do with it?", by CAP's Joseph Romm, 10 February 2009 http://gristmill.grist.org/story/2009/2/10/14180/3852

382 "Soros flunky runs Obama's pro-UN policy", *Accuracy in Media*, 11 January 2009

Chapter Eighteen

Priests Of The New Green Religion

"The United Nations is the chosen instrument of God; to be a chosen instrument means to be a divine messenger carrying the banner of God's inner vision and outer manifestation. One day, the world will...treasure and cherish the soul of the United Nations as its very own with pride, for this soul is all-loving, all-nourishing, and all-fulfilling"

– Sri Chinmoy, UN Interfaith Meditation Group head[383]

Last chapter we looked at the financial motivations for manipulating the climate change scare, but this time around we examine some of the true believers – those who bought the lie lock, stock and two smokin' barrels – and you'd be surprised not only at what they say, but who they are.

Cable TV magnate Ted Turner is one of these new anointed. He told a television audience last year that we're all in for a roasting if climate change isn't stopped immediately:[384]

"We'll be eight degrees hotter in 30 or 40 years and basically none of the crops will grow...Most of the people will have died and the rest of us will be cannibals," the 70 year old founder of CNN told his interviewer, PBS' Charlie Rose.

When he said "roasting", it was a bit of a shock to realise he meant it literally, with vegetables and gravy.

"Civilization will have broken down. The few people left will be living in a failed state – like Somalia or Sudan – and living conditions will be intolerable."

383 Speech in 1984, cited in *Now Is The Dawning of the New Age New World Order*, by Dennis L Cuddy Ph.D., Hearthstone Publishing, 1991, p269

384 "Ted Turner: Global warming could lead to cannibalism", *Atlanta Journal Constitution*, 3 April 2008 http://www.ajc.com/metro/content/news/stories/2008/04/03/turner_0404.html

As the *Atlanta Journal-Constitution* noted, Turner had a solution: population control.

"We're too many people; that's why we have global warming," he said. "Too many people are using too much stuff."

As America's biggest individual landowner, with holdings of 7,700 square kilometres, Ted would know. The fewer people that are left, the lower the likelihood of finding those pesky trespassers on your nature trails.

"On a voluntary basis, everybody in the world's got to pledge to themselves that one or two children is it."

You could write Turner off as a grumpy old man except, like his friend George Soros, he's a wealthy grumpy old man who has a heck of a lot of influence in media and political circles. His US$1 billion United Nations Foundation is a major lobby group.

Whilst he told PBS that he's a self-confessed "foot-in-the-mouth disease" sufferer, Turner, apparently without any trace of irony added, "I've gotten a lot better, though. It's been a long time since anybody caught me saying something stupid."

Maybe, maybe not. Try this one for size:

"A total world population of 250-300 million people, a 95 percent decline from present levels, would be ideal," Turner told *Audubon* magazine in 1996.[385]

But how to get rid of so many peasants?

"[Cannibalism is a] radical but realistic solution to the problem of overpopulation," South African zoologist and anthropologist Lyall Watson was quoted as saying,[386] this time evoking cannibalism as a lifestyle choice for committed Greens, who of course prefer not to dine on battery hens.

The idea of eating their fellow human, primitive as it sounds, is not a huge ideological leap for some of the numbnuts on the loopy left who have the ear of those in power. They don't see any value in humanity, and see global warming as a chance to spread their beliefs.

"We, in the green movement, aspire to a cultural model in which killing a forest will be considered more contemptible and more criminal than the sale of 6-year-old children to Asian brothels," remarked the late Carl Amery, German environmentalist writer, in

385 *You Don't Say?*, Fred Gielow, 1999, page 189, referencing *Audubon*
386 *The Financial Times*, 15 July 1995

a callous display of what's really driving his agenda.[387]

"The world has cancer, and that cancer is man," wrote the Rockefeller Foundation's former spokesman Merton Lambert,[388] in another example of an elite anti-civilisational worldview.

David Brower, the founder of the harmless-sounding environmental group Friends of the Earth, nonetheless evokes principles Adolf Hitler and Josef Mengele would have been proud to introduce as part of the Nazi empire, had they won the war:

"Childbearing [should be] a punishable crime against society, unless the parents hold a government license ... All potential parents [should be] required to use contraceptive chemicals, the government issuing antidotes to citizens chosen for childbearing."[389]

Twenty years ago, in 1989, Canada's *Harrowsmith* magazine ran an article about environmentalist and global warming priest David Suzuki, who was echoing the views of 'population explosion' Paul Erhlich:

"In one of his last columns in the *Globe*, Suzuki quoted Ehrlich's view of public apathy about the perils of economic growth ... A few weeks later, when the *Star* began to publish the column, Ehrlich was featured in it regularly. "Ehrlich concludes that it would be a dangerous miscalculation to look to technology for the answer to [environmental problems]. Scientific analysis points toward the need for a quasi-religious transformation of contemporary culture." ... and three weeks after that [Suzuki wrote], "Stanford University ecologist Paul Ehrlich reminds us that ... we face a 'billion environmental Pearl Harbors all at once.' " On December 2, Suzuki wrote, "We no longer have the luxury of time ... when people like Paul Ehrlich of Stanford University ... tell us we only have a decade to turn things around." And in his Christmas column on December 13, Suzuki wrote, "As eminent ecologist Paul Ehrlich says, 'the solution to ecocatastrophe is quasi-religious.'"

"Quasi-religious?" And you thought I was kidding when I called this chapter "Priests of the New Green Religion".

"The responsibility of each human being today is to choose between the force of darkness and the force of light," exclaimed the Secretary-General of the UN Conference on Environment and Development, Maurice Strong, to the UN's Earth Summit in 1992.

387 *Trashing the Planet,* by Dixy Lee Ray and Lou Guzzo, Regnery Publishing, 1990, page 169.
388 *Harpeth Journal*, 18 December 1962
389 *Trashing The Planet*, by Dixy Lee Ray and Louis Guzzo, Regnery Publishing, 1990, p.166

"We must therefore transform our attitudes and values, and adopt a renewed respect for the superior law of Divine Nature."

Divine Nature? This is a UN official using this language!

This collective liberal guilt about humanity, and the not so subtle move by UN officials to endorse Gaia worship as the planet's 'official' religion, appears to be the prime driving force – outside money and power – behind the push to legitimise global warming theory as "settled" science. It is as if a raft of disparate causes have found their common touchstone in global warming theory, and are spending billions to convince the public to bow to the environmental Trojan Horse.

The danger is that in many parts of the western world, governments have given in to pressure and actually funded green groups to give presentations to every classroom of schoolchildren about caring for the environment. Sounds great in principle, but with their real worldview unmasked here and elsewhere, what spin do you think vulnerable children will be subject to?

In the US, it's NASA's James Hansen who's provided input to the "National Teach-in on Global Warming Solutions. In the classroom material, he warns impressionable minds that "continued growth of greenhouse gas emissions, for just another decade", makes "the tipping level for catastrophic effects" virtually certain.

Scary stuff for a room full of 11 year olds, but he then tells them:

"We stand at a unique moment in human history. The window for action on global warming is measured in months, not years. Decisions that we make – or fail to make – in 2009 will have profound impacts not only for our children and grandchildren, but for every human being that will ever inhabit the face of this earth from now until the end of time."

Again, cue majestic pipe organ and perhaps the distant refrain of Dr Phibes' maniacal "mwah-ha-ha-ha" laugh.

Environmental NGO's and lobby groups generally put a congenial face towards the TV cameras for public consumption, but are much more radical underneath.

Take PETA (People for the Ethical Treatment of Animals), who have swung in heavily to persuade people that greenhouse emissions are killing the planet. But when you support PETA, you are also supporting beliefs like these:[390]

"Mankind is the biggest blight on the face of the earth."

---

390 "Animal Rights Versus Human Rights," by Edwin A. Locke, Ph.D., *eco-logic Powerhouse*, July 2005, page 21.

"I do not believe that a human being has a right to life."

"I would rather have medical experiments done on our children, than on animals."

Similar sentiments in this next one, as well:[391]

"If you haven't given voluntary human extinction much thought before, the idea of a world with no people in it may seem strange. But, if you give it a chance, I think you might agree that the extinction of Homo sapiens would mean survival for millions, if not billions, of Earth-dwelling species. . . . Phasing out the human race will solve every problem on earth, social and environmental."

Of course, if the kind of Wild Greens who advocate "voluntary human extinction" practised what they preached, there'd be no Wild Greens left to write about it.

What they really mean is they want ordinary families and kids to become extinct, leaving space for the Green elite to run the planet and enjoy exclusive bird-watching excursions whilst feasting on the bones of six year olds who'd earlier been sold to Asian brothels.

Now, again, you may dismiss all this as fantasy, and to an extent it is – right up to the moment that people with this kind of worldview start getting funding cheques from a newly enriched United Nations thanks to compulsory emissions trading fees or transaction taxes imposed. Because when these people get access to real money and political power, that's when life will become seriously interesting on Planet Earth, and not in a nice way.

Laugh and scoff about the 'New World Order' all you like, but if Copenhagen goes through at the end of this year you'll effectively be funding their daft schemes, and you won't be laughing for very long.

Part of the reason is that environmental politics was long ago captured by the hard left, who saw it as an opportunity to change the world in the name of a warm, fuzzy cause.

"I think if we don't overthrow capitalism, we don't have a chance of saving the world ecologically," Earth First! Member Judi Bari said in a 1992 interview.[392]

"I think it is possible to have an ecologically sound society under socialism. I don't think it's possible under capitalism."

---

391 "Voluntary Human Extinction," *Wild Earth*, Vol. 1, No. 2 (Summer 1991), page 72. Reprinted in *Animal Scam*, by Kathleen Marquardt with Herbert M. Levine and Mark LaRochelle, Regnery Publishing, 1993, page 175.

392 As quoted by William Williams, *State Journal-Register*, 25 June, 1992, and cited in *Environmental Overkill*, by Dixy Lee Ray with Lou Guzzo, 1993, page 203.

"The answer to global warming is in the abolition of private property and production for human need. A socialist world would place an enormous priority on alternative energy sources. This is what ecologically-minded socialists have been exploring for quite some time now," Columbia University's self-titled 'Unrepentant Marxist' Louis Proyect has said.

Yet you can't exactly sneak socialism in the front door, or can you?

"The American people will never knowingly adopt socialism," wrote the head of the US Socialist Party Norman Thomas fifty years ago. "But under the name of Liberalism, they will adopt every fragment of the socialist program until one day America will be a socialist nation without knowing how it happened."

The serendipity of both a global economic crash and the biggest conference on global warming since Kyoto coinciding in 2009 is a dream come true for the hard left, who sense the world is at an historic crossroads this year.

Again, you can look to past quotes to see the barely disguised excitement at the opportunity global warming provides.

"No matter if the science is all phony, there are collateral environmental benefits," Canada's Minister for the Environment told journalists 10 years ago.[393] "Climate change [ushers in] the greatest chance to bring about justice and equality in the world".

How exactly? Only if the real agenda behind tackling climate change is something much deeper and more fundamental than closing down smokestacks.

"We've got to ride the global warming issue," said Clinton advisor Timothy Wirth, now heading Ted Turner's United Nations Foundation. "Even if the theory of global warming is wrong, we will be doing the right thing in terms of economic policy and environmental policy."[394]

Again, how? Only if your agenda is to nobble the West and compulsorily transfer trillions of dollars and hundreds of millions of jobs from householders in the US, UK, Europe and Australasia into the Third World.

But if you are looking for honesty on this from leaders in the mainstream media, you may be out of luck.

"I would freely admit that on [global warming] we have crossed the

---

393 Reported in *Calgary Herald*, 14 December, 1998

394 *National Review*, January 27, 1997, page 10.

boundary from news reporting to advocacy," remarked *Time* magazine science columnist Charles Alexander to a conference audience.[395]

Another journalist firmly in the AGW camp, *Boston Globe* editor Ross Gelbspan, took a swipe at scientific sceptics at a Washington DC speech in 2000.

"These self-proclaimed 'greenhouse sceptics' would normally not be worthy of much attention. There are only about a dozen visible ones[396] versus a consensus of more than 2,000 scientists from 100 countries. But, with extraordinary access to the media thanks to their corporate sponsors, they have been able to create the general perception that the issue is hopelessly stuck in uncertainty.

"Not only do journalists *not* have a responsibility to report what sceptical scientists have to say about global warming, they have a responsibility not to report what these scientists say [fullstop]," exclaimed Gelbspan.[397]

In the same speech, Gelbspan tried to tell his audience that outbreaks of "frost" in equatorial Papua New Guinea were the fault of "global warming":

"In the fall of 1997, for instance, following a four-month spell of drought and frost, 700,000 Papua New Guineans left their homes and began wandering the countryside in search of food, water and warmth. And officials said they could not control the situation. Fortunately, other countries came to their aid."

The Papuan weather patterns of drought and frost (in the highlands), are firmly associated with El Niño, not increasing carbon levels, and occur frequently. It should have been a dead giveaway that frost on the equator could not be blamed on Earth turning into a hothouse.

Another journalist to concede that he and his colleagues have 'gilded the lily' is the late Environmental Protection Agency spokesman Jim Sibbison.

"We routinely wrote scare stories about the hazards of chemicals, employing words like 'cancer' and 'birth defects' to splash a little cold water in reporters' faces...Our press reports were more or less true...Few handouts, however, can be completely honest, and ours

395 "Prophets, False Prophets, and Profiteers," by Paul Driessen, *eco-logic Powerhouse*, February 2005, page 9
396 Well, actually the sceptics list now totals 32,000 and counting
397 "Prophets, False Prophets, and Profiteers," by Paul Driessen, *eco-logic Powerhouse*, February 2005, page 9

were no exception…we were out to whip the public into a frenzy about the environment."[398]

"It may seem stranger than fiction," writes *The New American's* William Norman Grigg, "but it's a documentable fact: the eco-socialist movement is financed by the super-rich as part of a comprehensive agenda for global control."[399]

Whether the super-rich get their wish remains to be seen, however. UN IPCC chairman Rajendra Pachauri has told a London newspaper President Obama could face a new civil war over climate change:[400]

"He [Obama] is not going to say [at Copenhagen] 'by 2020 I'm going to reduce emissions by 30%'. He'll have a revolution on his hands. He has to do it step by step," said Pachauri this year.

Interesting statement. There's no suggestion of Obama backing away from the end target, just a suggestion that he'll be massaging the spin to make sure he passes the hurdles one at a time. In which case, you can guarantee climate scare stories will start coming out your ears as part of a public softening up.

It's probably no coincidence that Pachauri announced, on the same day, he'd taken a position heading Yale University's Climate and Energy Institute, which will push global warming science at the highest levels in the US.[401]

"We are fortunate to attract one of the world's foremost climate change experts to lead this ambitious new institute," Yale president Richard Levin announced. The university, perhaps most famously known as the head office of the "Skull n Bones Society", said of Pachauri: "No one has a more comprehensive grasp of the science and policy of climate change or has done more to bring attention to this urgent issue."

Except, perhaps, for Pachauri failing to acknowledge the cooling years in his infamous Sydney speech.

For anyone who still believes scientists get funded on merit, Yale makes it clear Pachauri's new institute will be greasing the palms of scientists willing to tell him what he wants to hear:

"The new institute will provide seed grants, support postgraduate

398 *Environmental Overkill,* by Dixy Lee Ray with Lou Guzzo, 1993, page 165.

399 "Behind the Environmental Lobby," by William Norman Grigg, *The New American*, April 4, 2005, page 17

400 "UN climate chief: US carbon cuts could spark 'revolution'," *The Guardian*, 11 March 2009, http://www.guardian.co.uk/environment/2009/mar/11/us-carbon-cuts

401 "Yale to pick Pachauri brain", *The Calcutta Telegraph*, 11 March 2009

study, sponsor conferences and workshops, and foster interdisciplinary research spanning from basic atmospheric science to public policy."

Make no bones about it, they're aiming to beat you down so much with terrifying bought and paid for "scientific" studies that the grey matter inside your own skull turns to mush and you simply surrender, pay the new taxes and sell your children to them for medical experiments.

In a fascinating postscript to revelations that Yale is signing up to heavily promote Rajendra Pachauri's UN climate agenda, it's worth turning to the end of this year's mini-Copenhagen climate conference, where some 2,500 researchers and hangers on had gathered to compare notes on climate change, and organisers issued a statement containing "six key messages".

Britain's *Guardian* newspaper reported it under the headline, "Six ways to save the world: scientists compile list of climate change clinchers":[402]

"Scientists at the international congress in Copenhagen have prepared a summary statement of their findings for policy makers. This was handed today to the Danish prime minister, Anders Fogh Rasmussen. Ahead of the UN Climate Change Conference in December he will formally hand this statement over to officials and heads of state at the conference. The full conclusions from the 2,500 scientific delegates from 80 countries that have attended the three-day meeting this week will be published in full in June 2009."

Now, I think most readers would agree there is little room for ambiguity there. The 2,500 scientists at the conference have issued a summary of their key findings for policymakers to take on board ahead of December's main Copenhagen Treaty talks.

The BBC reported likewise:

"The worst-case scenarios on climate change envisaged by the UN two years ago are already being realised, say scientists at an international meeting. In a statement in Copenhagen on their six key messages to political leaders, they say there is an increasing risk of abrupt or irreversible climate shifts…"

Now, although the original news release contained a small disclaimer down the bottom that it may not represent the views of everyone, organisers were not exactly beating down the doors of newsrooms around the world to disavow them of the notion that the

402 *The Guardian*, 12 March 2009, http://www.guardian.co.uk/environment/2009/mar/12/copenhagen-summary

"six key messages" were carefully considered, 'consensus' opinions from scientists.

So quickly did this myth spread that the scientists had issued six key messages, that one of the scientists who actually attended at Copenhagen and chaired a session, Professor Mike Hulme, felt compelled to disassociate himself from the misinformation campaign.

"The six key messages are not the collective voice of 2,500 researchers," he wrote in a blog post,[403] "nor are they the voice of established bodies such as the World Meteorological Organisation. Neither are they the messages arising from a collective endeavour of experts, for example through a considered process of screening, synthesizing and reviewing of the knowledge presented in Copenhagen this week. *They are instead a set of messages drafted largely before the conference started by the organizing committee*, sifting through research that they see emerging around the world and interpreting it for a political audience."

Talk about "busted". News coverage right around the world in newspapers, radio and TV centred on these six political bombshell messages, as if delivered from on high by 2,500 scientists, when in fact the scientists had nothing to do with it.

So who, exactly, is responsible and what was the process behind this sham?

"The Copenhagen Climate Change Conference was no IPCC," says Hulme. "This was not a process initiated and conducted by the world's governments, there was no systematic synthesis, assessment and review of research findings as in the IPCC, and there was certainly no collective process for the 2,500 researchers gathered in Copenhagen to consider drafts of the six key messages or to offer their own suggestions for what politicians may need to hear. The conference was in fact convened by no established academic or professional body.

"Unlike the American Geophysical Union, the World Meteorological Organisation or the UK's Royal Society – who also hold large conferences and who from time-to-time issue carefully worded statements representing the views of professional bodies – this conference was organized by the International Alliance of Research Universities (IARU), a little-heard-of coalition launched in January

403 http://sciencepolicy.colorado.edu/prometheus/what-was-the-copenhagen-climate-change-conference-really-about-5055#comments

2006 consisting of ten of the world's self-proclaimed elite universities, including of course the University of Copenhagen.

"IARU is not accountable to anyone and has no professional membership. It is not accountable to governments, to professional scientific associations, nor to international scientific bodies operating under the umbrella of the UN. The conference statement therefore simply carries the weight of the Secretariat of this ad hoc conference, directed and steered by ten self-elected universities."

And who, exactly, are these ten "self-selected" universities who have stepped semi-covertly into global warming politics, as opposed to science? Well, what a surprise to find Rajendra Pachauri's Yale University listed in the chosen ten:[404]

Australian National University
ETH Zürich
National University of Singapore
Peking University
University of California, Berkeley
University of Cambridge
University of Copenhagen
University of Oxford
University of Tokyo
Yale University

One wonders, having nailed their colours well and truly to the mast, whether any of these cathedrals of learning would provide funding to any researcher whose line of inquiry might challenge the global warming religion? Would one of their own studies that cast doubt on global warming ever be published, or would it be swept under the carpet or thrown in the file marked "censored"?

If global warming believers can rant till the cows come home about the oil industry funding some sceptics years ago, what is the public to make of the much bigger agenda by global warming alarmists to control science and financially strangle any scientists who dare to oppose?

The Green priesthood has seized the altars of education, and swiped the proceeds of the collection plate. God help real science, and the rest of us.

404 http://climatecongress.ku.dk/newsroom/congress_key_messages/

Chapter Nineteen

Even Sillier Things Said In The Name Of Science

"An enormous propaganda campaign [is] saturating the American public with the idea that our environmental problems are too immense to be dealt with by our present system of independent, sovereign nation states. Thus we increasingly find ourselves confronting such pre-fabricated slogans as 'Global Problems Require Global Solutions', 'Global Problems Require Global Governance', and 'Think Globally, Act Locally'"

– ***William Jasper, 2001***[405]

Part of the problem with global warming is the enormous spin-off industry it's created in ridiculous statements and deeds.

It's bad enough when scientists make things up, but that behaviour has only encouraged such a spillover into laymen's turf that daft claims about global warming are at serious risk of becoming an Olympic sport.

Take President Obama's new special advisor on climate change, Carol Browner. *Dow Jones* news agency reported this in February:[406] "President Barack Obama's climate czar said Sunday the Environmental Protection Agency will soon issue a rule on the regulation of carbon dioxide, finding that it represents a danger to the public."

A gas essential for life, critical to the growth of rainforests and food crops, and life in the ocean, not to mention something our own bodies create when we breathe out, is deemed "a danger to the public".

What next, a tax on breathing? Don't laugh, it might yet happen.

Britain's chief scientific advisor Sir David King was reported in

405 *The UN Exposed*, William F Jasper, John Birch Society, 2001, p106
406 "US Climate Czar: CO2 Regulation Ruling To Come Soon", *Dow Jones* newswire, 22 February 2009

one newspaper saying that "Antarctica is likely to be the world's only habitable continent by the end of this century if global warming remains unchecked, because Earth was entering 'the first hot period' for 60 million years, when there was no ice on the planet and 'the rest of the globe could not sustain life'."[407]

King says the newspaper misquoted him in the interview based on his testimony to a House of Commons Select Committee in 2004. On that occasion, he said it this way: "Fifty-five million years ago was a time when there was no ice on the earth; the Antarctic was the most habitable place for mammals, because it was the coolest place, and the rest of the earth was rather inhabitable because it was so hot. It is estimated that it [the carbon dioxide level] was roughly 1,000 parts per million then, and the important thing is that if we carry on business as usual we will hit 1,000 parts per million around the end of the century."

As you've seen, the latest research suggests there has been ice on the planet always, and life as we know it did not cease to exist 60 million years ago, nor did all creatures great and small evacuate to Antarctica or die. Nor is this current warmish period the hottest in a thousand years, let alone 60 million years.

Another to wax lyrical about Armageddon is Gaia theorist James Lovelock.

"Before this century is over, billions of us will die, and the few breeding pairs of people that survive will be in the Arctic where the climate remains tolerable."[408]

In as much as few of the seven billion people alive today will survive to be a hundred, Lovelock is entirely correct in predicting that "billions of us will die" by the end of the century. His prognosis however that only a "few breeding pairs" will survive, and fly north to the Arctic to live, is science fantasy. It was hotter than it is now a thousand years ago, yet we are not all descended from Eskimos as a result of that temperature spike.

Elsewhere, it's the news media working themselves into a lather, as they report the vague ramblings of anyone who mumbles the words "climate change" during any otherwise incoherent sentence.

Drawing comparisons with an older generation who used to blame

407 *The Independent*, 2 May 2004

408 "James Lovelock: The Earth is about to catch a morbid fever", *The Independent*, 16 January 2006 http://www.independent.co.uk/opinion/commentators/james-lovelock-the-earth-is-about-to-catch-a-morbid-fever-that-may-last-as-long-as-100000-years-523161.html

weather changes on atmospheric nuclear testing, *Spiked Online's* John Brignell gently pokes the borax at the modern equivalent:

"Mind you, even in those far off innocent days they did not fly into a panic, as now, over a mild October. They just enjoyed it. They even had a term for it – Indian Summer. What a fine example of ratchet reporting we have seen in recent weeks, with almost every British newspaper showing horror pictures of... late flowering gardens."

Brignell has very kindly compiled a list of some of the weirder headlines on global warming, some of which I'll highlight here. For example, the Aussie cold spell blamed on warming:

"Australia is in the grip of a nationwide cold snap – and paradoxically, it could be another result of global warming. Last summer was the hottest on record. But last month many parts of Australia reported record or near-record cold nights. The average minimum temperature was 1.69 degrees below the long-term average, making it the second-coldest June since 1950."[409]

Be frightened of the warm because it will make you cold. Yeah, that makes sense.

While popular wisdom has been pushing the use of windfarms to generate electricity, other scientists are now warning windfarms could generate more global warming, not less, by upsetting essential wind currents and harming local and regional weather patterns, at a scale "somewhere between the environmental costs of deforestation and global warming".[410]

One Australian report suggests global warming is the cause of transsexual tuatara lizards – rising temperatures are turning girl eggs into boys:

"We're on a flight back in time. Across New Zealand's Cook Strait is the hideaway of a creature preserved since the Jurassic. This is Stephens Island, home of the prehistoric tuatara," reported the ABC.[411]

"But it's not the past we're here for. Unbelievably, it's in this lost world that scientists are detecting the early warning signs of a future global catastrophe. It may mean the end of this island's legendary inhabitants, and many others worldwide.

"When scientists first came here in search of tuatara, uncovering

409 "Cold spell's weird cause", *Sydney Morning Herald*, 4 July 2006, http://www.smh.com.au/news/national/cold-spells-weird-cause/2006/07/03/1151778873599.html

410 "Windmills to Change Local and Global Climates", *LiveScience*, 9 November 2004 http://www.livescience.com/environment/041109_wind_mills.html

411 "Tuatara", ABC 'Catalyst', 25 March 2004, http://www.abc.net.au/catalyst/stories/s1073835.htm

evidence of a looming disaster was the last thing on their minds."

Who knew? According to this explosive new evidence, tuatara eggs incubated at 21°C in the lab produce "one hundred percent females", while an increase to just 22°C produces "a hundred percent males". The warmer it gets, so the theory goes, tuatara will have to turn gay to get any action, and then they'll die out.

The tuatara colonies are showing significantly more males (65% to 35%) than females are being hatched, ergo, global warming proved in a nutshell (or eggshell).

Except for one minor problem. Or maybe two. The tuatara live somewhere between 150 and 300 years, and one recently became a father at the ripe young age of 111. Climate swings, as we've seen, come and go, with the Pacific Decadal Oscillation having a major impact on the islands where the tuatara live. Temperatures in any given year can vary by a number of degrees compared with the same period a year earlier. Birth rates in a warmish period like today might reflect a male weighting, but 30 years from now they might not. And conceivably, there are tuatara somewhere today born during the Little Ice Age – so they've seen an awful lot of climate change and survived happily.

In New York, global warming has taken the rap for "dull looking trees" in autumn,[412] while across the Atlantic British journalists were asking the question, "Are these [radiant] colours caused by global warming?"

"The answer is probably yes, says Nick Collinson, conservation policy adviser at the Woodland Trust. The Tree Council this week said global warming caused this season's russet reds to deep golden yellows."[413]

One day it's leaves on the trees, but the next it might be no trees at all – one science graduate has even gone so far as to suggest global warming could cause Planet Earth to vaporise in a nuclear explosion.[414]

"Consequences of global warming are far more serious than previously imagined," warns Dr Tom Chalko. "The REAL danger for our entire civilization comes not from slow climate changes, but from overheating of the planetary interior."

Struggling to see how the planet's interior 6,000 kilometres under-

412 "A Season A Tad Off Color, And Here's Why", *New York Times*, 16 October, 2005 http://www.nytimes.com/2005/10/16/nyregion/nyregionspecial2/16weleaf.html

413 http://www.guardian.co.uk/science/2004/nov/18/thisweekssciencequestions1

414 "Can Earth explode as a result of Global Warming?" Dr Tom J. Chalko, MSc, PhD, http://nujournal.net/core.pdf

ground can be affected by the weather? Well, apparently the warm days can transfer heat to the interior deep below with catastrophic effects.

"Overheating the centre of the inner core reactor due to the so-called greenhouse effect on the surface of Earth may cause a meltdown condition, an enrichment of nuclear fuel and a gigantic atomic explosion."

Sigh.

Hedging its bets on the significance of climate change, Britain's *Guardian* newspaper bravely asked the oxymoronic [or should that be carbon dioxymoronic?] question, "Will global warming trigger a new ice age?"[415]

The poor journalists at the *Guardian* had been scared sillier by one of the global warming computer models predicting a breakdown in the Gulf Stream that traditionally warms Europe. They overlooked the very simple fact that there isn't enough accessible fresh water close to the Atlantic Ocean capable of interfering with the Gulf Stream, even if Greenland's ice caps drain away. Another garbage in/garbage out computer programme causing needless anxiety.

Sceptical climatologist Pat Michaels summed up the problem nicely:[416]

"When the temperature is warm, the alarmists blame it on global warming. When the temperature is cold, the alarmists blame it on global warming. When it rains, global warming is to blame. When it doesn't rain, global warming is to blame. This is not science, this is the alarmists simply lobbying for attention and lobbying for large government research grants."

One thing global warming is no longer being blamed for, however, is the disappearance of the world's frogs. Around a third of the world's amphibian species are currently under threat, but the hardest hit are those in Central America and Australia.

The obvious cause of their demise is an organism known as the chytrid fungus, and up until now global warming has been blamed for rising temperatures to an optimal level for the fungus. The equally obvious response to that claim – that temperatures can fluctuate by several degrees because of entirely natural causes and that this is more likely a culprit than a half degree average increase in temperature over a century, appears lost on the global warming industry.

415 *The Guardian*, 13 November 2005

416 "Global Warming Simultaneously Baking, Freezing Europe, Alarmist Activists Claim", Heartland Institute, Feb 2006, http://www.heartland.org/policybot/results/18445/Global_Warming_Simultaneously_Baking_Freezing_Europe_Alarmist_Activists_Claim.html

However, the latest studies[417] showed temperature increases do not correlate with amphibian deaths, so much so that global warming "evidence linking it with the declines is weak", according to a University of South Florida study in 2008.

Instead, a farm chemical called atrazine may play a far larger role in killing off frogs.[418]

There's room for massive scepticism too on a claim that 85% of the Amazon will disappear because of global warming this century. In another garbage in/garbage out computer modelling scenario, researchers at the UK Met Office Hadley Centre punched in the numbers and told the mini-Copenhagen conference:[419]

"The impacts of climate change on the Amazon are much worse than we thought," said Hadley's Vicky Pope. Essentially, a temperature increase of two degrees Celsius above pre-industrial temperatures (tail-end of Little Ice Age) would kill 20-40% of the Amazon jungle this century, while a 3°C rise would take out 75% and 4°C would destroy 85%.

"The forest as we know it would effectively be gone," said Pope.

But is Hadley Centre right? Based on past performance one is tempted to laugh out loud, but let's take a more serious look.

For a start, the Amazon jungle survived the medieval warm period with no adverse effects. Secondly, increasing CO_2 levels lead to increased vegetation growth, because CO_2 is a food to jungles (that's one of the reasons we use greenhouses in agriculture, to increase the yield from plants as a result of higher CO_2). So it's counterintuitive for Hadley to claim that rising CO_2 levels are bad for forests per se.

Thirdly, the Hadley computer model used suggests the increasing temperature will reduce rainfall, and it's the reduced rainfall that will dry the Amazon out and kill it. We've already seen scientific studies in this book that have measured far higher levels of precipitation (rain and snowfall) than computer models had allowed for, so it is by no means certain that the Hadley Centre predictions on rainfall changes will be the correct ones.

---

417 "Amphibian Extinctions: Is Global Warming Off the Hook?", Brian Handwerk, *National Geographic News*, 1 December 2008, http://news.nationalgeographic.com/news/2008/12/081201-global-warming-frogs.html

418 "Study traces frog population decline to weed killer", National Public Radio, 29 October 2008

419 "Amazon could shrink by 85% due to climate change, scientists say", *The Guardian*, 11 March 2009, http://www.guardian.co.uk/environment/2009/mar/11/amazon-global-warming-trees

I'm an investigative journalist, not a climate scientist. I'm only reporting the conclusions of others and drawing inferences from their work. But it seems I was on the right track.

The icing on the cake in destroying this "key message" from mini-Copenhagen came from a global warming believer in the *New York Times*. Reporter Andy Revkin's *DotEarth* blog for the *NYT* describes apocalyptic claims like this one as "climate porn" and warns it will backfire when the public realise they're being played for suckers.[420]

> On Thursday, an e-mail message was distributed to a host of Amazon forest experts and to a journalist by Yadvinder Malhi, an Oxford University biologist who is focused on the Amazon and climate. He questioned the Amazon findings presented at the meeting, and decried the resulting media coverage:
>
> "I must say I find it frustrating that the gloomiest take on news gets such a big profile. This is based on *one* model, and that model has flaws (especially its temperature sensitivity that seems too great (David Galbraith's work), and its rainfall that seems too low (our PNAS paper PDF[421])). The danger is that that such apparent bad news makes all the efforts to conserve the Amazon forests worthless (why bother saving them if they are already doomed?), and encourages disengagement and hopelessness rather than action. If that conclusion was based on solid empirical science then so be it, but when such a story goes out on a pure model study (not yet peer-reviewed) with significant imperfections, it may do a lot of damage in the real world."
>
> A colleague of Dr. Malhi who attended the meeting responded by saying several scientists there were engaged in "damage control."

It is certainly true that development deforestation of the Amazon (including Soros and others clearing South American jungles to plant biofuel crops) will hasten the drying out and death of the forest, for the same reasons that Mt Kilimanjaro is suffering, but according to the alarmist Hadley Centre deforestation is not the primary problem, CO_2 is.

Australian climatologist William Kininmonth, formerly head of that country's National Climate Centre, disagrees. He's presented a paper at the Heartland climate conference in New York this year[422]

420 http://dotearth.blogs.nytimes.com/2009/03/12/copenhagen-summit-seeks-climate-action/?hp
421 http://www.eci.ox.ac.uk/news/press-releases/090210amazonia-dieback.pdf
422 "A Natural Limit to Anthropogenic Global Warming", William Kininmonth, paper presented to Heartland Conference on Climate, 10 March 2009, http://jennifermarohasy.com/blog/2009/03/redefining-the-limits-of-global-warming

suggesting the current computer models underestimate evaporation and subsequent rainfall, and consequently overestimate temperature increases. That's because evaporating moisture sucks heat back out of the atmosphere, as does condensing it into raindrops.

"Water vapour is important in regulating the magnitude of the enhanced greenhouse effect in two ways," he says: "increased water vapour in the atmosphere has an amplifying effect on the CO_2 forcing; and, more importantly, increased evaporation constrains the surface temperature rise. It is the evaporation that is dominant because the Earth's surface is more than 70 percent ocean and much of the remainder is covered by transpiring vegetation. A doubling of CO_2 concentration by the end of the century from current levels [from 385 ppm/v to 800 ppm/v] will cause a modest global temperature rise not exceeding 1^0C.

"The computer models on which the IPCC based its fourth assessment projections have been shown to significantly underestimate the rate of increase of evaporation with temperature. As a consequence, surface temperature rise from CO_2 forcing is grossly exaggerated. Suggestions that global temperature might pass a 'tipping point' and even go into a phase of 'runaway global warming' are an outcome of the flawed computer models and are not a realistic future scenario. The extensive oceans and the hydrological cycle are a natural constraint on global temperature and dangerous anthropogenic global warming is not a feasible outcome," concludes Kininmonth.

It's just more evidence, if ever it was needed, that the media and global warming believers have been too quick to blame everything under the sun, but not the sun itself, on or for global warming, and too quick to rely on simplistic computer models of what the climate might do in future.

Chapter Twenty

The 2010 Update

"Moving towards a green economy would also provide an opportunity to re-examine national and global governance structures"

– ***UN briefing paper for environment ministers, February 2010***

In the time since *Air Con* was first published in April 2009, affairs in the tides of men have ebbed and flowed with considerable drama. Significantly, the Copenhagen summit in December 2009 failed spectacularly to take any actual action on climate, although as predicted President Obama did nut out the skeleton of a global wealth transference idea and waved it round claiming "meaningfulness in our time", or words to that effect.

The hard ball on both world governance and climate change has, therefore, been kicked into touch for fine-tuning at Mexico late in 2010, giving the UN an extra year to firm up support for its agenda with a probable finalization in 2012.[423] Don't for a moment think that this is a victory for sceptics and we can all rest easy. The UN and its plutocrat supporters are like zombies that just keep on coming with their mindless droning soundbites of "it's worse than we thought". That's the crux of the whole problem – they haven't got an original thought between them. Climate change is merely a means to a very profitable end.

423 In late February 2010, both ***Fox News*** and ***TGIF Edition*** published stories based on a UN Environment Programme briefing paper for world environment ministers gathering in Bali that month. The briefing paper suggests the UN is pressing ahead with plans for a global environmental control agency capable of regulating all world economies in the name of fighting climate change. The UNEP has named 2012 as the year it will be confirmed, and the location will be Rio de Janeiro at a "sustainability summit" described as "Rio + 20" – a reference to the 1992 Earth Summit that kicked global warming into the mainstream. The briefing paper can be read here http://www.foxnews.com/projects/pdf/022510_greeneconomy.pdf and the covering story here http://www.foxnews.com/story/0,2933,587426,00.html

How profitable? Well, as Christopher Booker revealed this year[424], the WWF is caught in a US$60 billion conflict of interest:

"If the world's largest, richest environmental campaigning group, the WWF – formerly the World Wildlife Fund – announced that it was playing a leading role in a scheme to preserve an area of the Amazon rainforest twice the size of Switzerland, many people might applaud, thinking this was just the kind of cause the WWF was set up to promote. Amazonia has long been near the top of the list of the world's environmental concerns, not just because it includes easily the largest and most bio-diverse area of rainforest on the planet, but because its billions of trees contain the world's largest land-based store of CO_2 – so any serious threat to the forest can be portrayed as a major contributor to global warming.

"If it then emerged, however, that a hidden agenda of the scheme to preserve this chunk of the forest was to allow the WWF and its partners to share the selling of carbon credits worth $60 billion, to enable firms in the industrial world to carry on emitting CO_2 just as before, more than a few eyebrows might be raised. The idea is that credits representing the CO_2 locked into this particular area of jungle – so remote that it is not under any threat – should be sold on the international market, allowing thousands of companies in the developed world to buy their way out of having to restrict their carbon emissions.

"The net effect," wrote Booker, "would simply be to make the WWF and its partners much richer while making no contribution to lowering overall CO_2 emissions."

Likewise for those who wondered why BBC News has been so gung-ho on promoting man-made global warming as a problem, we now find out the BBC pension fund is one of the biggest investors in the European carbon markets.[425] The retirement savings of journalists and other BBC staff have been invested in the climate change industry, creating a conflict of interest between reporting the news, and financially benefiting from those news reports.

And yet, despite all that money slushing around and all those vested

424 "WWF hopes to find $60 billion growing on trees", by Christopher Booker, ***The Telegraph***, 20 March 2010, http://www.telegraph.co.uk/comment/columnists/christopherbooker/7488629/WWF-hopes-to-find-60-billion-growing-on-trees.html

425 "Look who's making money out of global warming scare", Ian Wishart's blog TBR.cc, 1 February 2010 http://briefingroom.typepad.com/the_briefing_room/2010/02/look-whos-making-money-out-of-global-warming-scare.html

interests, the Copenhagen climate talks suffered speed wobbles.

Firstly, the science has continued to stack up on the side of skeptics, not believers. As you've already seen, temperature increases in the noughties (2000-2009) have tapered off dramatically, despite a continuing rise in CO_2 levels.

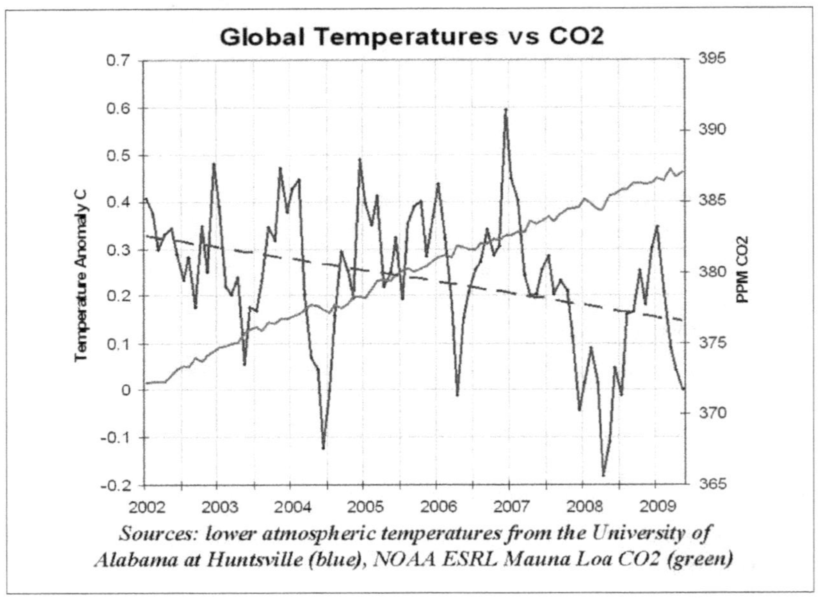

Sources: lower atmospheric temperatures from the University of Alabama at Huntsville (blue), NOAA ESRL Mauna Loa CO2 (green)

On any measure, such results fly in the face of greenhouse theory, which decrees that CO_2 is so strong that it overwhelms climate "noise" (natural fluctuations) and forces temperatures up regardless. Clearly, the latest figures show the central core of the IPCC argument is wrong – there is still no proven link between CO_2 levels in the wild, and matching movements in atmospheric temperatures. CO_2 levels have risen in the past ten years, but something stronger has allowed temperature increases to decelerate.

Proof, again, that climate is far more complex than the most sophisticated UN-approved computer models have allowed for.

Which, incidentally, is exactly what a peer-reviewed study published late January 2010 in the *Journal of Climate* admits.[426]

"Planet Earth has warmed much less than expected during the

426 "Why Hasn't Earth Warmed as Much as Expected?", Brookhaven Nat. Lab. News, 19 January 2010 http://www.bnl.gov/bnlweb/pubaf/pr/PR_display.asp?prID=1067

industrial era based on current best estimates of Earth's "climate sensitivity"– the amount of global temperature increase expected in response to a given rise in atmospheric concentrations of carbon dioxide (CO_2)," begins a news release on the study from Brookhaven National Laboratory.

"According to current best estimates of climate sensitivity, the amount of CO_2 and other heat-trapping gases added to Earth's atmosphere since humanity began burning fossil fuels on a significant scale during the industrial period would be expected to result in a mean global temperature rise of 3.8°F – well more than the 1.4°F increase that has been observed for this time span. Schwartz's analysis attributes the reasons for this discrepancy to a possible mix of two major factors: 1) Earth's climate may be less sensitive to rising greenhouse gases than currently assumed and/or 2) reflection of sunlight by haze particles in the atmosphere may be offsetting some of the expected warming.

" 'Because of present uncertainties in climate sensitivity and the enhanced reflectivity of haze particles', said Schwartz, 'it is impossible to accurately assign weights to the relative contributions of these two factors. This has major implications for understanding of Earth's climate and how the world will meet its future energy needs'."

In other words, just as you read at the start of *Air Con*, it's one thing to know how CO2 performs in a controlled lab experiment; it's an entirely different picture out in the real world with so many other complex factors to take into account.

A clue as to why temperatures might be stabilising emerged late January 2010, when the journal *Science* published new research showing water vapour plays a massive role in regulating Earth's temperature, also just as I suggested early in this book.

The study found water vapour content in the stratosphere increased sharply in the early 1990s, then fell sharply from around 2000 onwards. It's now estimated water vapour increases in the stratosphere – about 16 kilometres above the ground - caused around a third of what had been put down to human-caused warming in the 1990s, and likewise a big drop in water vapour – a 10% decrease – reduced the effects of global warming by 25% in the past ten years, as reported by NOAA.[427]

427 "Stratospheric water vapor is a global warming wild card", NOAA news release, 28 January 2010, http://www.noaanews.noaa.gov/stories2010/20100128_watervapor.html

"Observations from satellites and balloons show that stratospheric water vapour has had its ups and downs lately, increasing in the 1980s and 1990s, and then dropping after 2000. The authors show that these changes occurred precisely in a narrow altitude region of the stratosphere where they would have the biggest effects on climate."

As EcoFactory reported, "Susan Solomon, the respected climate scientist who lead the research... did point out that the research does allude to human emissions having a much smaller role in climate change than previously thought, and serves as a warning to climate modelers who 'over-interpret the results from a few years one way or another'."[428]

Yet despite the obvious trend in temperatures and new studies suggesting a cooler phase is underway, global warming belief websites and lobbyists have continued to deny that warming has in any way slowed down[429].

The UN IPCC's Rajendra Pachauri opened up the Copenhagen summit by utterly ignoring the temperature slowdown, and sea level rise slowdown, and instead parroting the UN IPCC's 2007 conclusion:

"Warming of the climate system is unequivocal as is now evident from observations of increases in global average air and ocean temperatures, widespread melting of snow and ice and rising global sea level."

That's how the global warming juggernaut operates, it simply ignores inconvenient evidence and keeps repeating convenient untruths in the hope that people won't realize they're being lied to.

But what did one of the world's leading climate scientists, Mojib Latif, tell a major international climate conference just a few weeks before Copenhagen?[430]

"Forecasts of climate change are about to go seriously out of kilter," reported the normally alarmist *New Scientist* magazine in early September. "One of the world's top climate modellers said Thursday

---

428 "NOAA, NASA: Water Vapor Largely Responsible for Global Warming," by Nate Kharrl, 29 January 2010 http://www.ecofactory.com/news/noaa-nasa-water-vapor-largely-responsible-global-warming-012910

429 In the interests of fairness, I would point out that Australian warming believer Tim Flannery has now conceded the obvious in the wake of Climategate: "We're dealing with an incomplete understanding of the way the earth system works... When we come to the last few years when we haven't seen a continuation of that (warming) trend we don't understand all of the factors that create earth's climate...So when the computer modelling and the real world data disagree you've got a very interesting problem... Sure for the last 10 years we've gone through a slight cooling trend." – ABC TV interview Australia, 23 November 2009, http://www.abc.net.au/lateline/content/2008/s2751390.htm

430 http://www.investigatemagazine.com/climatereality.pdf

we could be about to enter one or even two decades during which temperatures cool."

Eh? How come we didn't hear about this from Pachauri at Copenhagen?

"People will say this is global warming disappearing," Mojib Latif was quoted as telling 1500 delegates to a climate pow-wow in Geneva. "I am not one of the sceptics," insisted Latif of the Leibniz Institute of Marine Sciences at Kiel University, Germany. "However, we have to ask the nasty questions ourselves or other people will do it."

It turns out that Latif, and others, now believe a chunk of the warming of the last three decades (remembering that the UN IPCC has previously gone on record as saying it can only detect human influence on climate in the last three decades) appears to have been *wrongly* attributed to humans, and instead was really natural.

"Some of the climate scientists gathered in Geneva," writes Fred Pearce in the *New Scientist* article, "admitted that... natural variability is at least as important as the long-term climate changes from global warming.

"Latif predicted that in the next few years a natural cooling trend would dominate over warming caused by humans. The cooling would be down to cyclical changes to ocean currents and temperatures in the North Atlantic, a feature known as the North Atlantic Oscillation (NAO)."

The article signalled a major "break" with what it calls "climate-change orthodoxy" by highlighting Latif's comments that ocean oscillations "were probably responsible for some of the strong global warming seen in the past three decades".

"But how much?", Latif is quoted asking. "The jury is still out... The NAO is now moving into a colder phase."

The idea that oceanic oscillations are a major climate driver is, of course, a central theme of this book.

As if to prove Latif absolutely right, when the Copenhagen conference was visited by Al Gore and President Obama, Europe was hit by a massive early snowstorm, so bad that environmentally friendly electric trains failed, the Channel Tunnel was closed for days, and carbon-burning *steam locomotives* had to be used to rescue stranded electric train passengers.[431]

---

431 http://wattsupwiththat.com/2009/12/25/steam-train-rescues-stranded-passengers-in-britain-where-electric-trains-failed/

Across the US, record low temperatures were hit, and nearly 900 snowfall records were either broken or tied across different cities in the US alone. The blizzards led to a white Christmas, and totally confounded the climate agencies whose computer models had predicted warmer than average temperatures between October and December, and a record warm year in 2010.

Arctic sea ice extent hit a record high in April 2010 as a result.

Mojib Latif, for his part, has now conceded global warming may be taking a breather for "two decades or longer", as the *Daily Mail* reports:[432]

"Prof Latif, who leads a research team at the renowned Leibniz Institute at Germany's Kiel University, has developed new methods for measuring ocean temperatures 3,000ft beneath the surface, where the cooling and warming cycles start.

"He and his colleagues predicted the new cooling trend in a paper published in 2008 and warned of it again at an IPCC conference in Geneva last September.

"Last night he told *The Mail on Sunday*: 'A significant share of the warming we saw from 1980 to 2000 and at earlier periods in the 20th Century was due to these cycles – perhaps as much as 50 per cent.

" 'They have now gone into reverse, so winters like this one will become much more likely. Summers will also probably be cooler, and all this may well last two decades or longer.

" 'The extreme retreats that we have seen in glaciers and sea ice will come to a halt. For the time being, global warming has paused, and there may well be some cooling'."

Another plank of the global warming scare myth to come tumbling down is the claim that rising sea levels caused by CO2 emissions are wiping out low-lying tropical islands. As an author, I was hit with this back in May 2009 when *Air Con* was published when New Zealand's TVNZ network ran a documentary[433] promoting global warming belief and using footage of the sea over-running homes on the Takuu islands as proof that sea level rise "is the ugly face of global warming".

432 http://www.dailymail.co.uk/sciencetech/article-1242011/DAVID-ROSE-The-mini-ice-age-starts-here.html#ixzz0cloanDaV To be fair to Latif, he claims the ***Daily Mail*** beat up his story by using his study as evidence against climate change. Latif is a die-hard warmist, and no-one is accusing him of losing faith in the overall theory of warming, but his comments speak for themselves – cold periods can outforce CO2.

433 http://tvnz.co.nz/sunday-news/the-story-takuu-2687185/video?vid=2698367

What the TV network failed to disclose (a common media trait when it comes to climate stories), is that Takuu, off the coast of Papua New Guinea, sits on the edge of one of the world's most active continental plate collision zones, as veteran Pacific affairs journalist Michael Field had reported in 2006:[434]

"This dress rehearsal for sea level rise is *not* the result of global warming but rather of a grand, continental clash of the Australian and Pacific tectonic plates...It has created the most rapidly expanding active rift system on Earth, renowned for its heavy magnitude earthquakes."

Indeed, as one major newspaper reported in 2000, Takuu and its neighbouring islands are "sinking" rapidly:[435]

"The islands are just 12 ft above sea level, and they are sinking 11.8 inches a year."

That was in 2000. Do the maths and work out for yourself how many feet above sea level Takuu would have been by 2009 at that rate (I'll save you the trouble: three feet, or about a metre, which is exactly where they now are and dropping fast).

If you remember your high school geography and geology classes, you might recall learning about the 'life cycle of a coral atoll'. This is the well-documented phenomenon where undersea volcanoes punch through the ocean surface creating an island peak, which is eventually eroded away leaving the crater rim, upon which coral grows to form atolls. As sea levels gradually rise, the coral grows with them,[436] but when sea levels fall it leaves more coral exposed to the elements. After a long period of time, through a combination of erosion from ocean waves and storms, and also the weight of the volcanic seamount sinking back into the earth's crust, the islands gradually disappear.

So there is nothing scientifically remarkable about coral atolls being threatened by sea water. However, there's another extra ingredient they don't tell you about on the news either: human stupidity.

It seems many Pacific Islanders, and through Asia and elsewhere, adopted the dubious practice of "dynamite fishing" after World

434 "Takuu's Tragedy Unfolding", by Michael Field, *Islands Business*, http://www.islandsbusiness.com/islands_business/index_dynamic/containerNameToReplace=MiddleMiddle/focusModuleID=15847/overideSkinName=issueArticle-full.tpl

435 "1000 flee as sea begins to swallow New Guinea islands", by Kathy Marks, *NZ Herald*, 30 November 2000, http://www.nzherald.co.nz/world/news/article.cfm?c_id=2&objectid=162583

436 "Floating Islands" by Willis Eschenbach, 27 January 2010, http://wattsupwiththat.com/2010/01/27/floating-islands/

War 2, where explosives are set off in the lagoons to stun fish, which can then easily be harvested while floating on the surface. The dynamite, however, weakened the delicate coral reef structures, sometimes blowing holes in the reef as well as killing the coral via the shockwaves. The equally dubious practice of chemical fishing has been documented in a number of locales as having killed coral also. Islanders then harvested broken or dead coral to use on island roads, without thinking about the consequences.

It could have been the plot of a Wyllie Coyote performance in a *Roadrunner* cartoon.

One group of islands pleading the effects of sea level rise is Kiribati, in the South Pacific, yet the Kiribati government website lists dynamite fishing as an approved activity:[437]

"Some other areas in which foreign investment is allowed but subject to restrictions include:

> 1. Fisheries sector: drifter fishing, dynamite fishing, chemical fishing, conservation areas, coral exportation."

Not only are they blowing up their coral, they are even exporting it! The coral, you see, acted as a barrier to ocean swells. Without the barrier, the atolls became much more vulnerable to ocean surges and storms scouring away their beaches, and their erosion happened much faster. Which is also what has occurred on Takuu.

I was briefed[438] after *Air Con* was first published by a former UN Development Programme official who'd worked on the Takuu problem. He explained that the UN had been well aware of the real cause of Takuu's fate, but that UNDP officials from New York had arrived on the island "several years ago, telling islanders to refer to global warming as the cause in public, so as to qualify for emergency climate adaptation funding" from developed countries. It is likely the UNDP has been pitching the same fraud at Tuvalu, Kiribati, the Maldives and elsewhere.

One final aspect on coral atolls. The reefs are home to diverse fish life, including parrotfish that play a key role in breaking down dead coral to form sand. It's that sand that's washed up onto atoll beaches and which helps maintain land area. But dynamite fishing

437 http://www.spto.org/spto/export/sites/spto/investment/kiribati.shtml
438 At a public meeting in the NZ town of Kerikeri, in front of 250 others, 5 August 2009

tipped the ecological balance, stripping the atolls of these crucial creatures. Fewer parrotfish means less coral sand being produced, which means less sand on the beaches, and means more erosion. You get the picture.

It is true that humans have played a part in hastening the submergence of tropical atolls under the sea. It just isn't true that CO_2 had anything whatsoever to do with it. Significantly, TVNZ lost a broadcasting standards case brought by complainants, after adjudicators ruled the Takuu footage had been "inaccurate" and "sensational".[439]

The scare stories about global warming and coral atolls don't stop there, however. The UN IPCC has tried to drum up panic by suggesting rising CO_2 absorption in the oceans is affecting the pH balance, making the seas more "acidic" and killing coral and other marine life. But the latest scientific studies are actually suggesting other factors may be at play.

"An international team of scientists has solved a mystery that has puzzled marine chemists for decades," explained the University of Miami in 2009.[440] "They have discovered that fish contribute a significant fraction of the oceans' calcium carbonate production, which affects the delicate pH balance of seawater. The study gives a conservative estimate of three to 15 percent of marine calcium carbonate being produced by fish, but the researchers believe it could be up to three times higher."

Now this is an exceptionally important study. Rising CO_2 levels in the oceans cause chemical reactions that can start to harm shellfish and other marine life by dissolving their shells and exoskeletal structures. In theory. However, it appears fish can and do use the surplus CO_2 in the oceans, along with calcium-rich surface waters, to create calcium carbonates, which help keep the oceans alkaline.

If you've owned a fish tank you'll know fish quickly create their ideal pH level, and the same thing applies on a much bigger scale in the oceans.

"Until now," continues the University of Miami study, "scientists

439 "The Authority finds that this part of the programme was inaccurate because it implied unequivocally that the situation in Takuu was caused solely by global warming. It considers that the use of the footage was sensational, particularly as it was used as the opening and closing images for the item, and the impression that was created as a result demonstrated a lack of care and rigour in dealing with such a complex issue as global warming." – Broadcasting Standards Authority ruling, http://www.bsa.govt.nz/decisions/2009/2009-063.html

440 "Fishdunnit! Mystery solved", University of Miami news release, based on a study published in *Science*, 16 January 2009, http://www.rsmas.miami.edu/pressreleases/20090116-flounder.html

believed that the oceans' calcium carbonate, which dissolves in deep waters making seawater more alkaline, came from marine plankton. The recent findings published in *Science* explain how up to 15 percent of these carbonates are, in fact, excreted by fish that continuously drink calcium-rich seawater. The ocean becomes more alkaline at much shallower depths than prior knowledge of carbonate chemistry would suggest which has puzzled oceanographers for decades. The new findings of fish-produced calcium carbonate provides an explanation: fish produce *more soluble* forms of calcium carbonate, which probably dissolve more rapidly, before they [are able to] sink into the deep ocean."

This is important, because the UN IPCC study teams have claimed the calcium carbonates produced by plankton (which sink to the ocean floor) take millions of years to re-balance the oceans, but this new study shows fish produce a much more rapidly acting form of the alkaline that can benefit the upper layers of the oceans immediately.

"The digestive systems of fish play a vital role in maintaining the health of the oceans and moderating climate change," reported Reuters news agency[441] on the peer-reviewed study.

"Bony fish produced a large portion of the inorganic carbon that helps maintain the oceans' acidity balance and was vital for marine life, they said.

"The world's bony fish population, estimated at between 812 million and 2 billion tons, helped to limit the consequences of climate change through its effect on the carbon cycle.

" 'This study is really the first glimpse of the huge impact fish have on our carbon cycle – and why we need them in the ocean', researcher Villy Christensen and colleagues wrote.

"Calcium carbonate is a white, chalky material that helps control the acidity balance of sea water and is essential to the health of marine ecosystems and coral reefs."

The probable reason for decreasing alkalinity, then, is overfishing, not CO_2 emissions.[442] If we strip-mine the seas of alkaline-producing

441 "Fish digestions help keep oceans healthy", *Reuters* 16 January 2009, http://spoonfeedin.wordpress.com/2009/01/16/science-fish-digestions-help-keep-the-oceans-healthy/

442 More evidence that this is the case is provided by Australian geology researcher Tim Casey, who found that decreasing alkalinity in the oceans was not being matched in fresh water systems like rivers and lakes. If CO2 were the culprit, acidification should be happening in all open air water sources across the planet. The fact that it isn't suggests something other than atmospheric CO2 is to blame, and marine overfishing or submarine volcanism are far more likely explanations. See Casey's paper: http://carbon-budget.geologist-1011.net/

fish, we should hardly act surprised when we find less alkalinity in the seawater after a few decades of bad fishing practice.[443]

Then there's this recent bombshell published in the journal *Geology*: that increasing levels of CO_2 in the ocean are actually helping some shellfish thrive:[444]

> In a striking finding that raises new questions about carbon dioxide's (CO2) impact on marine life, Woods Hole Oceanographic Institution (WHOI) scientists report that some shell-building creatures – such as crabs, shrimp and lobsters – unexpectedly build more shell when exposed to ocean acidification caused by elevated levels of atmospheric carbon dioxide (CO2).
>
> Because excess CO2 dissolves in the ocean – causing it to "acidify" – researchers have been concerned about the ability of certain organisms to maintain the strength of their shells. Carbon dioxide is known to trigger a process that reduces the abundance of carbonate ions in seawater – one of the primary materials that marine organisms use to build their calcium carbonate shells and skeletons.
>
> The concern is that this process will trigger a weakening and decline in the shells of some species and, in the long term, upset the balance of the ocean ecosystem.
>
> But in a study published in the Dec. 1 issue of *Geology*, a team led by former WHOI postdoctoral researcher Justin B. Ries found that seven of the 18 shelled species they observed actually built more shell when exposed to varying levels of increased acidification. This may be because the total amount of dissolved inorganic carbon available to them is actually increased when the ocean becomes more acidic, even though the concentration of carbonate ions is decreased.
>
> "Most likely the organisms that responded positively were somehow able to manipulate...dissolved inorganic carbon in the fluid from which they precipitated their skeleton in a way that was beneficial to them," said Ries, now an assistant professor in marine sciences at the University of North Carolina. "They were somehow able to manipulate CO2...to build their skeletons."

In truth, it's a simple reminder of something the climate changers either forget or deliberately ignore when crafting their dumbed-down scary soundbites: when one species can no longer take the heat in

443 Additionally, if coral atolls are being severely overfished that could explain why decreasing alkalinity might be impacting coral structures.

444 "In CO2-rich Environment, Some Ocean Dwellers Increase Shell Production", Woods Hole Oceanographic Institute, Dec 1 2009, http://www.whoi.edu/page.do?pid=7545&tid=282&cid=63809&ct=162

the kitchen, another one rises up from the shadows swiftly to take its place that's more resilient and even thrives in the new conditions.

The moral of the story? Life appears far more adaptable than you hear about on the TV news. The next time you hear a Greenpeace lobbyist, or a TV reporter for that matter, sensationally warning of the dangers of ocean acidification, you can be forgiven if you choose to roll all over the floor in fits of laughter.[445]

In addition to CO_2, warming oceans, too, are accused of killing marine life irreparably and destroying the marine food chain as we know it. This seems more than a trifle far-fetched, given that marine life has survived periods where tropical waters were hotter than spa pools,[446] but the coral bleaching of the Great Barrier Reef is a good case study of how the alarmists got their predictions badly wrong.

"Marine scientists say they are astonished at the spectacular recovery of certain coral reefs in Australia's Great Barrier Reef Marine Park from a devastating coral bleaching event in 2006," researchers announced the very week *Air Con* was first published in 2009.[447]

"That year high sea temperatures caused massive and severe coral bleaching in the Keppel Islands, in the southern part of the GBR. The damaged reefs were quickly smothered by a single species of seaweed – an event that can spell the total loss of the corals.

"However, a lucky combination of rare circumstances meant the reefs were able to achieve a spectacular recovery, with abundant corals re-established in a single year, says Dr Guillermo Diaz-Pulido, from the ARC Centre of Excellence for Coral Reef Studies (CoECRS) and the Centre for Marine Studies at The University of Queensland.

"Dr Diaz-Pulido explains that the rapid recovery is due to an exceptional combination of *previously-underestimated* ecological mechanisms. [my emphasis]

" 'Three factors were critical. The first was exceptionally high regrowth of fragments of surviving coral tissue. The second was an unusual seasonal dieback in the seaweeds, and the third was the presence of a highly competitive coral species, which was able to outgrow the seaweed'."

445 For an absolute treasure trove of scientific studies revealing the ocean acidification scare is nothing more than a Sea Con, see the 2010 report from Science and Public Policy Institute: http://scienceandpublicpolicy.org/images/stories/papers/originals/acid_test.pdf

446 Far more likely than outright extinction is simply a shift in habitat zones – fish preferring colder waters will venture further toward the poles, and corals will spring up in areas they are comfortable in.

447 "Spectacular Recovery From Coral Bleaching At Great Barrier Reef Marine Park In Australia", *ScienceDaily* Apr. 24, 2009, http://www.sciencedaily.com/releases/2009/04/090423100817.htm

In other words, Nature is much more resilient than the climate scientists and green lobbyists have allowed for. To recap, ocean 'acidification' turns out not to be matched in freshwater systems, which it would be if it was genuinely caused by CO_2 in the atmosphere, and the acid/alkaline balance of the oceans turns out to be heavily dependent on how many fish exist in a cubic kilometre of water. We all know we have heavily overfished the oceans. And then we see predictions of coral bleaching shatter as reefs bounce back and adjust. Everywhere we look, the global warming scare stories just don't stack up under proper scrutiny.

I had some personal experience of the misinformation being pushed out there in the name of science, when I attended a phone-in news conference with a group of climate scientists from New Zealand's National Institute of Water and Atmospheric research, NIWA, in November.[448] One of those scientists, a former Pachauri offsider named Andy Reisinger, made the claim that up to 30% of species were expected to become extinct as a result of global warming this century. It's a load of old cobblers, as anyone who has studied evolutionary biology in detail knows. Far from being slow to adapt, species show remarkable ability to switch on and off genetic traits that allow them to rapidly meet environmental changes head on.[449]

I challenged Reisinger at the news conference by quoting an Oxford University study published in *Science* on the issue, and NIWA became so embarrassed at being caught out that my microphone was muted and I was prevented from asking any further questions.[450] But the Oxford study was clear:[451]

"Alarming predictions that climate change will lead to the extinction of hundreds of species may be exaggerated, according to Oxford scientists.

"They say that many biodiversity forecasts have not taken into

448 Incidentally, NIWA have been caught using highly dodgy techniques to manipulate raw temperature data, in a manner that seems to be widespread in the climate clique. Suggested reading can be found *here*: http://briefingroom.typepad.com/the_briefing_room/2010/02/breaking-news-niwa-reveals-nz-original-climate-data-missing.html and *here*: http://briefingroom.typepad.com/the_briefing_room/2010/03/fisking-ken-perrott-over-niwa-defence.html

449 A classic example is the good old stickleback fish, populations of which have evolved separately in North America. In every population, geneticists found, the same genes had switched on to allow rapid adaptation to local conditions. "Stickleback Genomes Shining Bright Light on Evolution", *ScienceDaily*, Feb. 25, 2010 http://www.sciencedaily.com/releases/2010/02/100225214757.htm

450 "Climate scientists run from challenging questions", TBR.cc, http://briefingroom.typepad.com/the_briefing_room/2009/11/nz-climate-scientists-run-from-challenging-questions.html

451 "Experts say that fears surrounding climate change are overblown", Hannah Devlin, *Times of London*, 6 November 2009, http://www.timesonline.co.uk/tol/news/science/article6905082.ece

account the complexities of the landscape and frequently underestimate the ability of plants and animals to adapt to changes in their environment.

" 'The evidence of climate change-driven extinctions have really been overplayed', said Professor Kathy Willis, a long-term ecologist at the University of Oxford and lead author of the article.

"Professor Willis warned that alarmist reports were leading to ill-founded biodiversity policies in government and some major conservation groups. She said that climate change has become a 'buzz word' that is taking priority while, in practice, changes in human use of land have a greater impact on the survival of species."

The problem seems to arise from dodgy garbage-in/garbage-out computer models, again.

"The article, published today in the journal *Science*, reviews recent research on climate change and biodiversity, arguing that many simulations are not sufficiently detailed to give accurate predictions.

"In one analysis of the likelihood of survival of alpine plant species in the Swiss Alps, the landscape was depicted with a 16km by 16km (10 miles by 10 miles) grid scale. This model predicted that all suitable habitats for alpine plants would have disappeared by the end of the century. When the simulation was repeated with a 25m by 25m (82ft by 82ft) scale, the model predicted that areas of suitable habitat would remain for all plant species."

Hardly rocket science, but apparently too difficult for the climate science people at NIWA to get their heads around, and they were so desperate that other journalists taking part in the carefully staged news conference should not hear about this that they cut my microphone and prevented me from completing my question. Never let the facts get in the way of a good scare story.[452]

"Other studies comparing predictions of extinction rates with actual extinction rates have come to similar conclusions," reported the *Times of London*. "According to a high-profile paper published in the journal *Nature* in 2004, up to 35 per cent of bird species would be extinct by 2050 due to changes in climate. To be on track to meet this figure, Professor Keith Bennett, head of geography at Queen's University Belfast, calculated that about 36 species would have to

452 In fact, a similar fate befell Irish journalist and documentary maker Phelim McAleer, who was pulled aside by armed UN security men and his microphone cut when he tried to ask inconvenient questions of Al Gore on Climategate, in December 2009. http://www.youtube.com/watch?v=fooYtalS9Gc&feature=player_embedded

have become extinct *each year* between 2004 and 2008. In reality, three species of bird became extinct [in that time]."

As Bennett later made the point, birds confronted with changing weather don't simply point their toes to the sky and drop backwards, dead, out of trees. Instead, he said quite simply, "they move."

Ouch. But despite this embarrassing slap-down for the global warming juggernaut, it just kept repeating its fabrications over and over, as this comment from Rajendra Pachauri's Copenhagen speech shows:

"Approximately 20% to 30% of species assessed so far [are] at increased risk of extinction if increases in global average warming exceed 1.5 to 2.5 degrees C", said Pachauri.

In the face of the studies on fish and coral reefs, and the comments from Oxford about the inaccuracies of current climate models used for species calculations, is the UN position still credible in *any* way?

Perhaps the biggest blow to the UN's scare stories is another peer-reviewed study published late 2009, which showed ocean temperatures were as warm a thousand years ago, during the Medieval Warm Period, as they are today. In other words, if today's warming is "calamitous" then the warming a thousand years ago must have been equally dangerous, and yet life survived and here we all are.

"A new 2,000-year-long reconstruction of sea surface temperatures (SST) from the Indo-Pacific warm pool (IPWP) suggests that temperatures in the region may have been as warm during the Medieval Warm Period as they are today," researchers reported.[453]

"The IPWP is the largest body of warm water in the world, and, as a result, it is the largest source of heat and moisture to the global atmosphere, and an important component of the planet's climate. Climate models suggest that global mean temperatures are particularly sensitive to sea surface temperatures in the IPWP. Understanding the past history of the region is of great importance for placing current warming trends in a global context.

"The study is published in the journal *Nature*."

The sea continues to be a battleground in the global warming propaganda war, but as you can see the believers are losing ground. Some of you may recall that just on the eve of December's Copenhagen conference, a new "study" was released claiming that sea

453 http://www.investigatemagazine.com/climatereality.pdf

levels would rise by two metres this century, far higher than the UN IPCC had predicted.

That study was conducted by the University of Potsdam's Stefan Rahmstorf, a climate activist who blogs at the discredited RealClimate site.

Rahmstorf's conclusions flew utterly in the face of what you've read in *Air Con*, but you may not have heard the latest – fellow scientists claim Rahmstorf got his maths, and therefore his study, badly wrong – as the *Times of London* reported early 2010.[454]

"It predicted an apocalyptic century in which rising seas could threaten coastal communities from England to Bangladesh and was the latest in a series of studies from Potsdam that has gained wide acceptance among governments and environmental campaigners.

"Besides underpinning the Copenhagen talks, the research is also likely to be included in the next report of the Intergovernmental Panel on Climate Change. This would elevate it to the level of global policy-making.

"However, the studies, led by Stefan Rahmstorf, professor of ocean physics at Potsdam, have caused growing concern among other experts. They say his methods are flawed and that the real increase in sea levels by 2100 is likely to be far lower than he predicts."[455]

There was further embarrassment when a prestigious study in *Nature Geoscience*, trumpeted as backing up UN IPCC predictions of sea level rise, had to be withdrawn from publication because of inaccuracies its methodology.[456]

Adding insult to injury for the IPCC, its claims that 20 million people in Bangladesh would be displaced by rising sea levels have been shot down by Bangladeshi scientists who say the UN failed to properly factor in the effect of estuaries in building up sediment:[457]

"DHAKA – Scientists in Bangladesh posed a fresh challenge to the UN's top climate change panel Thursday, saying its doomsday forecasts for the country in the body's landmark 2007 report were overblown.

"The Intergovernmental Panel on Climate Change (IPCC), already

---

454 "Climate change experts clash over sea-rise 'apocalypse'," *Times of London*, 10 January 2010, http://www.timesonline.co.uk/tol/news/environment/article6982299.ece

455 "Climate scientists withdraw journal claims of rising sea levels", *The Guardian*, 21 February 2010, http://www.guardian.co.uk/environment/2010/feb/21/sea-level-geoscience-retract-siddall

456 "Climate scientists withdraw journal claims of rising sea levels", *The Guardian*, 21 February 2010, http://www.guardian.co.uk/environment/2010/feb/21/sea-level-geoscience-retract-siddall

457 Challenge to IPCC's Bangladesh climate predictions", by Shafiq Alam, *Agence France Presse*, 23 April 2010

under fire for errors in the 2007 report, had said a one-metre (three-foot) rise in sea levels would flood 17 percent of Bangladesh and create 20 million refugees by 2050.

"The claim helped create a widespread consensus that the low-lying country was on the 'front line' of climate change, but a new study argues the IPCC ignored the role sediment plays in countering sea level rises.

"IPCC chairman Rajendra Pachauri defended his organisation's Bangladesh predictions Thursday, warning that 'on the basis of one study one cannot jump to conclusions'.

" 'The IPCC looks at a range of publications before we take a balanced view on what's likely to happen', he told AFP by telephone.

"But IPCC's prediction did not take into account the one billion tonnes of sediment carried by Himalayan rivers into Bangladesh every year, which are crucial in countering rises in sea levels, the study funded by the Asian Development Bank said.

" 'Sediments have been shaping Bangladesh's coast for thousands of years,' said Maminul Haque Sarker, director of the Dhaka-based Center for Environment and Geographic Information Services (CEGIS), who led research for the study.

"Previous "studies on the effects of climate change in Bangladesh, including those quoted by the IPCC, did not consider the role of sediment in the growth and adjustment process of the country's coast and rivers to the sea level rise," he told AFP.

"Even if sea levels rise a maximum one metre in line with the IPCC's 2007 predictions, the new study indicates most of Bangladesh's coastline will remain intact, said Sarker."

As if to prove the point, another peer reviewed study of shorelines in the Mediterranean has just found massive fluctuations in sea levels within the last two thousand years, that had nothing to do with CO_2:[458]

"During the Hellenistic period, the sea level was about 1.6 meters lower than its present level; during the Roman era the level was almost similar to today's; the level began to drop again during the ancient Muslim period, and continued dropping to reach the same level as it was during the Crusader period; but within about 500 years it rose again, and reached some 25 centimeters lower than

458 "The sea level has been rising and falling over the last 2,500 years", University of Haifa news release, 26 January 2010, http://www.eurekalert.org/pub_releases/2010-01/uoh-tsl012610.php

today's level at the beginning of the 18th century."

There were no major glacial melts, yet sea levels fluctuated by 1.6 metres over the space of just a few hundred years.

Like I said at the start of this chapter, the science is increasingly giving support to skeptics who say modern warming is part of a mostly natural climate cycle to be adapted to, not feared.

But it was science in a rather different light that lit the fuse under global warming theory late 2009, when thousands of emails and documents from one of the world's top climate research centres were released on the internet in the firestorm that quickly became known as "Climategate".

Chapter Twenty One

The Climategate Caper

"The fact is that we can't account for the lack of warming at the moment and it is a travesty that we can't."

– ***Kevin Trenberth, climate scientist, Climategate email***

On or around November 13 (the most recent of the leaked emails dated from November 12) 2009, someone – probably an insider at the University of East Anglia's Climatic Research Unit (CRU) – logged in to the CRU computer servers for the last time and downloaded more than 160 megabytes of files relating to climate change research. Cybercrime experts suspect an insider because of the sheer volume of files, and because the information appears to have been carefully selected, rather than the work of a 'smash and grab' hacker. Nonetheless, whoever it was probably was not working alone.

On November 17, hackers did indeed strike the RealClimate website, closely allied to the CRU scientists and partially funded by entities associated with one George Soros,[459] uploading a zipped file of the emails and data. RealClimate discovered the hack within minutes, and shut the link down.

As the CRU's director Phil Jones later confided to me in the world exclusive interview that confirmed the authenticity of the Climategate emails,[460] RealClimate failed to tell CRU what was out in the marketplace.

"Real Climate were given information, but took it down off their

459 RealClimate's website is funded by PR firm Fenton Communications, whose clients according to Discover The Networks include Marxist organizations and radical environmental groups, as well as Soros' Open Society Institute. Soros has also donated funding to Environmental Media Services, the Fenton-created entity that actually pays RealClimate's web costs. http://www.discoverthenetworks.org/groupProfile.asp?grpid=6958 See also: http://briefingroom.typepad.com/the_briefing_room/2009/11/look-whos-paying-global-warmings-top-bloggers.html?cid=6a00d8341c51bc53ef0120a663bbc4970b

460 http://briefingroom.typepad.com/the_briefing_room/2009/11/hadleycru-says-leaked-data-is-real.html

site and told me they would send it across to me. They didn't do that. I only found out it had been released five minutes ago."

That interview, on the morning of November 20, GMT, set the fate of Jones and the Climategate story. It was the first – and for several days the only – official media confirmation from Jones that the climate emails appearing on the internet were genuine, rather than a practical joke.[461]

The sequence of events after RealClimate took down the link is interesting.

Late on the morning of November 20, NZ time, US climate change blogger Anthony Watts posted a message about a breaking story – reports that one of the world's leading climate research centres, based in Britain, had been hacked into and its data leaked onto the internet.

By chance, staff at New Zealand's *Investigate* magazine stumbled on Watts' post literally only two minutes after it hit the web. The story described how someone claiming to have obtained data from the University of East Anglia's CRU (climatic research unit), had posted a link on another website to a very large 61mb file containing the leaked data.

Investigate checked the second website, The Air Vent, only to find the link had already been removed, but using Google cache we retrieved a version of the page that still contained the link to a Russian webserver carrying the allegedly hacked data.

We made an editorial decision to re-post that link on our own website, TBR.cc, as part of a story about the fact of the hacking. At the same time, we downloaded a copy of the data to the *Investigate* office.

This is how news broke on TBR:

"A MAJOR STORY IS BREAKING in climate science, after hackers posted a 61 megabyte data file on a Russian server that appears to be confidential emails and climate data hacked from the UK Met Office Hadley Centre.[462]

"The data raises major questions about the role of scientists in what appears to be a *deliberate* conspiracy to mislead the public..."

We outlined a couple of the emails, before commenting:

461 http://wattsupwiththat.com/2009/11/19/breaking-news-story-hadley-cru-has-apparently-been-hacked-hundreds-of-files-released/

462 The initial reports on WattsUpWithThat had named the Hadley Centre but in fact, as quickly became clear, it was East Anglia University's Climatic Research Unit involved. The confusion arose because the data produced by CRU is fed into what is known as the HadCRUT3 dataset.

"If there's an innocent explanation, I'll be interested in hearing it. In the meantime I've sent an email to Phil Jones asking if this email is genuine.

"For those interested, the large file can be downloaded here.

"UPDATE: Am busy on the TGIF deadline so have only generally perused the leaked emails. It appears to be a collection that might have been prepared for a possible FOIA (freedom of information) request and were in the process of being scrutinized. The tone of many is quite waspish, although like others the email above seems too damning to be true. Surely they weren't that stupid to commit such comments to writing back in 1999?

"UPDATE 2: One of the emails refers to stacking the peer-review process to ensure scientific papers by the likes of NZ's Chris de Freitas don't make it past review into the IPCC's 2007 AR4.

> *The other paper by MM is just garbage – as you knew. De Freitas again. Pielke is also losing all credibility as well by replying to the mad Finn as well – frequently as I see it. I can't see either of these papers being in the next IPCC report.* ***K and I will keep them out somehow – even if we have to redefine what the peer-review literature is !***

"Shocking. Lends authenticity to the documents as well – a US or European based hacker would not be likely to pluck de Freitas' name out of thin air if they were making something up."

What followed next became the stuff of internet legend. *Investigate's* coverage, directing people both to the main file and to ongoing coverage of the breaking story from WattsUpWithThat.com, was picked up and magnified by other blog sites. Within an hour, half a dozen websites around the world were carrying links to the download of CRU emails and documents, and hits on both TBR.cc and Watts were going through the roof as the world began waking up to the contents of the leaked emails.

Commenters on both websites had taken the opportunity to download and open the files, and begin posting incriminating emails and documents up in the comments thread of the respective websites. In a sense, what would have taken one newsroom a week to do – trawl through 3,000 emails and more than a thousand documents – was being outsourced to tens of thousands of volunteers around the world.

One volunteer with time on his hands used his computer expertise to turn the email files into a searchable online database that readers could navigate by entering keywords and search phrases.

Despite the internet catching fire with news of the devastating leak, coverage was confined to the blogosphere. No mainstream media journalist had apparently felt sufficiently motivated to pick up a phone and check out the story.

Instead, that task fell to *Investigate's TGIF Edition*. At 9pm, NZ time on Friday November 20, we finally made contact with CRU director Phil Jones, who starred in some of the most incriminating emails. It was 8am in the UK and he had only walked into his office five minutes earlier.

"A 62 MEGABYTE ZIP FILE, containing around 160 megabytes of emails, pdfs and other documents, has been confirmed as genuine by the head of the University of East Anglia's Climate Research Unit, Dr Phil Jones", *TGIF* reported a short time later in a world exclusive.

"In an exclusive interview with *Investigate* magazine's *TGIF Edition*, Jones confirms his organization has been hacked, and the data flying all over the internet appears to have come from his organisation.

" 'It was a hacker. We were aware of this about three or four days ago that someone had hacked into our system and taken and copied loads of data files and emails'."

" 'Have you alerted police?' "

" 'Not yet. We were not aware of what had been taken'."

"Jones says he was first tipped off to the security breach by colleagues at the website RealClimate.

"Real Climate were given information, but took it down off their site and told me they would send it across to me. They didn't do that. I only found out it had been released five minutes ago."

The files were first released from a Russian fileserver site by an anonymous tipster calling him or herself "FOIA", in an apparent reference to the US Freedom of Information Act. The zip file contains more than a thousand documents sitting in a "FOIA" directory, and it prompted speculation that the information may have been in the process of being compiled for consideration of an information act request.

Jones, however, says the files were not contained in a "FOIA" directory at the Climate Research Unit.

"No. Whoever is responsible has done that themselves."

"I'm not sure what we're going to do. I'll have to talk to other people here. In fact, we were changing all our passwords overnight and I can't get to my email, as I've just changed my password. I've gone into the Climate Audit website because I can't get into my own email.

"It's completely illegal for somebody to hack into our system."

In one email dating back to 1999, Jones appears to talk of fudging scientific data on climate change to "hide the decline":

> From: Phil Jones
> To: ray bradley ,mann@[snipped], mhughes@[snipped]
> Subject: Diagram for WMO Statement
> Date: Tue, 16 Nov 1999 13:31:15 +0000
> Cc: k.briffa@[snipped],t.osborn@[snipped]
>
> Dear Ray, Mike and Malcolm,
>
> Once Tim's got a diagram here we'll send that either later today or first thing tomorrow. I've just completed Mike's Nature trick of adding in the real temps to each series for the last 20 years (ie from 1981 onwards) and from 1961 for Keith's to hide the decline. Mike's series got the annual land and marine values while the other two got April-Sept for NH land N of 20N. The latter two are real for 1999, while the estimate for 1999 for NH combined is +0.44C wrt 61-90. The Global estimate for 1999 with data through Oct is +0.35C cf. 0.57 for 1998.Thanks for the comments, Ray.
>
> Cheers, Phil
> Prof. Phil Jones
> Climatic Research Unit

TGIF asked Jones about the controversial email discussing hiding "the decline", and Jones explained he was not trying to mislead.

"No, that's completely wrong. In the sense that they're talking about two different things here. They're talking about the instrumental data which is unaltered – but they're talking about proxy data going further back in time, a thousand years, and it's just about how you add on the last few years, because when you get proxy data you sample things like tree rings and ice cores, and they don't always have the last few years. So one way is to add on the instrumental data for the last few years."

Jones told *TGIF* he had no idea what he meant by using the words "hide the decline".

"That was an email from ten years ago. Can you remember the exact context of what you wrote ten years ago?"

"The other emails are described by skeptic commentators as "explosive", one talks of stacking the peer-review process to prevent qualified skeptical scientists from getting their research papers considered."

Investigate's confirmation that the data was genuine, and not someone dummying up emails as a prank, set off a chain reaction across the world. In their "autopsy" of Climategate, the global warming PR agency Hoggan and Associates wrote:[463]

"The *TGIF* article is referred to as proof of the validity of the stolen emails in most early stories... Six hours after Wishart confirmed the leak in *TGIF*, James Delingpole published the story in a *Telegraph* blog, with the meme 'the final nail in the coffin of anthropogenic global warming'. Minutes later the BBC posts its first story at 2:13pm, UMT time. Four hours after that, Fox News started posting it online in the US with the hook 'skeptics see smoking gun in leaked climate science emails.' At 1:48pm on Friday the *Drudge Report* first found the story, spreading it to an even broader network."

As climatologist Pat Michaels was quoted by the *Washington Post* in the days following, this wasn't "a smoking gun, it's a mushroom cloud".

The first northern hemisphere journalist to jump on the confirmation was the *Telegraph's* James Delingpole:[464]

"If you own any shares in alternative energy companies I should start dumping them NOW. The conspiracy behind the Anthropogenic Global Warming myth (aka AGW; aka ManBearPig) has been suddenly, brutally and quite deliciously exposed after a hacker broke into the computers at the University of East Anglia's Climate Research Unit (aka CRU) and released 61 megabytes of confidential files onto the internet.

"When you read some of those files – including 1079 emails and 72 documents – you realise just why the boffins at CRU might have preferred to keep them confidential..."

463 "Climategate – An Autopsy" by Morgan Goodwin, 30 March 2010, http://www.desmogblog.com/climatgate-autopsy

464 "Climategate: the final nail in the coffin of 'Anthropogenic Global Warming'?" by James Delingpole, *Telegraph*, 20 November 2009, http://blogs.telegraph.co.uk/news/jamesdelingpole/100017393/climategate-the-final-nail-in-the-coffin-of-anthropogenic-global-warming/

So what was the sequence of events leading up to the confirmation? US media website *The Examiner* soon established a more detailed timeline.

"THE TIMELINE BEGINS ON November 17, when the user named "FOIA" left this comment at The Air Vent site:

'We feel that climate science is, in the current situation, too important to be kept under wraps.

'We hereby release a random selection of correspondence, code, and documents. Hopefully it will give some insight into the science and the people behind it.

'This is a limited time offer, download now:'

"He then continued with a link to a Russian anonymous FTP account. (That account no longer works, but this Examiner was able to obtain the archive from it when a correspondent alerted him to it.)

"This is consistent with Phil Jones' statement to Ian Wishart of *Investigate* magazine, dated November 20. Jones said that he had known about a security breach of his organizations computers "three or four days ago," having heard about the matter first from the administrators of RealClimate.org. Concerning RealClimate's immediate reaction, Jones said:

" 'Real Climate were given information, but took it down off their site and told me they would send it across to me. They didn't do that. I only found out it had been released five minutes ago.'

"RealClimate's own statement says this:

" 'We were made aware of the existence of this archive last Tuesday morning when the hackers attempted to upload it to RealClimate, and we notified CRU of their possible security breach later that day.'

"This indicates that the tipster first tried to submit his material to RealClimate.org, and when the administrators refused to accept it, he then established his Russian anonymous FTP account and submitted the link in his comment to The Air Vent.

"The Air Vent's administrator, Jeff Id, was out-of-contact when the comment was posted. No one said another word about it until, two days later, the user named Steven Mosher alerted The Blackboard. Initially he left only a link to the original post, not a specific comment link. But apparently Lucia, the Blackboard administrator, followed the link and examined the files for herself. She was, however, reluctant to publish the link, but another user, Jean S, published it for her. In the process, she said this:

" 'Seems to me that someone has hacked UAH computers. All e-mails seem to contain at least an address ending uea.ac.uk. Also all the files seem to be UAH-related. At least some of the material has to be real, there are just so many small details that were just impossible to fake (for instance under briffa-treering-external/timonen there are some file names only a Finn would use).'

"She might be referring to file names like "kilpisj" and "hossapal", and extensions like "tuc". The file names fail to translate when subjected to Google's Translate routines.

"At the same time, Steven Mosher published an alert to Climate Audit. Then within hours, Anthony Watts at Watts Up With That published his own brief commentary. Shortly after that, this Examiner made his initial report, which is, as far as this Examiner has been able to determine, the first report by a professional or semi-professional journalist of this whole affair. Ian Wishart, editor and publisher of *Investigate*, also took note of the story at the same time and published his own initial blog entry, in which he announced that he had sent an e-mail to Phil Jones requesting an interview.

"In all that time, the original poster of the Russian FTP link never made another comment in any forum. As discussed above, this is not typical of a hacker. A hacker would be boasting about his act, and loudly. Instead, his file sat in that anonymous FTP account for more than forty-eight hours, and the poster never made any further attempt to publicize his find. Hence the conclusion, by this Examiner and a host of other commenters, including IP security professionals, that this unknown user was one who had had access to CRU computers, in accordance with his duties at the CRU, concludes the Examiner report."

The Daily Mail in the UK, incidentally, reports police are now working on the theory it was indeed an inside job, rather than a hack.

In the meantime, after being taken by surprise by *Investigate TGIF's* Friday phone call and its revelations reverberating around the world, Jones and his CRU team apparently went into bunker mode, and refused to answer any further media calls, according to the *Guardian's* George Monbiot: [465]

465 "Pretending the climate email leak isn't a crisis won't make it go away", George Monbiot, *The Guardian*, 25 November 2009, http://www.guardian.co.uk/environment/georgemonbiot/2009/nov/25/monbiot-climate-leak-crisis-response

"The university knew what was coming three days before the story broke. As far as I can tell, it sat like a rabbit in the headlights, waiting for disaster to strike. When the emails hit the news on Friday morning, the university appeared completely unprepared. There was no statement, no position, no one to interview. Reporters kept being fobbed off while CRU's opponents landed blow upon blow on it.

"When a journalist I know finally managed to track down Phil Jones, he snapped "no comment" and put down the phone. This response is generally taken by the media to mean "guilty as charged". When I got hold of him on Saturday, his answer was to send me a pdf called "WMO statement on the status of the global climate in 1999". Had I a couple of hours to spare I might have been able to work out what the heck this had to do with the current crisis, but he offered no explanation."

This was confirmed by the *Times of London*, who also reported Jones going to ground as a result of the *Investigate* call.[466]

"Several of the e-mails were allegedly written by Phil Jones, head of the Climate Research Unit. One, dated November 16, 1999, contained a sentence about temperatures and referred to a "trick" that could be used to "hide the decline".

"The university declined to confirm whether or not this e-mail was genuine. The spokesman said that Prof Jones would not be commenting but confirmed that he had spoken early yesterday to Ian Wishart, a climate sceptic who reported their conversation on his website.

"Mr Wishart quoted Prof Jones as denying that he had manipulated data and saying that he could not remember writing the words "hide the decline".

It was selective reporting by the *Times* of the Wishart/Jones conversation, partly because in failing to report my conversation in context it left open the question of genuineness, but also because of the implication that I was merely a skeptic, rather than a professional working journalist whose newspaper *TGIF Edition* had broken the story. By downplaying it from a media interview to merely a conversation with a "skeptic", the *Times* was creating further room for doubt – as if its environment reporter didn't want to believe Climategate was real. Regardless, it soon became clear the *TGIF* digital newspaper's story was entirely right.

466 "Sceptics publish climate e-mails 'stolen from East Anglia University'" by Ben Webster, *Times of London*, 21 November 2009, http://www.timesonline.co.uk/tol/news/environment/article6926325.ece

The Guardian's George Monbiot – an absolute believer in human-caused warming – commenting on the uncooperative University of East Anglia, lamented the lack of a Jones defence:

"By then he should have been touring the TV studios for the past 36 hours, confronting his critics, making his case and apologising for his mistakes. Instead, he had disappeared off the face of the Earth. Now, far too late, he has given an interview to the Press Association, which has done nothing to change the story."

On the deeper issue of the significance of the emails and their revelations, Monbiot had this to say:

"It's no use pretending that this isn't a major blow.[467] The emails extracted by a hacker from the climatic research unit at the University of East Anglia could scarcely be more damaging. I am now convinced that they are genuine, and I'm dismayed and deeply shaken by them.

"Yes, the messages were obtained illegally. Yes, all of us say things in emails that would be excruciating if made public. Yes, some of the comments have been taken out of context. But there are some messages that require no spin to make them look bad. There appears to be evidence here of attempts to prevent scientific data from being released, and even to destroy material that was subject to a freedom of information request.

"Worse still, some of the emails suggest efforts to prevent the publication of work by climate sceptics, or to keep it out of a report by the Intergovernmental Panel on Climate Change. I believe that the head of the unit, Phil Jones, should now resign. Some of the data discussed in the emails should be re-analysed."

One can only agree with Monbiot on this. So what do the emails say, and how devastating are they?

"HIDE THE DECLINE"

The first to gain attention in the *TGIF* story was the now infamous "Hide the decline" email.

From: Phil Jones

To: ray bradley ,mann@[snipped], mhughes@[snipped]

Subject: Diagram for WMO Statement

Date: Tue, 16 Nov 1999 13:31:15 +0000

Cc: k.briffa@[snipped],t.osborn@[snipped]

467 "The Knights Carbonic", George Monbiot, *The Guardian*, 23 November 2009

Dear Ray, Mike and Malcolm,

Once Tim's got a diagram here we'll send that either later today or first thing tomorrow. I've just completed Mike's *Nature* trick of adding in the real temps to each series for the last 20 years (ie from 1981 onwards) amd [sic] from 1961 for Keith's to hide the decline. Mike's series got the annual land and marine values while the other two got April-Sept for NH land N of 20N. The latter two are real for 1999, while the estimate for 1999 for NH combined is +0.44C wrt 61-90. The Global estimate for 1999 with data through Oct is +0.35C cf. 0.57 for 1998. Thanks for the comments, Ray.

Cheers

Phil

Prof. Phil Jones

Climatic Research Unit

TGIF asked Jones about the controversial email discussing hiding "the decline", and Jones explained he was not trying to mislead.

"No, that's completely wrong. In the sense that they're talking about two different things here. They're talking about the instrumental data which is unaltered – but they're talking about proxy data going further back in time, a thousand years, and it's just about how you add on the last few years, because when you get proxy data you sample things like tree rings and ice cores, and they don't always have the last few years. So one way is to add on the instrumental data for the last few years."

Jones told *TGIF* he had no idea what me meant by using the words "hide the decline".

"That was an email from ten years ago. Can you remember the exact context of what you wrote ten years ago?"

Except, despite Jones' protestations, it soon turned out that "hide the decline" had a very crafty meaning indeed. The proxy data that Jones referred to as "Mike's *Nature* trick" was in fact temperatures as determined by tree rings. The "decline" turned out to be the inconvenient fact that tree-ring data after 1960 showed a fall in temperatures, which strongly conflicted with the already-discussed and highly-suspect surface temperature station records. Those surface records taken by weather stations were showing a strong uptick in temperatures, and Mann and co had a problem – how to combine both sets of data in a graph to show global warming.

The "trick" involved burying the proxy downtrend line on the graph, underneath a blur of ink showing numerous upward swinging surface temperatures. Technically, the downtrend was there in the graph; you just couldn't see it.[468] The overall impression left by the trick was the false idea that all the temperature data agreed that modern warming was unprecedented and heading relentlessly upward.

Proof of this deviousness was found in some of the documents released in the Climategate haul, with this next passage referring specifically to "artificially adjusting" the figures in the "hide the decline" email above:

"Uses 'corrected' MXD – but shouldn't usually plot past 1960 because these will be artificially adjusted to look closer to the real temperatures."[469]

The significance of this is that tree ring proxies are what Mann used to try and negate evidence of the Medieval Warm Period. If, in fact, tree rings record lower than actual temperatures, it means the MWP was probably even hotter than currently estimated.

As I've said elsewhere in this book, if they have to lie, how much confidence can you have in their overall argument?

Scientists are supposed to report the facts, without fear or favour. These guys, the top climate scientists in the world, were fudging the facts so the public, politicians and even other scientists who were not part of the inner sanctum couldn't see the truth.

Why would they do such a thing? Professional pride is one possible motive, money is another. Phil Jones, for instance, was disclosed in the Climategate haul as the recipient of some US$22 million in mostly taxpayer-funded grant money to continue his work on climate change.[470]

Michael Mann's Pennsylvania State University banked an extra $30 million in grants *per year* after he came on board in 2005.[471]

Australian skeptic and author Joanne Nova had earlier in 2009 crunched the numbers and discovered climate change was a US$79 billion dollar industry – and that was just US grants![472] That amount of funding to "prove" warming dwarfs the $20 million or so that "Big Oil" is accused of funding some skeptical groups with.

468 http://wattsupwiththat.com/2009/11/26/mcintyre-data-from-the-hide-the-decline/

469 http://climateaudit.org/2009/11/22/these-will-be-artificially-adjusted/

470 The grants spreadsheet, and indeed links to all the Climategate documents, can be found here: http://motls.blogspot.com/2009/11/hacked-hadley-cru-foi2009-files.html

471 "The Economics of Climate Change", *Wall Street Journal* online, 30 November 2009 http://online.wsj.com/article/SB10001424052748703499404574559491076961008.html

472 http://scienceandpublicpolicy.org/originals/climate_money.html

"IT'S A TRAVESTY"[473]

Another of the emails to garner deserved attention is a debate between the University of Colorado's Kevin Trenberth and Penn State's Michael Mann:

> From: Kevin Trenberth <trenbert@xxxxxxxxx.xxx>
> To: Michael Mann <mann@xxxxxxxxx.xxx>
> Subject: Re: BBC U-turn on climate
> Date: Mon, 12 Oct 2009 08:57:37 -0600
> Cc: Stephen H Schneider <shs@xxxxxxxxx.xxx>, Myles Allen <allen@xxxxxxxxx.xxx>, peter stott <peter.stott@xxxxxxxxx.xxx>, "Philip D. Jones" <p.jones@xxxxxxxxx.xxx>, Benjamin Santer <santer1@xxxxxxxxx.xxx>, Tom Wigley <wigley@xxxxxxxxx.xxx>, Thomas R Karl <Thomas.R.Karl@xxxxxxxxx.xxx>, Gavin Schmidt <gschmidt@xxxxxxxxx.xxx>, James Hansen <jhansen@xxxxxxxxx.xxx>, Michael Oppenheimer <omichael@xxxxxxxxx.xxx>
>
> Hi all
> Well I have my own article on where the heck is global warming? We are asking that here inBoulder where we have broken records the past two days for the coldest days on record. Wehad 4 inches of snow. The high the last 2 days was below 30F and the normal is 69F, and itsmashed the previous records for these days by 10F. The low was about 18F and also arecord low, well below the previous record low. This is January weather (see the Rockiesbaseball playoff game was canceled on saturday and then played last night in below freezingweather).
>
> Trenberth, K. E., 2009: An imperative for climate change planning: tracking Earth's globalenergy. Current Opinion in Environmental Sustainability, 1, 19-27, doi:10.1016/j.cosust.2009.06.001. [1][PDF] (A PDF of the published version can be obtained from the author.)
>
> *The fact is that we can't account for the lack of warming at the moment and it is a travesty that we can't. The CERES data published in the August BAMS 09 supplement on 2008 shows there should be even more warming: but the data are surely wrong. Our observing system is inadequate.* [my emphasis]

Trenberth makes a valid point. Despite all the slogans at the climate change protest marches, warming has effectively stopped over the past decade, and Trenberth is candidly admitting that he and his colleagues simply don't know why, which may of course come back

473 http://www.eastangliaemails.com/emails.php?eid=1048&filename=1255352257.txt

to that problem highlighted in the Brookhaven study I mentioned at the start of this chapter.

If they can't explain why temperatures have decelerated despite higher levels of CO_2, what faith can the public have that any of their scaremongering based on "boys-toys" computer systems is actually real, as opposed to nerd-fantasy?

With the vast sums of money being doled out to scientists with a warming agenda, one might argue that the real winners are the people designing the "scientific" versions of Nintendo that the climate modellers play on.

This suspicion grows all the stronger, when you consider another of the Climategate documents – the "Harry" file.

"HARRY – READ ME.TXT"[474]

Harry (believed to be CRU's IT go-to guy Ian 'Harry' Harris) was a computer programmer tasked with helping compile the world temperature database as used by the UN IPCC. Unfortunately, as he found, the world temperature data (despite being publicly hailed as pristine) was a dog's breakfast behind closed doors, as these selected highlights from his file notations show.

There were missing data problems:

"Bear in mind that there is no working synthetic method for cloud, because Mark New lost the coefficients file and never found it again (despite searching on tape archives at UEA) and never recreated it."

There was bad documentation of what had been done with the data in earlier years (who had fiddled with it and why):

"So.. we don't have the coefficients files (just .eps plots of something). But what are all those monthly files? DON'T KNOW, UNDOCUMENTED. Wherever I look, there are data files, no info about what they are other than their names. And that's useless.. take the above example, the filenames in the _mon and _ann directories are identical, but the contents are not. And the only difference is that one directory is apparently 'monthly' and the other 'annual' – yet both contain monthly files."

On the issue of trying to match temperature computer records with what had been previously published as official temperatures last century, Harry found:

474 http://www.anenglishmanscastle.com/HARRY_READ_ME.txt

"These are very promising. The vast majority in both cases are within 0.5 degrees of the published data. However, there are still plenty of values more than a degree out."

Given that the UN IPCC is measuring temperatures to 0.1 degrees and claims global temperatures rose 0.7°C last century, errors of more than a degree don't provide great confidence for the public in the official records.

After further experimentation, Harry throws his hands in the air, hampered by the horrific state of the records:

"It's botch after botch after botch…

"..Knowing how long it takes to debug this suite - the experiment endeth here. The option (like all the anomdtb options) is totally undocumented so we'll never know what we lost."

Lost?

So just how good *are* the weather records?

"It [the program] wouldn't work, and on investigating I found 200-odd stations with zero precipitation [rainfall] for the entire 1901-2006 period!"

If you believe 200 locations around the world on the UN database received no rainfall in 105 years, you'll believe anything.

In the attempt to import US temperature data, there were 210 "duplicates" where one year's data had been copied into the system twice in place of the real data, meaning exact temperature details for those stations in those years were wrong.

"In fact, on examination the US database record is a poor copy of the main database one, it has more missing data and so forth. By 1870 they have diverged, so in this case it's probably OK.. but what about the others? I just do not have the time to follow up everything. We'll have to take 210 year-repetitions as 'one of those things'," complained Harry in his notations.

"So, uhhhh.. what in tarnation is going on? Just how off-beam are these datasets?!!"

Remember, you can rest assured that all of this behind the scenes angst is irrelevant, because the UN IPCC's Rajendra Pachauri has told reporters that the Climategate documents don't cast *any* doubt on the data behind global warming at all.

On further investigation, Harry found even more "duplicates" corrupting the temperature data:

"So 200 duplication events are unique to the older database, and

2572 are unique to the new database - with 1809 common to both. A quick look at the 2572 'new' ones showed a majority of those with the first WMO [World Meteorological Organisation station ID code] as -999: this is the key. The databases do not have any records with WMO=-999 as far as I know, so something is going on.. With huge reluctance, I have dived into 'anomdtb' - and already I have that familiar Twilight Zone sensation."

Soon after, a very chilling revelation about the true state of the CRU database:

"Wrote 'makedtr.for' to tackle the thorny problem of the tmin [minimum temp] and tmax [figure it out] databases not being kept in step. Sounds familiar, if worrying. Am I the first person to attempt to get the CRU databases in working order?!!"

All the way through, Harry is having to write new software and "adjust" data to make it fit what his bosses expect to see. And this, just to remind you, is the data being used to tell the world that global temperatures are rising uncontrollably. This is the data that underpins all that "settled science".

When Harry moved on to tackle Australia's temperature station records, he found more dismal errors, including temperatures wrongly attributed to other cities and locations:

"..so roughly 100 don't match. They are mostly altitude discrepancies, though there are an alarming number of name mismatches too."

When Harry dug deeper, and wrote software to dump stations that did not have "normal" data, he got the shock of his life – more than 9,000 temperature stations dropped out of the records, and only 5,000 remained.

"I suspect the high percentage lost reflects the influx of modern Australian data. Indeed, nearly 3,000 of the 3,500-odd stations with missing WMO codes were excluded by this operation. This means that, for tmn.0702091139.dtb, 1240 Australian stations were lost, leaving only 278.

"This is just silly. I can't dump these stations, they are needed to potentially match with the bulletin stations."

His proposed solutions, however, failed to work:

"Neither give me results that are anything near reality. FFS."

As just one example, he found temperatures being recorded at one Australian weather station three decades before it even came into existence!

"Now looking at the dates.. something bad has happened, hasn't it. COBAR AIRPORT AWS cannot start in 1962, it didn't open until 1993!"

Then, a further sideswipe at CRU's "flagship" temperature calculating program:

"Back to the gridding. I am seriously worried that our flagship gridded data product is produced by Delaunay triangulation - apparently linear as well. As far as I can see, this renders the station counts totally meaningless. It also means that we cannot say exactly how the gridded data is arrived at from a statistical perspective - since we're using an off-the-shelf product that isn't documented sufficiently to say that.

"Confidence in the fidelity of the Australian station in the database drastically reduced. Likelihood of invalid merging of Australian stations high. Let's go..

"I'm quickly realising that the Australian stations are in such a state that I'm having to constantly refer to the station descriptions on the BOM [Bureau of Meteorology] website, which … takes time.. time I don't have!

"…Getting seriously fed up with the state of the Australian data. so many new stations have been introduced, so many false references.. so many changes that aren't documented."

Some time later, as Harry begins to check other countries, the news is just as bleak:

"I am very sorry to report that the rest of the databases seem to be in nearly as poor a state as Australia was. There are hundreds if not thousands of pairs of dummy stations, one with no WMO and one with, usually overlapping and with the same station name and very similar coordinates. I know it could be old and new stations, but why such large overlaps if that's the case? Aarrggghhh! There truly is no end in sight.

"I honestly have no idea what to do here. And there are countless others of equal bafflingness."

As his desperation gets worse, Harry begins to joke about taking shortcuts:

"It's not documented, but then, none of the process is so I might as well bluff my way into it!

"So.. should I really go to town (again) and allow the Master database to be 'fixed' by this program? Quite honestly I don't have time – but it just shows the state our data holdings have drifted into. Who added those two series together? When? Why? Untraceable, except anecdotally.

"It's the same story for many other Russian stations, unfortunately – meaning that (probably) there was a full Russian update that did no data integrity checking at all. I just hope it's restricted to Russia!!

"What the hell is supposed to happen here? Oh yeah – there is no 'supposed,' I can make it up. So I have : –)"

Reduced virtually to tears (he sometimes writes the notation *cries* alongside some of the worst offences), Harry ends up having to provide false codes to rogue weather stations:

"You can't imagine what this has cost me - to actually allow the operator to assign false WMO codes!! But what else is there in such situations? Especially when dealing with a 'Master' database of dubious provenance (which, er, they all are and always will be).

"False codes will be obtained by multiplying the legitimate code (5 digits) by 100, then adding 1 at a time until a number is found with no matches in the database. THIS IS NOT PERFECT but as there is no central repository for WMO codes – especially made-up ones – we'll have to chance duplicating one that's present in one of the other databases."

"This still meant an awful lot of encounters with naughty Master stations, when really I suspect nobody else gives a hoot about. So with a somewhat cynical shrug, I added the nuclear option - to match every WMO possible, and turn the rest into new stations (er, CLIMAT excepted). In other words, what CRU usually do. It will allow bad databases to pass unnoticed, and good databases to become bad, but I really don't think people care enough to fix 'em, and it's the main reason the project is nearly a year late."

You can almost sympathise with Harry when he says:

"Gotta love the system! Like this is ever going to be a blind bit of use... Oh, sod it. It'll do. I don't think I can justify spending any longer on a dataset, the previous version of which was completely wrong (misnamed) and nobody noticed for five years."

If you think I've been too harsh calling the computer models "garbage in/garbage out", you'll be interested to know Harry found he could only get results comparable to the official published figures under certain circumstances:

"So, we can have a proper result, but only by including a load of garbage!"

"OH F*** THIS. It's Sunday evening, I've worked all weekend, and just when I thought it was done I'm hitting yet another problem that's based on the hopeless state of our databases. There is no uniform data integrity, it's just a catalogue of issues that continues to grow as they're found."

Harry would, I'm sure, have been delighted that his Climatic Research Unit computer system routinely signed off, in true gamer-style, with the words:

"Thanks for playing! Byeee!"

Now, here's the frightening thing. IPCC boss Rajendra Pachauri has this year vehemently defended the University of East Anglia and its datasets:[475]

"These scientists are highly reputed professionals, whose contributions over the years to scientific knowledge are unquestionable.

"It is also a well-established fact that the IPCC relies on datasets - not from any single source – but from a number of institutions in different parts of the world. *Significantly, the datasets from East Anglia were totally consistent with those from other institutions*, [my emphasis] on the basis of which far-reaching and meaningful conclusions were reached in the AR4."

If CRU's datasets are the same as the rest of those used by the UN IPCC and "totally consistent with" as claimed by Pachauri himself… well, you can do the math on how good the UN IPCC science is.

New Zealand's Climate Change Minister Nick Smith is another singing from Pachauri's hymnbook, telling one citizen who wrote to him:[476]

"I note your concern over the leaked emails which you suggest questions the reliability of climate change science. I am advised that a careful review of the content of the emails shows that they do not undermine the scientific case that climate change is real – or that human activities are almost certainly the cause. That scientific case is supported by multiple, robust lines of evidence…"

Tell that to the Oxburgh inquiry into Climategate which although a whitewash nonetheless found the CRU's knowledge of statistical analysis was seriously faulty.[477]

"With very noisy data sets a great deal of judgement has to be used. Decisions have to be made on whether to omit pieces of data

475 "Climate change has no time for delay or denial", Rajendra Pachauri, *The Guardian*, 4 January 2010, http://www.guardian.co.uk/environment/cif-green/2010/jan/04/climate-change-delay-denial
476 Letter from Smith to a constituent, dated 23 December, and released by ClimateRealists.org.nz
477 http://www.uea.ac.uk/mac/comm/media/press/CRUstatements/SAP

that appear to be aberrant. These are all matters of experience and judgement. The potential for misleading results arising from selection bias is very great in this area.

"It is regrettable that so few professional statisticians have been involved in this work because it is fundamentally statistical. Under such circumstances there must be an obligation on researchers to document the judgemental decisions they have made so that the work can in principle be replicated by others.

"CRU accepts with hindsight that they should have devoted more attention in the past to archiving data and algorithms and recording exactly what they did.

"We cannot help remarking that it is very surprising that research in an area that depends so heavily on statistical methods has not been carried out in close collaboration with professional statisticians. Indeed there would be mutual benefit if there were closer collaboration and interaction between CRU and a much wider scientific group outside the relatively small international circle of temperature specialists."

As UN IPCC author Richard Tol commented in April 2010:[478]

"The Oxburgh report confirms that the CRU is disorganised and not competent in statistical methods. As most of what they do is database management and statistical analysis, this is a harsh verdict."

There have now been two inquiries – into both Michael Mann and CRU – that have found climate scientists don't have a clue about statistical analysis, which calls into question the worth of climate predictions and projections that they make.

To the world's governments, and mainstream media, it is as if none of the evidence you've read here exists. They are hellbent on ushering in this new global system under the guise of "climate change".

Take Al Gore. Questioned in a *Slate* magazine interview about Climategate, the man tipped to become the world's first carbon billionaire displayed either total incompetence on the subject, or he lied, when he told the interviewer the most "recent" of the Climategate emails was 10 years old:[479]

Interviewer: "How damaging to your argument was the disclosure

---

478 "Global warming: Examiner exclusive interview with Richard Tol: IPCC was 'captured'", by Thomas Fuller, 21 April 2010, http://www.examiner.com/x-9111-Environmental-Policy-Examiner~y2010m4d21-Global-warming-Examiner-exclusive-interview-with-Richard-Tol-IPCC-was-captured

479 http://wattsupwiththat.com/2009/12/08/al-gore-cant-tell-time-thinks-most-recent-climategate-email-is-more-than-10-years-old/

of e-mails from the Climate Research Unit at East Anglia University?"

Gore: "To paraphrase Shakespeare, it's sound and fury signifying nothing. I haven't read all the e-mails, but the most recent one is more than 10 years old. These private exchanges between these scientists do not in any way cause any question about the scientific consensus. But the noise machine built by the climate deniers often seizes on what they can blow out of proportion, so they've thought this is a bigger deal than it is."

In fact, the "most recent" emails date from November 12, 2009, just *a week* before they were publicly released on November 19 GMT. Mind you, this is the same Al Gore who told a US TV interviewer the temperature at the centre of the earth was "millions of degrees"[480], when it's actually only a few thousand degrees in temperature. How many mistakes can a climate change guru make before his followers wake up and realize he doesn't know what he's talking about?[481]

But if the politicians are still trying to convince the public that Climategate emails "do not undermine the scientific case that climate change is real", let's examine some more of those emails.

"THIS IS ALL GUT FEELING, NO SCIENCE"[482]

CRU director Phil Jones let the cat out of the bag in one email diatribe about the existence or non-existence of the Medieval Warm Period, which we touched on earlier in this book. Publicly, Michael Mann and his colleagues have tried to put out pseudo-scientific reasons as to why the MWP isn't relevant, but the following email shows the real basis of their claim:

> "Bottom line – their [sic] is no way the MWP (whenever it was) was as warm globally as the last 20 years. There is also no way a whole decade in the LIA period was more than 1 deg C on a global basis cooler than the 1961-90 mean. *This is all gut feeling, no science*, [my emphasis] but years of experience of dealing with global scales and varaibility. [sic]
>
> Must got [sic] to Florence now. Back in Nov 1.
>
> Cheers
>
> Phil

480 http://www.youtube.com/watch?v=ag2AWst3Qv4

481 Perhaps that's why former President Bill Clinton gently mocked his former VP at a speech in Washington the night before the official start of Spring, "otherwise known to Al Gore as proof of global warming.". "Clinton Returns to Washington, Needling Himself, Obama and the Press", WSJ Blogs, 21 March 2010

482 http://www.eastangliaemails.com/emails.php?eid=440&filename=1098472400.txt

So this is the "settled science" the politicians believe the public should trust? A mere hunch? Based on "years of experience" delving into CRU's global databases? Jones should probably have a word to Harry about the "varaibility" lurking in the official temperature records.

"THE IMPRESSION OF GLOBAL WARMING WILL BE MUTED"[483]

Jones, incidentally, touches on one of the other frauds used by CRU and the UN IPCC – the insistence of measuring their temperature increases against a baseline of 1961-1990. As anyone familiar with statistics knows, you can make them lie to say anything. In the case of climate, if you pick a really cold period to provide your baseline (such as 1961-1990), then it's easy to claim temperatures are rising because anything warmer than a mini ice-age will show a strong warming trend. And that's exactly why they did it, as another of the emails shows:

> Neil
>
> There is a preference in the atmospheric observations chapter of IPCC AR4 to stay with the 1961-1990 normals. This is partly because a change of normals confuses users, e.g. *anomalies will seem less positive than before if we change to newer normals, so the impression of global warming will be muted.*[my emphasis]

Just how cold was it in the sixties? Britain's *Independent* rates one of the winters thus:[484]

"1962-63 The coldest winter for 200 years left the sea frozen in places. Wales and South-west England had snowdrifts up to 6m deep and the mercury dropped to -22.C in Braemar, Scotland, on 18 January. The cold spell lasted into early March."

So 1963 was the coldest winter since the Little Ice Age, and even the sea froze up. Some areas of Britain were buried under six metre high snow drifts.[485] Had the UN IPCC used the warm 1911-1940 period as a baseline instead, modern temperatures would barely be

483 http://www.eastangliaemails.com/emails.php?eid=426&filename=1105019698.txt

484 "Heavy snowfall sees Met Office put Britain on high alert", *The Independent*, 6 January 2010, http://www.independent.co.uk/news/uk/home-news/heavy-snowfall-sees-met-office-put-britain-on-high-alert-1858997.html

485 "Britain braced for heaviest snowfall in 50 years", *The Telegraph*, 5 January 2010 http://www.telegraph.co.uk/topics/weather/6937854/Britain-braced-for-heaviest-snowfall-in-50-years.html

showing an increase at all. And of course, the central issue in the email above is whether the IPCC should move to a modern, 1971-2000 baseline. This latter period was much warmer than 1961-1990, so of course the UN and Jones don't want to measure against it.

"HIDE THE DECLINE #2"[486]

You'll recall how Michael Mann, Jones and others have used a "trick" to "hide the decline". Well, another email by climate scientist Mick Kelly discloses a similar willingness to hide the truth about the current slowdown in global warming from the public and politicians, except his trick was simply deleting the inconvenient stuff:

> From: Mick Kelly mick.tiempo@xxxxxxxxx.xxx
> To: <P.Jones@xxxxxxxxx.xxx>
> Subject: RE: Global temperature
> Date: Sun, 26 Oct 2008 09:02:00 +1300
>
> Yeah, it wasn't so much 1998 and all that that I was concerned about, used to dealing with that, but the possibility that we might be going through a longer – 10 year – period of relatively stable temperatures beyond what you might expect from La Nina etc.
>
> Speculation, but if I see this as a possibility then others might also.
>
> Anyway, I'll maybe cut the last few points off the filtered curve before I give the talk again as that's trending down as a result of the end effects and the recent cold-ish years.
>
> Enjoy Iceland and pass on my best wishes to Astrid.
>
> Mick

How many public lectures or political briefings have been done where inconvenient evidence has been "cut" so no-one would know?

"DISHONEST PRESENTATIONS"[487]

NASA scientist Tom Wigley may have had exactly these kinds of briefings in mind when he wrote this email to Michael Mann:

"The Figure you sent is very deceptive...In my (perhaps too harsh) view, there have been a number of dishonest presentations of model results by individual authors and by IPCC."

Hardly confidence-inspiring, is it?

486 http://www.eastangliaemails.com/emails.php?eid=927&filename=1225026120.txt
487 http://www.eastangliaemails.com/emails.php?eid=1057&filename=1255553034.txt

"A GLOBALIZATION AGENDA"[488]

The same Mick Kelly mentioned a moment ago was the recipient of an email from Greenpeace's Paul Horsman in 2000, where both agreed the UN IPCC was using climate change for a globalization agenda:

> Hi Mick,
>
> It was good to see you again yesterday - if briefly. One particular thing you said – and we agreed – was about the IPCC reports and the broader climate negotiations were working to the globalisation agenda driven by organisations like the WTO. So my first question is do you have anything written or published, or know of anything particularly on this subject, which talks about this in more detail?
>
> My second question is that I am involved in a working group organising a climate justice summit in the Hague and I wondered if you had any contacts, ngos or individuals, with whom you have worked especially from the small island States or similar areas, who could be invited as a voice either to help on the working group and/or to invite to speak?
>
> All the best,
>
> Paul

Three years earlier, a 1997 email discloses CRU scientists moving beyond the science and becoming global change agents – activists – tasked with helping dummy up a list of "scientists" who believe in global warming ahead of the original Kyoto protocol summit. Apparently, experience and qualifications didn't matter:

> From: Joseph Alcamo <alcamo@xxxxxxxxx.xxx>
> To: m.hulme@xxxxxxxxx.xxx, Rob.Swart@xxxxxxxxx.xxx
> Subject: Timing, Distribution of the Statement
> Date: Thu, 9 Oct 1997 18:52:33 0100
> Reply-to: alcamo@xxxxxxxxx.xxx
>
> Mike, Rob,
>
> Sounds like you guys have been busy doing good things for the cause.I would like to weigh in on two important questions –
>
> Distribution for Endorsements –

488 Leaked emails, "greenpeace.txt",

> I am very strongly in favor of as wide and rapid a distribution as possible for endorsements. I think the only thing that counts is numbers. The media is going to say "1000 scientists signed" or "1500 signed". No one is going to check if it is 600 with PhDs versus 2000 without. They will mention the prominent ones, but that is a different story.
>
> Conclusion – Forget the screening, forget asking them about their last publication (most will ignore you.) Get those names!

Yet, believers jump all over lists of scientists quoted by sceptics with accusations of stacking! The hypocrisy knows no bounds.

On the issue of stage-managing this PR stunt, Joe Alcamo had a firm time-frame in mind with which to get maximum media attention:

> "Timing – I feel strongly that the week of 24 November is too late.
>
> 1. We wanted to announce the Statement in the period when there was a sag in related news, but in the week before Kyoto we should expect that we will have to crowd out many other articles about climate.
>
> 2. If the Statement comes out just a few days before Kyoto I am afraid that the delegates who we want to influence will not have any time to pay attention to it. We should give them a few weeks to hear about it.
>
> 3. If Greenpeace is having an event the week before, we should have it a week before them so that they and other NGOs can further spread the word about the Statement. On the other hand, it wouldn't be so bad to release the Statement in the same week, but on a different day. The media might enjoy hearing the message from two very different directions.
>
> Conclusion – I suggest the week of 10 November, or the week of 17 November at the latest.
>
> Best wishes,
>
> Joe Alcamo
>
> Prof. Dr. Joseph Alcamo, Director
>
> Center for Environmental Systems Research University of Kassel, Germany

This transition from objective scientists to social engineering activists is even better illustrated in one UN IPCC discussion document[489] contained in the emails and sent to climate scientists and other hangers on (including Joe Alcamo) explicitly talks of the IPCC's real agenda in selling the climate change message:

> Our approach has been to develop a set of four "scenario families".

489 http://www.eastangliaemails.com/emails.php?eid=54&filename=889554019.txt

The storylines of each of these scenario families describes a demographic, politico-economic, societal and technological future. Within each family one or more scenarios explore global energy industry and other developments and their implications for Greenhouse Gas Emissions and other pollutants. These are a starting point for climate impact modelling.

The scenarios we have built explore two main questions for the 21st century, neither of which we know the answer to:

• Can adequate governance – institutions and agreements – be put in place to manage global problems?

• Will society's values focus more on enhancing material wealth or be more broadly balanced, incorporating environmental health and social well-being.

The way we answer these questions leads to four families of scenarios:

• Golden Economic Age (A1): a century of expanded economic prosperity with the emergence of global governance

• Sustainable Development (B1): in which global agreements and institutions, underpinned by a value shift, encourages the integration of ecological and economic goals

• Divided World (A2): difficulty in resolving global issues leads to a world of autarkic regions

• Regional Stewardship (B2): in the face of weak global governance there is a focus on managing regional/local ecological and equity

Easy to see the preferred option. Given a list containing the choices, "Golden Age", "Sustainable", "Divided World" or mere "Regional Stewardship", which would you instinctively generate the most warm fuzzies over?

"CLIMATE VARIABILITY: WE KNOW F*** ALL"[490]

Publicly, the climate scientists present a united front to the world. Everything is certainty, everything is doomsday on a popsicle stick. Behind the scenes, in their candid moments in the Climategate emails however, they can be seen quietly confessing to each other that really, they and the UN IPCC know diddly-squat about longer term climate trends, as this email from Edward Cook of Lamont-Doherty Earth Observatory in New York, to East Anglia's Keith Briffa in September 2003 discloses.

"Without trying to prejudice this work, but also because of what

490 http://www.eastangliaemails.com/emails.php?eid=356&filename=1062592331.txt

I almost think I know to be the case, the results of this study will show that we can probably say a fair bit about [less than] <100 year extra-tropical NH [Northern Hemisphere] temperature variability (at least as far as we believe the proxy estimates), but honestly know f***-all about what the [greater than] >100 year variability was like with any certainty (i.e. we know with certainty that we know f***-all)."

On the strength of that "certainty", the United Nations wants to turn the world economy upside down in a massive multi-trillion dollar wealth and jobs transfer and social engineering experiment, and they want *you* to believe the scientists know what they're doing.

They may have known "f***-all" about climate change, but that didn't stop the IPCC's top brains from trying to hide the little they did know:

"DELETE THIS ATTACHMENT"

From: Phil Jones <p.jones@uea.ac.uk>

To: "Michael E. Mann" <mann@meteo.psu.edu>, "raymond s. bradley" <rbradley@geo.umass.edu>

Subject: A couple of things

Date: Fri May 9 09:53:41 2008

Cc: "Caspar Ammann" <ammann@ucar.edu>

....

2. You can delete this attachment if you want. Keep this quiet also, but this is the person who is putting in FOI requests for all emails Keith and Tim have written and received re Ch 6 of AR4. We think we've found a way around this.

Followed a couple of weeks later by this:

From: Phil Jones <p.jones@uea.ac.uk>

To: "Michael E. Mann" <mann@meteo.psu.edu>

Subject: IPCC & FOI

Date: Thu May 29 11:04:11 2008

Mike,

Can you delete any emails you may have had with Keith re AR4? Keith will do likewise. He's not in at the moment – minor family crisis.

Can you also email Gene and get him to do the same? I don't have his new email address.

> We will be getting Caspar to do likewise.
> I see that CA claim they discovered the 1945 problem in the Nature paper!!
> Cheers
> Phil

Given that these were government scientists, paid by the taxpayer and accountable through the law to make documents available, conspiring to destroy publicly-owned documents has a decidedly unscientific and unethical ring to it.

To all those who claimed this was *not* a criminal offence, it turned out it was:[491]

"The university at the centre of the climate change row over stolen e-mails broke the law by refusing to hand over its raw data for public scrutiny," reported the *Times of London*.

"The University of East Anglia breached the Freedom of Information Act by refusing to comply with requests for data concerning claims by its scientists that man-made emissions were causing global warming.

"The Information Commissioner's Office decided that UEA failed in its duties under the Act but said that it could not prosecute those involved because the complaint was made too late, *The Times* has learnt. The ICO is now seeking to change the law to allow prosecutions if a complaint is made more than six months after a breach."

But perhaps the saddest thing of all about the Climategate emails (and I've really only scratched the surface with this selection), is the enormous damage to the credibility of science itself.

Essentially, the scandal blows massive holes in the quality of scientific peer review. All of us rely on scientists to work objectively towards discovering the truth about nature and the world around us. Scientific debate has always been robust – indeed it has to be and that's why it's a 'debate' – but when scientists conspire to prevent opposing viewpoints or research from being published, that's corruption under any definition. Debate is silenced, and the winners of the power struggle go on to win prestige, public accolades and fat, lucrative research contracts.

Can you imagine, in a university exam, arranging to have your paper marked by your best friend? No, well believe it or not that's what passes for "peer review" among global warming believers, as

491 "Scientists in stolen e-mail scandal hid climate data", *The Times*, 28 January 2010, http://www.timesonline.co.uk/tol/news/environment/article7004936.ece

this next leaked email shows.[492]

Look how casually CRU director Phil Jones sets himself up to "peer review" the work of his close colleagues – a blatant conflict of interest in my view:

> From: Phil Jones <p.jones@xxxxxxxxx.xxx>
>
> To: "Folland, Chris" <chris.folland@xxxxxxxxx.xxx>
>
> Subject: RE: FW: Temperatures in 2009
>
> Date: Wed Jan 7 12:51:51 2009
>
> Chris,
>
> Apart from contacting Gavin and Mike Mann (just informing them) you should appeal. In essence it means that Real Climate is a publication. **If you do go to GRL I wouldn't raise the issue with them. Happy to be a suggested reviewer if you do go to GRL.**
>
> Cheers
>
> Phil
>
> Chris,
>
> Worth pursuing - even if only GRL. Possibly worth sending a note to Gavin Schmidt at Real Climate to say what Nature have used as a refusal!
>
> Cheers
>
> Phil

How cosy it must be for the elite, knowing they will be favourably peer-reviewed by their friends so long as they keep paying homage to the human caused global warming scam.

Peer-review should be independent and objective. The scientist should never know, let alone be able to nominate, who reviews his or her papers. The lobbying, the schmoozing, the commonality of purpose, the use of their dominant clique to freeze out scientists they didn't agree with – this is why the Copenhagen Diagnosis is now worthless.

The practice of silencing critics, intimidating journal editors into not publishing something, and other scientific malpractice, reeks of fraud, and strongly suggests that the UN's climate science advisors have something to hide.

One scientific journal editor who's felt pressured to tow the party

492 http://www.eastangliaemails.com/emails.php?eid=952&filename=1231350711.txt

line is Sonja Boehmer-Christiansen, editor of *Energy & Environment*. Her testimony to Britain's Climategate inquiry is telling:[493]

"I WAS PEER REVIEWER FOR IPCC (Intergovernmental Panel on Climate Change)… Since 1998 I have been the editor of the journal, *Energy & Environment* (E&E) published by Multi-science, where I published my first papers on the IPCC. I interpreted the IPCC "consensus" as politically created in order to support energy technology and scientific agendas that in essence pre-existed the "warming-as -man-made catastrophe alarm."…

"3.2 Scientific research as advocacy for an agenda (a coalition of interests, not a conspiracy,) was presented to the public and governments as protection of the planet… CRU, working for the UK government and hence the IPCC, was expected to support the hypothesis of man-made, dangerous warming caused by carbon dioxide, a hypothesis it had helped to formulate in the late 1980s…

"3.3 ... In persuading policy makers and the public of this danger, the "hockey stick" became a major tool of persuasion, giving CRU a major role in the policy process at the national, EU and international level. This led to the growing politicisation of science in the interest, allegedly, of protecting the "the environment" and the planet. I observed and documented this phenomenon as the UK Government, European Commission, and World Bank increasingly needed the climate threat to justify their anti-carbon (and pro-nuclear) policies. In return climate science was generously funded and required to support rather than to question these policy objectives… Opponents were gradually starved of research opportunities or persuaded into silence. The apparent "scientific consensus" thus generated became a major tool of public persuasion…

"4.1 ... As editor of a journal which remained open to scientists who challenged the orthodoxy, I became the target of a number of CRU manoeuvres. The hacked emails revealed attempts to manipulate peer review to E&E's disadvantage, and showed that libel threats were considered against its editorial team…

"4.4 Most recently CRU alleged that I had interfered "maliciously" with their busy grant-related schedules, by sending an email to the UKCIP (Climate Impact Programme) advising caution in the use of CRU data for regional planning purposes. This was clearly reported

493 http://blogs.news.com.au/heraldsun/andrewbolt/index.php/heraldsun/comments/how_government_cash_created_the_climategate_scandal

to [CRU head Phil] Jones who contacted my Head of Department, suggesting that he needed to reconsider the association of E&E with Hull University. Professor Graham Haughton, while expressing his own disagreement with my views, nevertheless upheld the principle of academic freedom...

"4.5 The emails I have read are evidence of a close and protective collaboration between CRU, the Hadley Centre, and several US research bodies such as the Lawrence Livermore National Laboratory where former CRU students had found employment. Together they formed an important group inside IPCC Working Group 1, the science group...

"The CRU case is not unique. Recent exposures have taken the lid off similar issues in the USA, the Netherlands, Australia, and possibly in Germany and Canada... It is at least arguable that the real culprit is the theme- and project-based research funding system put in place in the 1980s and subsequently strengthened and tightened in the name of "policy relevance". This system, in making research funding conditional on demonstrating such relevance, has encouraged close ties with central Government bureaucracy. Some university research units have almost become wholly-owned subsidiaries of Government Departments. Their survival, and the livelihoods of their employees, depends on delivering what policy makers think they want. It becomes hazardous to speak truth to power...

"Postglacial climatic history is by no means well understood and the human contributions cannot yet be assessed."

Sonja Boehmer-Christiansen's experiences are not unique. And then there's the issue of the actual quality of 'peer review'.

If we are to believe the embattled Phil Jones, peer review doesn't actually involve any real checking, and is merely a rubber stamp. This comment from Phil Jones' testimony to the UK parliament inquiry on Climategate is fascinating:

"Prof Jones today said it was not 'standard practice' in climate science to release data and methodology for scientific findings so that other scientists could check and challenge the research. He also said the scientific journals which had published his papers had never asked to see it."

If they had never asked to see his data, how exactly did *Science* and *Nature's* "peer reviewers" know whether Jones had got it right?

Another newspaper reported it thus:[494]

"Previously sympathetic MPs were beginning to be more hostile. One asked why it was so unusual for somebody to replicate Jones' work from scratch. Jones said that during the peer review process, nobody had ever asked for raw data or methodology."

Never even *once* questioned about data or methodology? Is this the "no clothes" moment for peer reviewed climate science?

The climate scientists wanted us to trust them with the fate of the world's economies, and to pay massive new taxes for each household, on the basis of homework that was never even checked by Teacher?

You may well ask, how does all this impact on the prestigious UN IPCC reports? Good question. It impacts badly.

In a *Guardian* interview, Rajendra Pachauri tried to claim the UN IPCC review process was so robust that the biases and agendas displayed in Climategate could not affect the UN reports.[495]

"The processes in the IPCC are so robust, so inclusive, that even if an author or two has a particular bias it is completely unlikely that bias will find its way into the IPCC report," he said.

"Every single comment that an expert reviewer provides has to be answered either by acceptance of the comment, or if it is not accepted, the reasons have to be clearly specified. So I think it is a very transparent, a very comprehensive process which insures that even if someone wants to leave out a piece of peer reviewed literature there is virtually no possibility of that happening."

That's what Pachauri would have you, the public, believe. But here's what a scientist who actually worked as an IPCC coordinating lead author said publicly in a South African newspaper at the end of 2009:[496]

> Prof Bruce Hewitson (Uninformed vitriol, November 19) pontificates on Andrew Kenny's assessment (Ideology and money drive global-warming religion, November 16). Unfortunately for him, there has been a reformation. The time for pontification is over. The critics must be answered. Instead Prof Hewitson stood in his pulpit and preached the gospel according to St IPCC.
>
> He says he was a lead author for the IPCC (Intergovernmental Panel on Climate Change). That is not material – I was a coordinating lead author, but

494 "Climategate hits Westminster", *The Register*, 2 March 2010, http://www.theregister.co.uk/2010/03/02/parliament_climategate/page2.html

495 "Leaked emails won't harm UN climate body, says chairman", *The Guardian*, 29 November 2009, http://www.guardian.co.uk/environment/2009/nov/29/ipcc-climate-change-leaked-emails

496 Originally published in ***Business Day***, South Africa, re-quoted in "Climategate: What The Media Didn't Tell You", ***Investigate*** magazine, January 2010 issue

it gives me no mantle of infallibility. Instead, it gave me insight into the flaws behind the whole process.

The IPCC claims that it has thousands of scientists and almost as many reviewers of the scientists' work to produce their reports. There are two problems, however. In the scientific world I move in, "review" means that your work is scrutinised by several independent, anonymous reviewers chosen by the editor.

However, when I entered the IPCC world, the reviewers were there at the worktable, criticising our drafts, and finally meeting with all us coordinators and many of the IPCC functionaries in a draftfest.

The product was not reviewed in the accepted sense of the word – there was no independence of review, [all emphasis mine] and the reviewers were anything but anonymous. The result is not scientific.

The second problem is that the technical publication is not completed by the time the IPCC reports. Instead, it produces a Summary for Policy Makers. Writing the summary involves the coordinators, the reviewers and the IPCC functionaries as before, and also various chairmen.

The summary goes out in a blaze of publicity, but there is no means of checking whether it represents what the scientists actually said, because the scientific report isn't published for another four months or more.

In the Fourth Assessment, the summary was quietly replaced several months after it was first published because some scientists who were involved complained of misrepresentation.

In the early years of the IPCC, there was a slightly different process. The Summary for Policy Makers and the scientific reports were issued at the same time. In those years, however, the Summary for Policy Makers bore a warning that it was the last current word on the subject, whereas the scientific reports were correctly identified as being subject to continuing development.

Someone smelled a rat about the "last word" story, so the process was changed, and *now the summary is issued with no means of checking.*

It isn't necessary to list all the changes I have identified between what the scientists actually said and what the policy makers who wrote the Summary for Policy Makers said they said. *The process is so flawed that the result is tantamount to fraud. As an authority, the IPCC should be consigned to the scrapheap without delay.*

Dr Philip Lloyd, Pr Eng

In other words, contrary to what Pachauri says, the UN IPCC process is widely regarded as a joke, even by the scientists who worked on it.

But what was it Pachauri said again?

"The processes in the IPCC are so robust, so inclusive, that even if an author or two has a particular bias it is completely unlikely that bias will find its way into the IPCC report," he said.

"Every single comment that an expert reviewer provides has to be answered either by acceptance of the comment, or if it is not accepted, the reasons have to be clearly specified. So I think it is a very transparent, a very comprehensive process which insures that even if someone wants to leave out a piece of peer reviewed literature there is virtually no possibility of that happening."

If that was indeed true, and not just cynical spin designed to fool gullible journalists, then how on earth did *this* happen:[497]

"A WARNING THAT CLIMATE change will melt most of the Himalayan glaciers by 2035 is likely to be retracted after a series of scientific blunders by the United Nations body that issued it.

"Two years ago the Intergovernmental Panel on Climate Change (IPCC) issued a benchmark report that was claimed to incorporate the latest and most detailed research into the impact of global warming. A central claim was the world's glaciers were melting so fast that those in the Himalayas could vanish by 2035.

"In the past few days the scientists behind the warning have admitted that it was based on a news story in the *New Scientist*, a popular science journal, published eight years before the IPCC's 2007 report.

"It has also emerged that the *New Scientist* report was itself based on a short telephone interview with Syed Hasnain, a little-known Indian scientist then based at Jawaharlal Nehru University in Delhi.

"Hasnain has since admitted that the claim was 'speculation' and was not supported by any formal research.[498] If confirmed it would be one of the most serious failures yet seen in climate research. The IPCC was set up precisely to ensure that world leaders had the best possible scientific advice on climate change."

Pachauri, who had only a few weeks earlier labelled as "voodoo science" a claim by a glaciologist[499] that the UN IPCC was wrong on the 2035 figure, was faced with the reality that his much vaunted

---

497 "World misled over Himalayan glacier meltdown", by Jonathan Leake and Chris Hastings, *Sunday Times*, 17 January 2010, http://www.timesonline.co.uk/tol/news/environment/article6991177.ece

498 In fact, Hasnain says he only spoke of the glaciers starting to rapidly melt by around the middle of the century, and that he did not specifically mention 2035 or that the glaciers would be gone by then. This date appears to have been in *New Scientist's* story through misreading a 1996 study suggesting the Himalayas could melt by 2350.

499 "Glaciologist demands apology from Pachauri for 'voodoo' remark", *The Times of India*, 19 January 2010 http://timesofindia.indiatimes.com/home/environment/global-warming/Glaciologist-demands-apology-from-Pachauri-for-voodoo-remark/articleshow/5477796.cms

2007 IPCC report had plucked the Himalayan glacier melt story out of thin air, without *any* peer review at all. What's worse, scientists came out of the woodwork and said they had tried to warn their IPCC colleagues during the "robust" review process, and were ignored.[500]

And politicians, the media and the public are supposed to *trust* the UN IPPC?

But it wasn't just the Himalayan gaffe. An IPCC claim that most of the Netherlands was below sea level turned out to be false as well, along with scare stories about Bangladesh, and so did a claim that the Amazon jungle was in grave danger from climate change (as already refuted in *Air Con* several months earlier, incidentally), and the IPCC's claim of massive species extinction from climate change this century has also been rubbished.

How did many of these mistakes turn up in the IPCC report? Probably because its "robust" peer review process turned out to rely on non-scientific, non peer-reviewed articles written by biased outfits like WWF or Greenpeace. In fact, one analysis of AR4 found more than five thousand references to non peer-reviewed articles.[501]

Not that the lobby groups are under-funded in any way. While Exxon may have spent US$23 million on sceptics' research, take a look at how much money WWF has raked in as a result of the public scare campaign about climate change:

Year Income ($US)
2003 $370 million
2004 $469 million
2005 $499 million
2006 $550 million
2007 $663 million
TOTAL (2003-2007) $2,551,783,000

Once you add in 2008 and 2009 financial results, you will find WWF revenues have climbed to over $3.6 billion since 2003.

Greenpeace, incidentally, is raking in around US$200 million a year. Never let it be said that Big Green is comprised of poor, penniless activists. Big Green is big biz.

500 "UN climate report: Scientist warned glacier forecast was wrong", by Marlowe Hood, *Agence France-Presse*, January 18, 2010 http://www.canada.com/technology/climate+report+Scientist+warned+glacier+forecast+wrong/2455973/story.html

501 http://nofrakkingconsensus.blogspot.com/2010/04/ipcc-reliance-on-grey-literature-30.html

It would be remiss of me, however, to close this book without shifting focus one more time from Big Green to its inbred cousin, Daft Green.

Chapter Twenty Two

A Planetary Regime

"The Regime would have some power to enforce the agreed limits. If this could be accomplished, security might be provided by an armed international organization, a global analogue of a police force"

– John Holdren, now Obama administration Science Czar

You'll recall earlier in the book some discussion of the real agenda behind the climate change scare, and some of the more chilling claims from activists in support of that agenda. Well, a few months after *Air Con* was published someone tracked down a book co-written by John Holdren, now the Obama administration's 'Science Czar' where he laid out his vision for an environmentally sound future:502

"Adding a sterilant to drinking water or staple foods is a suggestion that seems to horrify people more than most proposals for involuntary fertility control. Indeed, this would pose some very difficult political, legal, and social questions, to say nothing of the technical problems. No such sterilant exists today, nor does one appear to be under development. To be acceptable, such a substance would have to meet some rather stiff requirements: it must be uniformly effective, despite widely varying doses received by individuals, and despite varying degrees of fertility and sensitivity among individuals; it must be free of dangerous or unpleasant side effects; and it must have no effect on members of the opposite sex, children, old people, pets, or livestock."

"A program of sterilizing women after their second or third child,

502 John Holdren, Obama's Science Czar, from his 1,051 page book 'Ecoscience', co-authored with Paul Erhlich in 1978. ISBN 978-0716700296

despite the relatively greater difficulty of the operation than vasectomy, might be easier to implement than trying to sterilize men.

"The development of a long-term sterilizing capsule that could be implanted under the skin and removed when pregnancy is desired opens additional possibilities for coercive fertility control. The capsule could be implanted at puberty and might be removable, with official permission, for a limited number of births.

"In today's world, however, the number of children in a family is a matter of profound public concern. The law regulates other highly personal matters. For example, no one may lawfully have more than one spouse at a time. Why should the law not be able to prevent a person from having more than two children?"

A little further on, Holdren writes of his vision for a new global governing ~~group~~ 'regime':

"Perhaps those agencies, combined with UNEP and the United Nations population agencies, might eventually be developed into a Planetary Regime—sort of an international superagency for population, resources, and environment. Such a comprehensive Planetary Regime could control the development, administration, conservation, and distribution of all natural resources, renewable or nonrenewable, at least insofar as international implications exist. Thus the Regime could have the power to control pollution not only in the atmosphere and oceans, but also in such freshwater bodies as rivers and lakes that cross international boundaries or that discharge into the oceans. The Regime might also be a logical central agency for regulating all international trade, perhaps including assistance from DCs to LDCs, and including all food on the international market.

"The Planetary Regime might be given responsibility for determining the optimum population for the world and for each region and for arbitrating various countries' shares within their regional limits. Control of population size might remain the responsibility of each government, but the Regime would have some power to enforce the agreed limits.

"If this could be accomplished, security might be provided by an armed international organization, a global analogue of a police force. Many people have recognized this as a goal, but the way to reach it remains obscure in a world where factionalism seems, if anything, to be increasing. The first step necessarily involves partial surrender of sovereignty to an international organization."

If I need to spell out the implications of Holdren's vision, I've failed as an author, and clearly the text of the draft treaty put forward for Copenhagen was wasted on you.

Interestingly in light of the "planetary regime" total control suggestion above, confidential Obama administration briefing papers for the 2010 climate progress conference in Bonn, Germany, contained this as their number one item on the agenda:[503]

"Reinforce the perception that the US is constructively engaged in UN negotiations in an effort to produce a global regime to combat climate change."

It's not the only time Obama has used the phrase. In a speech in Prague in 2009 on nuclear weapons becoming a global crisis, the President dog-whistled his solution:[504]

"All nations must come together to build a stronger, global regime."

Then in July 2009, a US Department of Defense report directly quoted Obama:

"In the 21st century, a strong and global regime is the only basis for security from the world's deadliest weapons."[505]

It appears that, whether it is climate change, nukes or the economy, the preferred and proffered answer is increasingly "a global regime", with all of the inherent 'Big Brother', we-know-best-what's-good-for-you, anti-democratic imagery that the word 'regime' brings with it.

The global regime concept is cunningly simple. If you can successfully make the case for a global regime in any area – say feeding the poor for example – then once the infrastructure and treaties are in place to give legitimacy to such a regime, it becomes a mere stroke of the pen to extend its jurisdiction to other areas. Pretty soon, the little global regime that began with warm fuzzies is truly global and truly a regime.

It is true that some problems require complex global negotiations, but the devil is in the detail. There's a very fine line between genuine efforts to fix problems, and creating something that turns into a Trojan Horse, welcomed with open arms by unsuspecting members of the public who may not realise they are trading away hard-won

503 "Confidential document reveals Obama's hardline US climate talk strategy", *The Guardian*, 12 April 2010 http://www.guardian.co.uk/environment/2010/apr/12/us-document-strategy-climate-talks

504 http://sweetness-light.com/archive/obama-must-come-together-for-global-regime

505 "President Barack Obama Says the U.S. and China Can Cooperate on Security Issues", by Gerry J. Gilmore

United States Department of Defense, 27 July 2009 http://www.2010military.com/military-news-story.cfm?textnewsid=3549

personal freedoms in response to a slick 'crisis' sales pitch.

One of the reasons Copenhagen failed abysmally is that the draft treaty included a clause establishing a new global "government", via the UN Framework Convention on Climate Change. The specific clause is Article 38 of Annex 1 to the draft Copenhagen document.[506]

"The scheme for the new institutional arrangement under the Convention will be based on three basic pillars: government; facilitative mechanism; and financial mechanism."

While you might initially regard the "government" word as innocuous, and referring to governments generally, the next sub-clause in the treaty showed it was to be a new "government" established primarily to implement world climate change policies and enforce them:

"The government will be ruled by the COP [Conference of the Parties] with the support of a new subsidiary body on adaptation, and of an Executive Board responsible for the management of the new funds and the related facilitative processes and bodies. The current Convention secretariat will operate as such, as appropriate."

It is believed to be the first time in history an international treaty negotiation process has purported to actually set up a world "government", to be overseen by representatives of the various national governments via the UN.

Naturally, when word of this hit the headlines just prior to Copenhagen the political upheaval was large enough to see the offending clause removed, for now. But as the UN's Ban Ki-moon noted in an interview, it will be back:[507]

"How can you scrap the role of the United Nations? The United Nations has global reach. The United Nations will be there and should be involved in this implementation process. One of the principles agreed upon is that all commitments should be reportable, measurable and verifiable. This is what has been agreed by both developed and developing countries.

"We will establish a global governance structure to monitor and manage the implementation of this. Experts from both worlds (first and developing) should participate."

And if you think all of this is happening by accident and mere coincidence, think again.

---

506 http://clareswinney.files.wordpress.com/2009/11/un-fccc-copenhagen-2009.pdf
507 "Ban Ki-moon interview", *Los Angeles Times*, 16 Dec 2009, http://www.latimes.com/news/nation-and-world/la-fg-climate-ban16-2009dec16,0,1781040.story

Back in 1991, the world-government organisation The Club of Rome published a book outlining its agenda for establishing global governance. One of their points was the need to create a post-democratic world.

"Democracy is no longer well suited for the tasks ahead. The complexity and the technical nature of many of today's problems do not always allow elected representatives to make competent decisions at the right time.

"The crucial need is to revitalize democracy and give it a breath of perspective that will enable it to cope with the evolving global situation. In other words, is this new world we find ourselves in governable? The answer is probably not with the existing structures and attitudes. Have we gathered the necessary means and wisdom to make decisions on the scale of the world *problematique* taking into account the exigencies of our time? There is an increasingly evident contradiction between the urgency of making some decisions and the democratic procedure founded on various dialogues, such as Parliamentary debate, public debate and negotiations with trade unions or professional organizations. The obvious advantage of this procedure is its achievement of consensus. Its disadvantage lies in the time it takes, especially at the international level."

In other words, the global governance types don't like the public and their representatives making considered decisions after long periods of debate.

"Time in these matters has acquired a deep ethical content. The costs of delay are monstrous in terms of human life and hardship, as well of resources. The slowness of decision in a democratic system is particularly damaging at the international level. When dictators attack and international policing is required, delays of decisions can be fatal."

It's the old, 'we must act now, There Is No Alternative (TINA)' principle. As I recall, that was the same line of reasoning that led to the invasion of Iraq.

The people who want global governance at all costs will just keep manufacturing crises or overhyping existing ones, until the will of the public to resist is worn down, and the Trojan Horse is finally welcomed in.

While the UN and others try and convince you Climategate and its fallout is a storm in a teacup and you can continue to trust the climate scientists, I think these words from Clive Crook in *The Atlantic* are worth reading.[508]

"In my previous post on Climategate I blithely said that nothing in the climate science email dump surprised me much. Having waded more deeply over the weekend I take that back.

"The closed-mindedness of these supposed men of science, their willingness to go to any lengths to defend a preconceived message, is surprising even to me. The stink of intellectual corruption is overpowering. And, as Christopher Booker argues, this scandal is not at the margins of the politicised IPCC [Intergovernmental Panel on Climate Change] process. It is not tangential to the policy prescriptions emanating from what David Henderson called the environmental policy milieu [subscription required]. It goes to the core of that process.

"…I'm also surprised by the IPCC's response. Amid the self-justification, I had hoped for a word of apology, or even of censure. (George Monbiot called for Phil Jones to resign, for crying out loud.) At any rate I had expected no more than ordinary evasion. The declaration from Rajendra Pachauri that the emails confirm all is as it should be is stunning. Science at its best. Science as it should be. Good lord. This is pure George Orwell. And these guys call the other side 'deniers'.

"While I'm listing surprises, let me note how disappointed I was by *The Economist's* coverage of all this. 'Leaked emails do not show climate scientists at their best," it observes. No indeed. I should say I worked at the magazine for years, admire it as much as ever, and rely on the science coverage especially. But I was baffled by its reaction to the scandal. "Little wonder that the scientists are looking tribal and jumpy, and that sceptics have leapt so eagerly on such tiny scraps as proof of a conspiracy," its report concludes. Tiny scraps? I detest anti-scientific thinking as much as *The Economist* does. I admire expertise, and scientific expertise especially; like any intelligent citizen I am willing to defer to it. But that puts a great obligation on science. The people whose instinct is to respect and admire science should be the ones most disturbed by these revelations. The scientists have let them down, and made the anti-science crowd look wise. That is outrageous," concludes Clive Crook.

508 "More on Climategate", by Clive Crook, *The Atlantic*, 30 November 2009 http://www.theatlantic.com/business/archive/2009/11/more-on-climategate/30945/

American climate scientist Dr Judith Curry also finds the official response, including the Oxburgh report exonerating the CRU team of everything but utter incompetence, outrageous:[509]

"Criticisms of the Oxburgh report that have been made include: bias of some of the members including the Chair, not examining the papers that are at the heart of the controversies, lack of consideration of the actual criticisms made by Steve McIntyre and others, and a short report with few specifics that implies a superficial investigation.

"When I first read the report, I thought I was reading the executive summary and proceeded to look for the details; well, there weren't any. And I was concerned that the report explicitly did not address the key issues that had been raised by the skeptics. Upon reading Andrew Montford's analysis, I learned: 'So we have an extraordinary coincidence – that both the UEA submission to the [UK Parliament's Science and Technology] Select Committee and Lord Oxburgh's panel independently came up with almost identical lists of papers to look at, and that they independently neglected key papers like Jones 1998 and Osborn and Briffa 2006.' I recall reading this statement from one of the blogs, which seems especially apt: the fire department receives report of a fire in the kitchen; upon investigating the living room, they declare that there is no fire in the house," notes a cynical Curry.

Indeed. Believe such official assurances at your peril.

So where does this leave global warming as we head into 2011? Climategate has shown the world that the UN's leading climate scientists were incompetent on statistical analysis, and that the UN side attempted to hide inconvenient studies challenging their claims, by pressuring scientific journal editors not to publish work by sceptical scientists.

In this way, the UN IPCC team was then able to dominate the political and media spheres with false claims about a "scientific consensus", backed up by assurances that the only "peer reviewed" studies in existence were those supporting belief in global warming.

But then Climategate showed us how deeply the peer review process had been corrupted, with UN climate scientists sending emails behind the scenes offering to volunteer themselves as 'anonymous' peer reviewers of studies done by their friends and colleagues.

509 http://www.collide-a-scape.com/2010/04/23/an-inconvenient-provocateur/

In short, you've been conned. The scientists involved seem to have been captured by the idea that the ends justify the means – that they are on a noble quest to save the planet and they are entitled to cut corners or manipulate the scientific process to support that quest.

In this sense, they are no longer scientists but global warming evangelists who have joined hands with a more cynical and corrupt political and big business lobby who've realised they can greatly increase control over the public and make a lot of money in the process. By the time the public realise global warming is not the threat it has been portrayed as, it will be too late; the global governance infrastructure will be in place creating a planetary empire that Darth Vader would envy.

Unless, of course, the public figure out how to make their politicians come to heel, which is where we now turn.

Epilogue

What Can I Do?

"The new world order will not be fully realised unless the United Nations and its Security Council create structures… which are authorised to impose sanctions and make use of other measures of compulsion"

– ***Mikhail Gorbachev, 1992***[510]

Presented with facts like those in this book, you are liable to experience a range of emotions. Firstly, incredulity and disbelief: surely this can't be true, after everything you've heard about global warming. Then comes anger, anger at being played for the fool by the news media, lobby groups and politicians. Frustration follows next: "what can I do? Someone has to do something!"

You're right. And as high priest Al Gore might say, that someone is you. Only your course of action is not about switching light bulbs or buying a hybrid car.

Instead, it's important to recognise what ordinary householders are up against: a collection of fabulously wealthy vested interests with a hidden agenda. Sounds unbeatable? Don't worry, they're not. In similar fashion to the biblical Goliath, Soviet communism fell to a stone from the Berlin Wall. History is full of examples of seemingly impossible odds being beaten by human ingenuity and passion.

Air Con has hopefully armed you with the data needed to see global warming for what it really is: a natural climate cycle unrelated in any significant way to human greenhouse gas emissions. Vested interests want you to buy their lie and feel sufficiently guilty that you'll agree to massive tax increases and reductions in jobs and standard of living.

510 "The River of Time and The Necessity of Action" speech by Mikhail Gorbachev, delivered at Fulton, Missouri, 6 May 1992

They want you to see it as a "global" problem requiring a "global" solution. They want you to think that only the West can pay for the massive amounts of clean technology the Third World needs to prevent pollution.

Ultimately, for reasons of "social justice" rather than environment, they want to vastly increase the powers and funding of the United Nations, and transfer wealth to developing nations, via selected multinational companies who will become extremely rich clipping the ticket in this New World Order.

You might agree with those aims, or you may disagree, but those are the facts; that's the agenda. You might want to call it "conspiracy theory" but, let's be honest, I've quoted them in their own words. I'd be the first to grant that flaky internationalists have been delivering earnest and boring speeches for years about the coming 'new world order' and nothing ever came of it. But this December, in Mexico, those same 'flakes' will be just a heartbeat away from signing a compulsory global deal to tax every household in the western world and spend the money on their own pet projects. It doesn't matter what you call it, the end result will be the same.

Surprisingly, however, your power to resist, however, is actually quite strong.

For a start, these people are getting at you via your kids. As Eilis O'Hanlon, a commentator for the *Irish Independent* newspaper, recently wrote, you can blame it on an old student newspaper headline that once warned students that every time they self-pleasured, "God kills a kitten":[511]

"The other day, I was driving to the shop when my six-year-old daughter suddenly announced that I shouldn't be taking the car because it kills polar bears. 'Polar bears?' I said.

"Polar bears, she assured me. Well, it makes a change from kittens.

"She'd heard about it in school. Too many cars equals global warming equals melting polar ice caps equals dead polar bears. St Thomas Aquinas couldn't have demonstrated a clearer link between sinfulness and suffering.

"And yet I couldn't help being reminded about those kittens. It's such a loaded bell to ring. If you're naughty, bad things happen to

511 "Ecological evangelists force-feed our children a diet of green pieties", Eilis O'Hanlon, *Sunday Independent*, 15 March 2009, http://www.independent.ie/world-news/ecological-evangelists-forcefeed-our-children-a-diet-of-green-pieties-1673517.html

cute little furry animals. Polar bears, kittens, meerkats: the principle is the same. It's all just emotional blackmail. Backing it up with scientific data about the effects of climate change on sea levels doesn't change the nature of the game. You haven't been able to convince people to change their ways by persuasion, so pile them up with guilt instead and hope that works. Or, better still, pile the kids up with it, and then send them home to nag their elders into green action, if only for a quiet life.

"Schools are by far the worst culprits when it comes to strident ecological evangelism," warns O'Hanlon.

"Teachers seem unable to get through a day without hammering at least one dodgy environmental slogan into captive pupils' heads. They're obsessed, even though most of them know less about the planet's ecosystem than I know about 17th-century Hungarian acrylics.

"Recently, I've heard from my own and other people's children how they were told in school that Googling uses as much energy as boiling a kettle (that one, lifted wholesale from some newspaper article, turned out not to be true, unsurprisingly).

"They've also been fed the lie that air travel is the biggest producer of CO_2 gases (it isn't); that ferries are an ecologically friendly way to go on holiday (they're not); and that you should always buy local because it's less damaging to the earth (not necessarily, you have to take account of production methods as well as air miles, so it varies).

"All of these statements were presented to them as incontestable facts, like gravity, when they're really nothing but political posturings hiding behind statistics, each one designed to ram home the self-hating message that industry and electricity and progress and big cities and capitalism and the wicked West are bad, bad, bad.

"Schoolchildren are now force-fed these pieties across the curriculum, in history, geography, English, science, sociology and politics classes. Even in maths they manage to find a green angle. There's no questioning, no debate. It's a catechism which children are expected to repeat back unthinkingly. The only difference is that, in the past, these rules were presented as God's will, and now they're touted as Mother Nature's.

"As for Al Gore's film about global warming, which contained so many scenes of intellectual self-love from the former US vice president that scores of kittens must have died during production, that's shown to kids so often they can probably recite it from memory.

"It'll only get worse in the coming weeks, as schools stockpile all those little niggly factlets which have emerged from the latest climate change conference in Copenhagen – you know, the one that thousands of scientists and politicians flew halfway across the world to get to, just so they could warn others about the evils of air travel – then bring them out, one by one, to fling triumphantly at the defenceless children behind their desks. The Christian Brothers in their heyday didn't have this much self-righteous zeal," concludes O'Hanlon.

If only the column was fictional. Alas, compulsory environmental and climate change classes now exist in most Western schools. As parents, you can ask your school to provide copies of the curricula items they teach on climate change and the environment. You can see whether it is balanced or simply propaganda. If a majority of parents at a school object, you may be able to get those classes modified or stopped.

Likewise at universities, where climate change papers have been slipped in as compulsory in some degree courses, you may have sufficient strength if enough fee-paying students band together, to demand that the university justify inclusion of the paper and its spin as part of your degree course.

Secondly, you can write letters to your local political representatives, in councils, state, federal or national governments, warning them that the science on human-caused global warming is far from settled and that there may be an electoral backlash if they vote to support costly initiatives at Mexico this year.

Thirdly, you can discuss the evidence against human-caused global warming with friends and neighbours, because the more people who wake up to what's in the pipeline the more chance there is of diverting an economic disaster.

Fourthly, you can begin to recognise spin when you see it. *The Guardian* newspaper, for example, recently published a report on the International Conference on Climate Change held by 700 sceptical scientists and interested parties in New York 2009 (the Heartland conference).

Suzanne Goldenberg wrote a fairly waspish piece which included negative stereotyping as a put-down:

"Attendees... are almost entirely white males, and many, if not most, are past retirement age. Only two women and one African-American man figure on the programme of more than 70 speakers."

Her words, as the *WattsUpWithThat* site noted[512], could equally be applied to attendees at the UN IPCC meetings, as the photo[513] above shows.

Goldenberg's column was nothing more than an attempt to marginalise sceptics based on appearance and race. Another marginaliser is Wikipedia, where the online encyclopedia's page on climate change is controlled by a diehard global warming believer who has locked the page so no contrary opinion can be entered.[514] In their discussion pages, the Wikipedia editors constantly ridicule scientists with opposing views. You may not be able to change Wikipedia, but you can have influence on the news media.

So fifthly, in many western countries, every time you see an unbalanced TV or radio news report on climate change, you can tie your local broadcaster up in red tape by mounting an official written complaint and citing some of the scientific studies in this book relevant to the particular complaint. Once a broadcaster has been forced to face the real scientific evidence a few times, rather

512 http://wattsupwiththat.com/2009/03/13/who-makes-up-the-ipcc/
513 http://changingclimates.colostate.edu/pastEvents/colloquium08/ChangingClimatesSept08_web.pdf
514 "Wikipropaganda: Spinning green", by Lawrence Solomon, *National Review Online*, http://article.nationalreview.com/?q=NjU1ZDBhOGExOWRlNzc5ZDcwOTUxZWM3MWU2Mjc5MGE=

than one of James Hansen's news releases, you can bet they will start to be more careful.

None of this is going to happen on a dime. The media are not going to wake up tomorrow and realise they've been played for suckers as part of a propaganda exercise to play you for suckers. Like the Titanic, turning around something as ingrained as global warming theory takes time.

Write letters to newspapers, ring talk radio, comment on blogs.

If a local charity you support (such as Greenpeace, Friends of the Earth or Oxfam to name just a few) is pushing anthropogenic global warming as an issue, write to the director and advise you are withdrawing funding and why. Many charities get away with radicalism because their members are not really aware of the wider issues. If you don't know what your charity's position is, Google its name (eg. "Greenpeace") along with the phrase "climate change" or "global warming".

Let the world hear your voice. Donate a copy of the book or one like it to your local school library or staffroom or your children's teachers – if you want to fight indoctrination you actually have to fight them where it's happening, not confine the battle to your armchair.

Finally, don't mistake me as saying that protecting the environment doesn't matter. It does. There are good and valid reasons to stop polluting streams, rivers and the air, and to stop squandering natural resources. However, none of those "good and valid reasons" include the UN's Trojan Horse, climate change.

If we develop cleaner-burning technology, that's a good thing. But let's not mask it in panic and paranoia about sea levels rising 10 metres in 90 years. And let's not use it as an excuse to raid your wallets in the name of climate change, when the real agenda is putting money into the pockets of vested interests like George Soros or the United Nations.

The 'ends' in this case do not justify the 'means' – theft of trillions of dollars and millions of jobs.

For a fraction of that amount, we could sink thousands of water wells deep into the Sahara desert and re-irrigate the land from the massive underground lakes that still exist[515], turning the Sahara green enough to feed all of Africa's starving millions. Wouldn't

515 "New technique dates Saharan groundwater as million years old", American Geophysical Union, 2 March 2004, http://www.sciencedaily.com/releases/2004/03/040302075003.htm

that be a better use of money than trading in carbon credits? And come to think of it, wouldn't a greener Sahara soak up more CO_2 from the atmosphere?

As this book makes clear, we are definitely seeing evidence of warming, in the sense of increased ice melt, but this is what we'd expect as earth emerges from the Little Ice Age. Yes, the warming which began in the 1800s has unleashed CO_2 and methane and water vapour, and yes, those could cause increased warming. Humanity's tiny contribution to overall greenhouse gas output in real terms – just 1.7% – is so low that on its own it is inconceivable that it's causing massive warming, or acidic oceans. The other 98.3% of emissions generated naturally will have a far higher impact, as common sense clearly shows.

The good thing about all this, if it is part of natural climate cycles as it appears to be, is that Earth and its lifeforms have survived and adapted through thousands of climate cycles in the past. We still have shellfish, we still have birds, we still have many creatures and organisms that date back to the Jurassic and beyond. The odds, measured against nature's evidence rather than flawed computer models, overwhelmingly favour life continuing as normal.

However, if we are really concerned about the impacts of natural warming, we can plan ahead quite easily and without spending trillions. City and state governments can identify areas vulnerable to inundation if the sea levels do rise by a metre over the next 200 years. They can, starting today, cancel the establishment of new towns or housing developments lower than three metres above sea level (allowing for a multi-century redundancy if we are heading into a Modern Warm Period). As part of organised town planning, when buildings in vulnerable areas reach the end of their useful lives property owners can be encouraged to shift to higher ground rather than re-building so close to the tideline.

Earth, and life, have frequently enjoyed in modern times climates far warmer than our current one. We have to learn to deal with it, rather than rage against it.

Perhaps most important of all, and the reason this book is crucial in getting the message to you and your friends, and through you the political establishment: if we are indeed in the early stages of a multi-century warm period, like the ones the Romans enjoyed and again later in 1000 AD, then spending trillions of dollars on reducing our relatively tiny emissions will break the back, and the bank, of this civilisation. It will have no measurable effect apart from financially benefiting some dubious organisations and individuals, and it will leave no money for adaptation to climate or for projects like geo-engineering the Sahara Desert.

In effect, a proposed worldwide carbon cap and trade emissions system is perhaps the quickest way to financially bomb humanity back to the Stone Age. But then again, based on what you've just read, maybe that's been the green movement's plan all along. They're not really motivated by saving human life – how can you argue on the one hand that world population needs to be cut by 90%, and on the other shed crocodile tears about how millions of Bangladeshi people and others will die if we don't tackle global warming? Those two statements are mutually exclusive, yet environmentalists are saying them. These are people who see the life of a tree as more significant than the life of a six year old girl sold to a paedophile "in an Asian brothel".

At Copenhagen this year, you are being asked to trust your and your family's future to environmentalists. And pay them to do this work via higher taxes and higher prices.

Do you *really* trust them with your best interests?

In the final analysis, you can measure the real agenda of the green movement with this comment from the US State Department's Richard Benedict:

"A global climate treaty must be implemented even if there is no scientific evidence to back the greenhouse effect."

After all you've read in this book, the glaring question in response to that is "why?"

We should all be very worried when officials tell us we must adopt "the solution" even though the crisis they've laid out does not actually exist.

One thing, very important to keep in mind as this book begins to create debate in the media and elsewhere: *Air Con* is more up to date on climate change science than the most recent UN IPCC report published at the end of 2007. Even more so when you consider the UN data cut-off point for that review was two years earlier still, in 2005. It is more up to date than the Copenhagen conference of December 2009.

Therefore, if you hear pressure groups or alarmists trying to use IPCC or mini-Copenhagen data to override *Air Con*, remember it's the IPCC's 2005 data that is well out of date, not this book, where data analysis was closed off at April 2010.

But don't think for a moment that global warming believers will wave the flag of surrender. They will push even harder this year to bamboozle the public with scare stories, such as this from Canadian environmentalist David Suzuki:[516]

"There has been no peer-reviewed scientific study that has called into question the conclusions of the IPCC."

As you've seen, this very book quotes a number of peer-reviewed scientific studies whose conclusions fly in the face of claims made in the UN IPCC reports.

"Why does the public often pay more attention to climate change deniers than climate scientists?" Suzuki complains. "Why do denial arguments that have been thoroughly debunked still show up regularly in the media?"

Well again, as you've seen, this book quotes scientific papers, most of them presented by scientists the UN IPCC has utilised, and allows readers to see for themselves what scientists are actually saying. "Thoroughly debunked"? I spent the entire book waiting to see literature debunking the arguments I've highlighted here. If I'd found it, the title of this book would not have been *Air Con*. I was waiting to be convinced by people like Suzuki or James Hansen, but it never happened. There were so many emotional pleas aimed at gullible nine-year-olds, and so few hard facts.

As if on cue, Al Gore turned up on the pages of Britain's *Guardian* newspaper in March 2009, lying like a flatfish again on the back of publicity generated by the mini-Copenhagen conference:

"They're seeing the complete disappearance of the polar ice caps right before their eyes in just a few years," he told the newspaper.[517]

For the record, have you seen any scientific study in this book or elsewhere that remotely suggests the South Pole will be free of ice in just a few years? Of course not. Not even James Hansen is claiming that we'll be able to grow bananas at the South Pole by 2020, because if that happened we'd be dealing with a sea level rise of around 90 metres or more in the next five years (anything hot enough to melt all the ice at the South Pole would probably melt all the ice on the planet). Yet because it's Al Gore, the media let him get away with outrageous statements. If George W. Bush had said it…

516 "Climate change a certain threat", David Suzuki with Faisal Moola, LondonTopic.ca, 8 March 2009, http://www.londontopic.ca/article.php?artid=13347

517 "World will agree new climate deal, says Al Gore", *The Guardian*, 14 March 2009, http://www.guardian.co.uk/world/2009/mar/14/al-gore-climate-change1

There will be those who call me a 'climate change denier'. Where in this book have I argued that climate change doesn't happen? They will call me a 'global warming denier', despite the fact I argue the planet is warming up. Some will attempt to cherry-pick their evidence to oppose this book, going back to some of the scientific studies I've quoted and picking out a conclusory paragraph or an opinion that supports global warming, then trumpeting that as if it shoots down the argument. It doesn't. For the most part when looking at studies, by believers or sceptics, I wasn't interested in opinions, just the facts. The fact that Greenland glaciers or Antarctic ice has stopped melting as fast is a relevant fact, regardless of how either side spins it. Additionally, many scientists have learnt to conclude their studies with a hymnal confession confirming their belief in global warming, so as to ensure they get continued funding.

But most of all, I've referenced the studies and provided hyperlinks in many cases so you can read them yourself. The science isn't settled, in fact it's tipping heavily to a natural cause.

Today, as I write this final page (2009 edition) and send the book for typesetting, a story is breaking across the US based on a leaked United Nations briefing document on climate change. It echoes exactly what I have warned you of:[518]

"A United Nations document on 'climate change' that will be distributed to a major environmental conclave next week envisions a huge reordering of the world economy, likely involving trillions of dollars in wealth transfer, millions of job losses and gains, new taxes, industrial relocations, new tariffs and subsidies, and complicated payments for greenhouse gas abatement schemes and carbon taxes – all under the supervision of the world body [the UN].

"Those and other results are blandly discussed in a discretely worded United Nations 'information note' on potential consequences of the measures that industrialized countries will likely have to take to implement the Copenhagen Accord, the successor to the Kyoto Treaty, after it is negotiated and signed by December 2009. The Obama administration has said it supports the treaty process if, in the words of a U.S. State Department spokesman, it can come up with an "effective framework" for dealing with global warming..."

Get your wallets out. Latest costings of Obama's proposed scheme

518 "U.N. 'Climate Change' Plan Would Likely Shift Trillions to Form New World Economy," *Fox News*, 27 March 2009, http://www.foxnews.com/story/0,2933,510937,00.html

suggest it will cost up to US$6,000 *per person* in taxes and charges over its first eight years.[519] For a family of five, that's extra charges of US$3,750 a year, and that's carbon reduction 'lite', not the full impact version the lobby groups are seeking. If carbon prices hit US$40 a tonne as predicted, that equates to NZ$560 per person per year, not including higher prices resulting from firms changing to new cleaner technology. The cost of this book pales into insignificance compared to what you might eventually be paying to turn the UN into a de-facto world government under the guise of combating 'climate change'.

I began this book with Nazi propagandist Josef Goebbels' famous quote about repeating the lie often enough that the people come to believe it. With world governments, the UN and global corporations lining up to make belief in the lie compulsory in December this year, it is fitting to end with the final line from Goebbels' quote:

"It thus becomes vitally important for the State to use all of its powers to repress dissent, for the truth is the mortal enemy of the lie, and thus by extension, the truth is the greatest enemy of the State."

And the truth, as always, really will set you free. Congratulations. You've taken the first step towards reclaiming your freedom from fear. Don't let it be the last such step you take. Tell your friends.

Further information: If you're a die-hard global warming believer after reading this book, then sadly there's nothing else I can do for you except to point you in the direction of www.greenpeace.org so you can support them. If you'd like to contribute to groups lobbying against the Copenhagen Treaty, visit www.climatescienceinternational.org

ALSO BY THIS AUTHOR:

Absolute Power: The Helen Clark Years
Eve's Bite
The Divinity Code

See your nearest bookstore, Amazon.com or IanWishart.com for details
For climate change updates, see twitter.com/investigatemag

519 Based on briefing given by senior White House aide Jason Furman in March 2009, warning that cost of President's cap and trade scheme was much more than the US$646 billion announced, and could be three times higher - US1.9 trillion divided across a population of 305 million. See http://icecap.us/index.php/go/political-climate/obama_climate_plan_could_cost_2_trillion/

T

WHAT OTHERS SAY:

"*Air Con* demonstrates, with hundreds of scientific references, that "global warming" was not, is not, and will not be a global crisis; that, even if *per impossibile* it might be, it is far more cost-effective to adapt as *and if* needed than to attempt to mitigate 'global warming' by cutting emissions of carbon dioxide; and that all attempts at mitigation would serve only to imprison the very poorest in their poverty, thereby perversely increasing world population and consequently the 'carbon footprint' of humankind, achieving an outcome precisely the opposite of that which was (however piously) intended.

"The UN, Mikhail Gorbachev, Jacques Chirac, and other world-government wannabes are plotting to establish nothing less than a global, bureaucratic-centralist dictatorship under the pretext that it is necessary to 'Save The Planet'. Ian Wishart's book demonstrates that there is not the slightest scientific reason for the new, quasi-religious belief that The Planet needs Saving. The new religion is merely an excuse for world government. World government will not, repeat not, be democratic government.

"The 'global warming' debate is not really a debate about climatology - it is a debate about freedom. It is the aim of the growing world-government faction among the international *classe politique* to take away our hard-won freedom and democracy forever. I commend this timely book, which makes the scientific arguments comprehensible to the layman. Those who read it will help to forestall the new Fascists and so to keep us free."

– Lord Christopher Monckton, Viscount of Brenchley, former adviser to Prime Minister Margaret Thatcher

"I started reading this book with an intensely critical eye, expecting that a mere journalist could not possibly cope with the complexities of climate science ... [But] He gives chapter and verse for almost everything he says and he has been far more far-ranging in searching the web than anyone else I know. The book is brilliant. The best I have seen which deals with the news item side of it as well as the science. He has done a very thorough job and I have no hesitation in unreserved commendation."

– Dr Vincent Gray, UN IPCC expert reviewer

"From the Paleozoic to the Intergovernmental Panel on Climate Change, investigative journalist Ian Wishart delves into the science and statistics of anthropogenic climate change, only to discover the not-so-hidden agenda underlying the global warming scare. *Air Con* is a thorough summary of the current state of the debate, the science, and the politics; it will be an important reference in any AGW skeptic's arsenal."

– Vox Day, columnist, WorldNetDaily

www.ingramcontent.com/pod-product-compliance
Lightning Source LLC
Chambersburg PA
CBHW071630270326
41928CB00010B/1863

* 9 7 8 0 9 5 8 2 4 0 1 6 1 *